Biology of Skates

Developments in Environmental Biology of Fishes 27

Series Editor

DAVID L.G. NOAKES

For further volumes:
http://www.springer.com/series/5826

David A. Ebert · James A. Sulikowski
Editors

Biology of Skates

Springer

Editors:

David A. Ebert
Moss Landing Marine Laboratories
Pacific Shark Research Center
Moss Landing, CA 95039, USA

James A. Sulikowski
University of New England
Marine Science Center
11 Hills Beach Rd.
Biddeford ME 04005, USA
jsulikov@hotmail.com

ISBN: 978-1-4020-9702-7   e-ISBN: 978-1-4020-9703-4

DOI: 10.1007/978-1-4020-9703-4

Library of Congress Control Number: 2008942792

© Springer Science+Business Media B.V. 2008

No part of this work may be reproduced, stored in a retrieval system, or transmitted in any form or by any means, electronic, mechanical, photocopying, microfilming, recording or otherwise, without written permission from the Publisher, with the exception of any material supplied specifically for the purpose of being entered and executed on a computer system, for exclusive use by the purchaser of the work.

Printed on acid-free paper

Springer.com

Contents

Preface: Biology of skates by *D.A. Ebert & J.A. Sulikowski* 1–4

Biodiversity and systematics of skates (Chondrichthyes: Rajiformes: Rajoidei) by *D.A. Ebert & L.J.V. Compagno* 5–18

Southern African skate biodiversity and distribution by *L.J.V. Compagno & D.A. Ebert* 19–39

Food habits of the sandpaper skate, *Bathyraja kincaidii* (Garman, 1908) off central California: seasonal variation in diet linked to oceanographic conditions by *C.S. Rinewalt, D.A. Ebert & G.M. Cailliet* 41–57

Food habits of the longnose skate, *Raja rhina* (Jordan and Gilbert, 1880), in central California waters by *H.J. Robinson, G.M. Cailliet & D.A. Ebert* 59–73

Dietary comparisons of six skate species (Rajidae) in south-eastern Australian waters by *M.A. Treloar, L.J.B. Laurenson & J.D. Stevens* ... 75–90

Comparative feeding ecology of four sympatric skate species off central California, USA by *J.J. Bizzarro, H.J. Robinson, C.S. Rinewalt & D.A. Ebert* 91–114

Standardized diet compositions and trophic levels of skates (Chondrichthyes: Rajiformes: Rajoidei) by *D.A. Ebert & J.J. Bizzarro* .. 115–131

Normal embryonic development in the clearnose skate, *Raja eglanteria*, with experimental observations on artificial insemination by *C.A. Luer, C.J. Walsh, A.B. Bodine & J.T. Wyffels* 133–149

Endocrinological investigation into the reproductive cycles of two sympatric skate species, *Malacoraja senta* and *Amblyraja radiata*, in the western Gulf of Maine by *J. Kneebone, D.E. Ferguson, J.A. Sulikowski & P.C.W. Tsang* 151–159

Morphological variation in the electric organ of the little skate (*Leucoraja erinacea*) and its possible role in communication during courtship by *J.M. Morson & J.F. Morrissey* 161–169

Reproductive biology of *Rioraja agassizi* from the coastal southwestern Atlantic ecosystem between northern Uruguay (34°S) and northern Argentina (42°S) by *J.H. Colonello, M.L. García & C.A. Lasta* 171–178

Profiling plasma steroid hormones: a non-lethal approach for the study of skate reproductive biology and its potential use in conservation management by *J.A. Sulikowski, W.B. Driggers III, G.W. Ingram Jr., J. Kneebone, D.E. Ferguson & P.C.W. Tsang* 179–186

Age and growth estimates for the smooth skate, *Malacoraja senta*, in the Gulf of Maine by *L.J. Natanson, J.A. Sulikowski, J.R. Kneebone & P.C. Tsang* 187–202

Age, growth, maturity, and mortality of the Alaska skate, *Bathyraja parmifera*, in the eastern Bering Sea by *M.E. Matta & D.R. Gunderson* 203–217

Age and growth of the roughtail skate *Bathyraja trachura* (Gilbert 1892) from the eastern North Pacific by *C.D. Davis, G.M. Cailliet & D.A. Ebert* 219–230

Age and growth of big skate (*Raja binoculata*) and longnose skate (*R. rhina*) in the Gulf of Alaska by *C.M. Gburski, S.K. Gaichas & D.K. Kimura* 231–243

Preface: Biology of skates

David A. Ebert · James A. Sulikowski

Originally published in the journal Environmental Biology of Fishes, Volume 80, Nos 2–3, 107–110.
DOI 10.1007/s10641-007-9244-3 © Springer Science+Business Media B.V. 2007

Synopsis Skates have become a concern in recent years due to the preponderance of these elasmobranchs that are caught as bycatch or as a directed fishery. This has raised concern because skates have life history characteristics that may make them vulnerable to over-exploitation. It was due to this increasing awareness and concern about these batoids that prompted us to organize an international symposium on the "*Biology of Skates*". The aims and goals of the symposium were to bring together an international group of researchers to meet, discuss, perhaps develop collaborations, and present their most recent findings. The symposium was held over two days, on 13–14 July, 2006, in conjunction with the 22nd annual meeting of the American Elasmobranch Society in New Orleans, LA. A total of 31 authors from four countries contributed 16 papers that appear in this special issue. The papers are broadly arranged into four separate categories: systematics and biogeography, diet and feeding ecology, reproductive biology, and age and growth. To the best of our knowledge this is the first dedicated book on the biology of skates. We hope that readers will find this issue of interest and that it helps encourage and stimulate future research into these fascinating fishes.

Keywords Rajidae · Symposium · Life history

Skates have become a concern in recent years due to the preponderance of these elasmobranchs to be landed as part of a directed fishery or indirectly as bycatch. This has raised concern because skates, like other chondrichthyans, have life history characteristics that make them vulnerable to over-exploitation. However, until recently this ubiquitous batoid group has received little study relative to the charismatic shark-like fishes that are the target of major fisheries worldwide. Furthermore, due to their demersal life-style on soft bottom substrates skates are especially vulnerable to trawl fisheries. Although, shark fisheries are now being scrutinized and managed more closely than ever, skates despite being one of the more common and visible components of bycatch fisheries are generally still overlooked. In Alaskan waters, for example, the biomass of skate discard has been estimated at over 25 million metric tons, while skate landings along the California coast have increased 10-fold in the past 2 decades. It has been well documented that trawl fisheries in the North

D. A. Ebert (✉)
Pacific Shark Research Center, Moss Landing Marine Laboratories, 8272 Moss Landing Road, Moss Landing, CA 95039, USA
e-mail: debert@mlml.calstate.edu

J. A. Sulikowski
Marine Science Center, University of New England, 11 Hills Beach Road, Biddeford, ME 04005, USA

Atlantic have impacted the abundance, population structure, and distribution of several skate species, causing several to be given commercially prohibited status. Given the increased awareness and concern globally, researchers have begun to pay closer attention to these enigmatic fishes and have begun to study their biology with a newfound sense of urgency.

It was due this increasing awareness and concern about these fishes that prompted us to organize an international symposium on the "*Biology of Skates*". The aims and goals of the symposium were to bring together an international group of researchers to meet, discuss, perhaps develop collaborations, and present their most recent findings. The symposium was broadly organized around four areas of research, systematics and biogeography, diet and feeding ecology, reproductive biology, and age and growth aspects of their biology. Our initial inquiry to gauge the interest of such a symposium was very well received as approximately 75 researchers from nearly 20 countries responded positively. The overwhelming response we received confirmed our belief that there was a high level of interest, and concern, for skates.

The symposium was held over 2 days, on 13–14 July 2006, in conjunction with the 22nd annual meeting of the American Elasmobranch Society in New Orleans, LA (Fig. 1). A total of 25 speakers, representing nine countries, presented papers during the symposium proper with another three papers and five posters presented during the general and posters sessions, respectively. The presenters included a nice mix of established professionals ($n = 12$) and young budding researchers ($n = 13$), about 10 of whom were presenting results from their masters theses or Ph.D. dissertations.

A total of 31 authors from four countries contributed 16 papers that appear in this special issue. The papers are broadly arranged into the same four separate categories as the symposium: systematics and biogeography, diet and feeding ecology, reproductive biology, and age and growth. The first paper by David Ebert and Leonard Compagno presents a historical perspective on skate systematics over the past 250 years, chronologically detailing the exponential growth in skate diversity, e.g. number of species and genera, and how a few individual researchers over the past 60 years have largely been responsible for this dynamic increase in skate numbers. Included in this paper, in an appendix, is a checklist of all living skate species. Leonard Compagno and David Ebert then present a

Fig. 1 Participants at the Biology of Skates symposium held in conjunction with the 22nd Annual American Elasmobranch Society meeting in New Orleans, 13–14 July, 2006. From left to right; Jeff Kneebone, Romney McPhie, Chanté Davis, Lisa Natanson, Chris Rinewalt, Oscar Sosa Nishizaki, Heather Robinson, Malcolm Francis, Carl Luer, Joe Bizzarro, Michelle Treloar, Dana Bethea, David Ebert, Beth Matta, James Sulikowski, Chris Gburski, Duane Stevenson, Jim Ellis, Jorge Colonello, Daniel Figueroa, Dave Kulka, Jason Morson. Absent from photograph: Marcelo de Carvalho, Peter Last, and John McEachran

regional account on the biodiversity of skates from their nearly 25 years of research on the southern African fauna. In his paper on the feeding ecology of the sandpaper skate, *Bathyraja kincaidii*, Christopher Rinewalt and co-authors demonstrate by use of a three-factor MANOVA significant differences in the diet of this skate by sex, maturity status, and oceanographic season using numeric and gravimetric measures of importance for the major prey categories. Heather Robinson and her co-authors report on a dietary shift from crustaceans to teleosts associated with depth and size in the longnose skate, *Raja rhina*, a moderately large eastern North Pacific species. The following two papers, by Michelle Treloar and Joe Bizzarro and their collaborators, each deal with the trophic ecology of a complex of skates from southeastern Australia and central California, USA, respectively. The final paper on diet and feeding ecology is by David Ebert and Joe Bizzarro, and examines the standardized diet composition and trophic levels of skates, finding that these fish occupy a similar tophic position to that of other upper trophic level marine predators. The five papers on skate reproductive biology cover a wide range of aspects, including one by Carl Luer and colleagues on embryonic development of the clearnose skate, *Raja eglanteria*, with observations on artificial insemination in this species, a technique which may have important implications on other captive elasmobranchs. Jeff Kneebone and his co-authors investigated the seasonal cycle of two common Gulf of Maine skates by examining seasonal changes in steroid hormone levels over the course of a year. Jason Morson and John Morrissey demonstrate that the electric organ discharge in little skates, *Leucoraja erinacea*, may play a role in communication during courtship. Jorge Colonello and his colleagues examined the seasonal reproductive cycle of the monotypic Rio skate, *Rioraja agassiz*, a common southwestern Atlantic species of the continental shelf, concluding that females have a partially defined reproductive cycle with two peaks seasons. James Sulikowski and collaborators conclude the reproductive part of this issue by presenting evidence that circulating levels of estradiol and testosterone concentrations can be used as a non-lethal alternative to study reproductive biology in male and female skates. The final portion of this special issue presents four excellent studies detailing the age and growth of five skate species. The first paper by Lisa Natanson and her co-authors presents age and growth estimates for the smooth skate, *Malacoraja senta*, in the Gulf of Maine using an innovative histological technique to enhance band clarity. This is followed by two papers presenting the first age and growth estimates for two North Pacific Ocean skate species. Beth Matta and Donald Gunderson in their paper on aspects of the age, growth, maturity, and mortality of the Alaska skate, *Bathyraja parmifera*, found that caudal thorns may represent a non-lethal method to estimate age and growth in this species. Conversely, Chante Davis and her co-authors in their study on the age and growth of the roughtail skate, *Bathyraja trachura*, a common mid to upper slope species, found that caudal thorns were not a useful method for estimating age in this species. These two studies illustrate the importance of analyzing and interpreting ageing structures, e.g. vertebrae and caudal thorns, on an individual species basis. This special issue concludes with an important paper by Christopher Gburski and his collaborators on the age and growth of two skates (*Raja binoculata* and *R. rhina*) that have been at the center of an emerging fishery in the Gulf of Alaska. Results from this study will be critical to developing demographic models for stock assessments in these two species. To the best of our knowledge this is the first dedicated book on the biology of skates. We hope that the readers find this issue of interest and that it helps encourage and stimulate future research into these fascinating fishes.

We as guest editors would like to thank all of the participants who presented papers and posters at the symposium and to Environmental Biology of Fishes (EBF) for publishing this special issue, especially the editor in chief, David Noakes, managing editor, Lynn Bouvier, and the staff at EBF for making this a smooth process. Funding for the "*Biology of Skates*" Symposium was provided by NOAA/NMFS to the National Shark Research Consortium and Pacific Shark Research Center and by the American Elasmobranch Society. Finally, we would like to thank the following referees for taking the time out of their busy schedules to expeditiously review manuscripts, in a timely manner, for this special issue. (*Number in parenthesis indicates more than one manuscript reviewed).

Referees	Institution
Cynthia Awruch	University of Tasmania, Australia
Ivy Baremore	NOAA Fisheries Services, Southeast Fisheries Science Center
Lewis Barnett	Moss Landing Marine Laboratories, Pacific Shark Research Center
Dana Bethea	NOAA Fisheries Service, Southeast Fisheries Science Center
Joseph Bizzarro (2)	Moss Landing Marine Laboratories, Pacific Shark Research Center
Juan Matias Braccini	University of Adelaide, Australia
Gregor Cailliet	Moss Landing Marine Laboratories, Pacific Shark Research Center
Enric Cortés	NOAA Fisheries Service, Southeast Fisheries Science Center
Paul Cowley	South African Institute of Aquatic Biodiversity, South Africa
Chanté Davis	Moss Landing Marine Laboratories, Pacific Shark Research Center
Marcelo de Carvalho	University of São Paulo, Brazil
Jim Ellis	CEFAS, Lowestoft Laboratory, England
Lara Ferry-Graham	Moss Landing Marine Laboratories
Malcolm Francis	National Institute of Water and Atmospheric Research, New Zealand
Christopher Gburski	National Marine Fisheries Service, Alaska Fisheries Science Center
Jim Gelsleichter	Mote Marine Laboratory
Kenneth Goldman	Alaska Department of Fish & Game
Dean Grubbs	Virginia Institute of Marine Science, College of William & Mary
Alan Henningsen	National Aquarium in Baltimore
Gerald Hoff	NOAA Fisheries Service, Alaska Fisheries Science Center
Brett Human	Marine Science and Fisheries Centre of Oman
Sarah Irvine	Deakin University, Australia
Hajime Ishihara	Taiyo Engineering Overseas, Japan
Jeff Kneebone	University of New Hampshire
David Koester	University of New England
David Kulka	Department of Fisheries & Oceans, Canada
Luis Lucifora	Dalhousie University, Canada

Referees	Institution
Mabel Manjaji-Matsumoto	University Malaysia Sabah, Malaysia
Beth Matta	NOAA Fisheries Service, Alaska Fisheries Science Center
Romney McPhie	Dalhousie University, Canada
Lisa Natanson	NOAA Fisheries Service, Northeast Fisheries Science Center
Christopher Rinewalt	Moss Landing Marine Laboratories, Pacific Shark Research Center
Heather Robinson	Moss Landing Marine Laboratories, Pacific Shark Research Center
Duane Stevenson	NOAA Fisheries Service, Alaska Fisheries Science Center
Adam Summers	Ecology and Evolutionary Biology, University California Irvine
Michelle Treloar (2)	CSIRO Marine Research, Australia
Paul Tsang	University of New Hampshire
William White (2)	CSIRO Marine Research, Australia

Biodiversity and systematics of skates (Chondrichthyes: Rajiformes: Rajoidei)

David A. Ebert · Leonard J. V. Compagno

Originally published in the journal Environmental Biology of Fishes, Volume 80, Nos 2–3, 111–124.
DOI 10.1007/s10641-007-9247-0 © Springer Science+Business Media B.V. 2007

Abstract Skates (Rajiformes: Rajoidei) are a highly diverse fish group, comprising more valid species than any other group of cartilaginous fishes. The high degree of endemism exhibited by the skates is somewhat enigmatic given their relatively conserved body morphology and apparent restrictive habitat, e.g. soft bottom substrates. Skates are primarily marine benthic dwellers found from the intertidal down to depths in excess of 3,000 m. They are most diverse at higher latitudes and in deepwater, but are replaced in shallower, warm temperate to tropical waters by stingrays (Myliobatodei). The number of valid skate species has increased exponentially, with more species having been described since 1950 ($n = 126$) than had been described in the previous 200 years ($n = 119$). Much of the renaissance in skate systematics has largely been through the efforts of a few individuals who through author–coauthor collaboration have accounted for 78 of the 131 species described since 1948 and for nine of 13 genera named since 1950. Furthermore, detailed regional surveys and accounts of skate biodiversity have also contributed to a better understanding of the diversity of the skates. A checklist of the living valid skate species is presented.

Keywords Skates · Classification · Arhychobatidae · Anacanthobatidae · Rajidae · Checklist

Introduction

Batoids (Chondrichthyes: Rajiformes), including the skates (suborder: Rajoidei), comprises more valid species ($n = 574$) than all of the other nine chondrichthyan orders combined ($n = 528$) (Fig. 1). They comprise more than twice the number of chondrichthyan species relative to the nearest group, the Carcharhiniformes ($n = 296$). Furthermore, among batoids, the skates (Rajoidei) are the most diverse group comprising more genera ($n = 27$) and species ($n = 245$) than any of the other nine batoid suborders (Fig. 2). The stingrays (Myliobatoidei), the other most diverse batoid suborder, are morphologically far more varied than skates and show a far richer diversity at the family level ($n = 10$) and nearly as many genera ($n = 24$). Overall, the total number of skate species represents over 22% of all known chondrichthyan species and about 43% of all batoids.

The high degree of biodiversity and endemism exhibited by skates is somewhat enigmatic given their relatively conservative dorso-ventrally flattened body morphology and apparent restrictive habitat

D. A. Ebert (✉)
Pacific Shark Research Center,
Moss Landing Marine Laboratories,
Moss Landing, CA 95039, USA
e-mail: debert@mlml.calstate.edu

L. J. V. Compagno
Shark Research Center, Iziko – Museums of Cape Town,
Cape Town 8000, South Africa

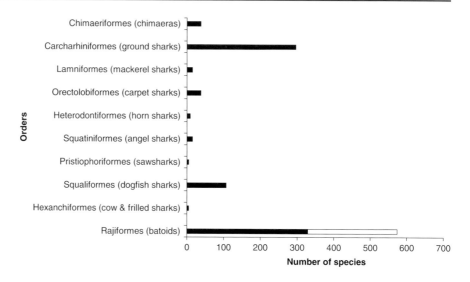

Fig. 1 The number of described valid Chondrichthyan species per order as compiled by the authors as of 2 March 2007. Solid bar is number of batoid species excluding skates and open bar represents number of skate species

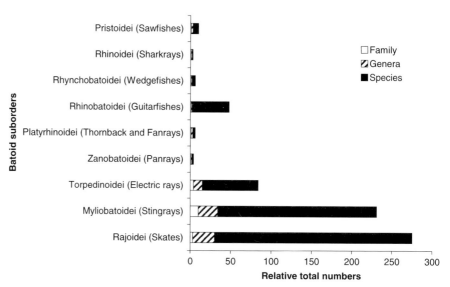

Fig. 2 The total number of families, genera, and species per batoid suborder as compiled by the authors as of 2 March 2007. The suborders Rhinoidei, Rhynchbatoidei, Rhinobatoidei, and Zanobatoidei each have only a single family and genus, while the suborders Pristoidei and Platyrhinoidei each have a monotypic family and two genera each

preference, e.g. soft bottom substrates. Skates are primarily marine benthic dwellers found from the intertidal down to depths in excess of 3,000 m. Skates are most diverse at higher latitudes and in deeper waters, but are generally replaced in shallower, warmer, temperate to tropical waters by Myliobatoids. Skates are generally found in shallower waters towards the poles, but occur deeper in warm temperate to tropical equatorial waters. They are conspicuously absent from brackish and freshwater environs, except for a single estuarine species found in Tasmania, Australia (species "L" in Last and Stevens 1994) and there are no known pelagic species within this group. Although long thought to be primarily inhabitants of soft bottom substrates, recent research by remote operated vehicles has revealed a somewhat diverse skate fauna inhabiting areas of rock cobble to high rocky relief. Several eastern North Pacific species once considered uncommon are in fact quite common in areas previously considered uninhabitable for skates (Kuhnz et al. 2006).

The number of valid skate species has increased exponentially over the past century (Fig. 3). Prior to 1900 only 62 skate species had been described, with another 57 species being described between 1900 and 1949 that brought the total number to 119 described species. However, since 1950 126 skate species have

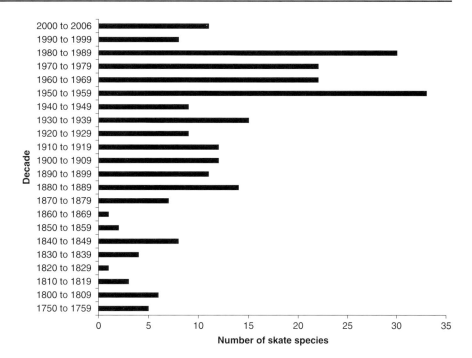

Fig. 3 Number of skate species described by decade from 1750 to 2006. Number of valid species compiled by authors as of 2 March 2007

been described. This rapid increase over the past nearly 60 years was due largely to the efforts of five individuals who through author–coauthor combinations have accounted for 78 (59.5%) of the 131 species described since 1948. At the forefront of this renaissance in skate systematics was the collaboration of Henry Bigelow and William Schroeder who accounted for the most new species descriptions with 29 (Bigelow and Schroeder 1948, 1950, 1951a, b, 1954, 1958, 1962, 1964, 1965). Also, contributing during the 1950's and 60's was Ishiyama (1952, 1955, 1958, 1967) who named 10 species and later in collaboration with his student Hajime Ishihara, who has contributed much to the systematics of skates in his own right, described another five species (Ishiyama and Ishihara 1977; Ishihara and Ishiyama 1985). Since the early 1970's two individuals, John McEachran and Mathias Stehmann, have each either authored or coauthored the descriptions of 18 and 16 species each, respectively. Other notable ichthyologists who have been involved in authoring–coauthoring the descriptions of five or more skate species includes (with number of species), V.N. Dolganov ($n = 7$), Samuel Garman ($n = 7$), David Starr Jordan ($n = 7$), John Norman ($n = 7$), Hajime Ishihara ($n = 6$), Peter Last ($n = 6$), Charles Gilbert ($n = 6$), Albert Günther ($n = 6$), P. Alexander "Butch" Hulley ($n = 5$), Carolus Linnaeus ($n = 5$), Johannes Müller and Fredrich Gustav Jacob Henle ($n = 5$), Cecil von Bonde and D.B. Swart ($n = 5$), and John Wallace ($n = 5$).

Paralleling the increase in skate species has been an increase in the number of recognized genera (Fig. 4). Of the 27 recognized skate genera, 21 have been erected since 1900 with 13 having been described since 1950 alone. As with the increase in new species descriptions there have been a few individuals who through author–coauthor combinations have been responsible for naming most of the skate genera. Ishiyama (1952, 1958) has authored the most with four (*Bathyraja, Notoraja, Okamejei, Rhinoraja*) generic descriptions, followed by Gilbert Whitley who described three (Whitley 1931, 1939; *Irolita, Pavoraja, Rioraja*), and Bigelow and Schroeder (1948, 1954) who collaborated in naming three (*Breviraja, Cruriraja, Pseudoraja*) genera. McEachran in collaboration with Leonard Compagno and Peter Last has also contributed to the descriptions of three (*Brochiraja, Fenestraja, Neoraja*) skate genera. Those having authored–coauthored the descriptions of two genera include Compagno (McEachran and Compagno 1982), Stehmann (1970; *Malacoraja, Rajella*), and A. W. Malm (Malm 1877; *Amblyraja, Leucoraja*).

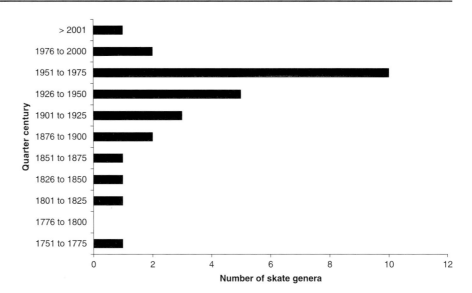

Fig. 4 Number of skate genera described by quarter century between 1750 and 2000. Prior to 1800 only a single genus was recognized and since 2000 one new genus has been erected. Number of valid genera compiled by authors as of 2 March 2007

An early chapter in expanding skate generic diversity was provided by W. Harold Leigh-Sharpe who did a comprehensive study of clasper morphology in cartilaginous fishes (Leigh-Sharpe 1920, 1921, 1922a–c, 1924a, b, 1926a–d) and provided several terms for skate claspers that are currently used by systematists. Leigh-Sharpe studied the claspers of 24 species of skates from the Western Atlantic, Eastern North Atlantic, Mediterranean Sea, Eastern North Pacific, and sub-Antarctic islands that were assigned to the genus *Raia* (=*Raja*). Leigh-Sharpe subdivided these species into 10 "pseudogenera" on clasper morphology, with their names based on Greek letters prefixed to *Raia* (e.g., *Alpharaia*). These pseudogenera are equivalent to subgenera in the International Code of Zoological Nomenclature and were formatted as genera and assigned species including designated types. Leigh-Sharpe's subgenera are synonyms of current genera as follows: *Etaraia* Leigh-Sharpe, 1924 is a possible senior synonym of *Rhinoraja* Ishiyama 1952; *Zetaraia* Leigh-Sharpe, 1924 is a possible senior synonym of *Bathyraja* Ishiyama, 1958 or represents a separate genus; *Thetaraia* Leigh-Sharpe, 1924 is a senior synonym of *Bathyraja* Ishiyama 1968; *Epsilonraia* Leigh-Sharpe, 1924 is a senior synonym of *Atlantoraja* Menni, 1972; *Alpharaia* Leigh-Sharpe, 1924 and *Kapparaia* Leigh-Sharpe, 1926 are junior synonyms of *Leucoraja* Malm, 1877; *Betaraia* Leigh-Sharpe, 1924 is a junior synonym of *Raja* Linnaeus 1758; *Gammaraia* Leigh-Sharpe, 1924 is a junior synonym of *Dipturus*

Rafinesque, 1810. *Deltaraia* Leigh-Sharpe, 1924 is a junior synonym of *Amblyraja* Malm, 1877; and *Iotaraia* Leigh-Sharpe, 1924 is a senior synonym of *Rostroraja* Hulley, 1972. Leigh-Sharpe's taxa were published in a decidedly mainstream journal and recognized as genera by Fowler (1941) and by Bigelow and Schroeder (1953) in synonymy of *Raja* Linnaeus, 1758. Most skate systematists have subsequently ignored, overlooked or dismissed these names which has resulted in synonyms and posed future problems that need to be resolved.

Much of the renaissance in skate systematics over the past 50 plus years has been largely due to an increase in regional studies. The western North Atlantic fauna was extensively examined by Bigelow and Schroeder who detailed this and the Gulf of Mexico regions (1948, 1950, 1951a, b, 1953, 1954, 1958, 1962, 1964, 1965). The western South Atlantic region has improved markedly in recent years with works by McEachran (1982, 1983), Menni and Stehmann (2000), Cousseau (Cousseau et al. 2000), Gomes (2002), and de Carvalho (2005). Stehmann has reviewed the eastern North Atlantic (Stehmann 1970; Stehmann and Bürkel 1984) and the West African continental slope from Morocco to South Africa (Stehmann 1995). Hulley (1970, 1972a, b) and Wallace (1967) contributed much new information to the southern African region, including both the southeastern Atlantic and southwestern Indian Oceans. Their research has been followed upon by Compagno et al. (1989, 1991) and Compagno and

Ebert (2007). Seret (1986, 1989a, 1989b) has reported on the deepwater skates from around Madagascar which included the descriptions of two new species. The Australasian region, including Australia, New Zealand, and New Caledonia, is undergoing an extensive revision that may eventually produce upwards of 30 or more new species (Last and Stevens 1994; Last and Yearsley 2002). The systematics of the western North Pacific, especially around Japanese waters and in the Bering Sea, has been studied in detail by Ishiyama (1958, 1967), Ishiyama and Ishihara (1977), Ishihara and Ishiyama (1985, 1986), Ishihara (1987), Dolganov (1985), and from around Taiwanese waters by Chen and Joung (1989). Recent efforts by Stevenson (2004) and Stevenson and Orr (2005) are beginning to elucidate the complex skate fauna of the Bering Sea and in Alaskan waters, a region very poorly known. The eastern North Pacific, although still lacking detailed systematic work, has improved markedly over the past 25 years (Dolganov 1983; Eschmeyer et al. 1983; Zorzi and Anderson 1988; Ebert 2003). The eastern South Pacific batoid fauna, including the skates, has been revised by Lamilla and Sáez (2003). Although the identification and systematics of regional skate faunas is improving there are still many areas where the fauna is still relatively unknown including the eastern Central Pacific, most of the Indian and Southern oceans, and the Caribbean Sea.

The number of valid described skate species as of this writing is approximately 245 (Appendix 1) with perhaps another 50–100 species still to be described. The higher classification (order, suborder, family) presented here follows Compagno (1999, 2001, 2005) in recognizing three families (Arhynchobatidae, Anacanthobatidae, Rajidae); these families are sometimes classified as subfamilies as per other recent classification schemes (de Carvalho 1996; McEachran and Ashliman 2004). The generic arrangement follows McEachran and Dunn (1998) with the addition of one new genus, *Brochiraja* Last and McEachran (2006). The genera *Bathyraja* and *Dipturus* have the most species with 46 and 31, respectively (Fig. 5). Seven genera (*Amblyraja, Anacanthobatis, Okamejei, Raja, Rajella, Rhinoraja*) have between 10 and 15 species, while five genera (*Arhynchobatis, Irolita, Pseudoraja, Rioraja, Rostroraja*) have only a single species each. Fifteen species are currently unassigned, 13 to two as yet undescribed genera, the Amphi-American and North Pacific Assemblage, and two western Pacific species

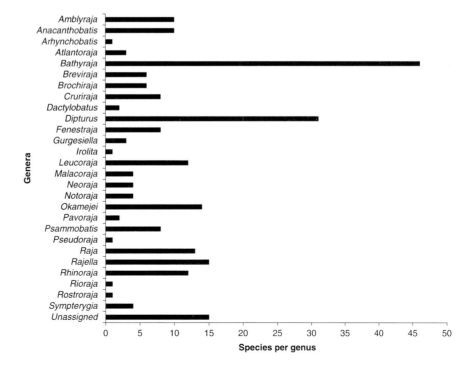

Fig. 5 Number of species per skate genera as compiled by the authors as of 2 March 2007

not yet assigned to a subgenus. The species checklist presented in Appendix 1 has been compiled by the authors as of 2 March 2007.

Acknowledgements Funding for this research was provided to DAE by NOAA/NMFS to the National Shark Research Consortium and Pacific Shark Research Center and by the American Elasmobranch Society. LJVC's research was supported by grants from the South African Foundation for Research Development (FRD), National Science Foundation (NRF) and Fisheries Development Corporation, and by the South African Museum and the former J.L.B. Smith Institute of Ichthyology (now South African Institute of Aquatic Biodiversity).

Appendix 1

A checklist of the known skate species.
Names of doubtful species are preceeded by a question mark; valid species that are doubtfully placed in agences have a question mark at the end of the generic name.

> Class Chondrichthyes—Cartilaginous fishes
> Subclass Elasmobranchii—Sharks and Rays
> Superorder Batoidea—Rays
> Order Rajiformes—Skates
> Suborder Rajoidei

Family Arhynchobatidae Fowler, 1934—Softnose Skates

This group in an alternate arrangement is ranked as subfamily Arhynchobatinae (e.g., McEachran and Dunn 1998) within a single family, Rajidae, for all skates. Worldwide this family has 12 genera and approximately 89 valid species and 25 undescribed and dubious species.

Genus *Arhynchobatis* Waite, 1909—Longtailed skates. Found only in New Zealand waters on insular shelves. A single species.

Arhynchobatis asperrimus Waite, 1909—Long-tailed skate

Genus *Atlantoraja* Menni, 1972—La Plata skates. Endemic to the western South Atlantic coast of South America, mostly on continental shelves. Three species, plus possibly one undescribed species.

Atlantoraja castelnaui (Ribero, 1904)—Spotback skate

Atlantoraja cyclophora (Regan, 1903)—Eyespot skate

Atlantoraja platana (Günther, 1880)—La Plata skate

Genus *Bathyraja* Ishiyama 1958—Softnose skates. This wide-ranging speciose and primarily deep-water genus is most diverse at high latitudes, on continental insular shelves and slopes. Forty-six species and at least four, possibly more, undescribed species.

Bathyraja abyssicola (Gilbert, 1896)—Deepsea skate

Bathyraja aguja (Kendall and Radcliffe, 1912)—Aguja skate

Bathyraja aleutica (Gilbert, 1895)—Aleutian skate

Bathyraja andriashevi Dolganov, 1985—Little-eye skate

Bathyraja bergi Dolganov, 1985—Bottom skate

Bathyraja brachyurops (Fowler, 1910)—Broadnose skate

Bathyraja caerulonigricans Ishiyama and Ishihara, 1977—Purpleblack skate

Bathyraja cousseauae de Astarloa and Mabragaña, 2004—Joined fins skate

Bathyraja diplotaenia (Ishiyama, 1952)—Dusky-pink skate

Bathyraja eatoni (Günther, 1876)—Eaton's skate

Bathyraja fedorovi Dolganov, 1985—Cinnamon skate

Bathyraja griseocauda (Norman, 1937)—Graytail skate

Bathyraja hesperafricana Stehmann, 1995—West African skate

Bathyraja irrasa Hureau and Ozouf-Costaz, 1980—Kerguelen sandpaper skate

Bathyraja ishiharai Stehmann, 2005—Abyssal skate

Bathyraja isotrachys (Günther, 1877)—Challenger skate

Bathyraja kincaidii (Garman, 1908)—Sandpaper skate

Bathyraja lindbergi Ishiyama and Ishihara, 1977—Commander Skate

Bathyraja longicauda (de Buen, 1959)—Slimtail skate

Bathyraja maccaini Springer, 1971—McCain's skate

Bathyraja maculata Ishiyama and Ishihara, 1977—Whiteblotched skate
Bathyraja mariposa Stevenson, Orr, Hoff, and McEachran, 2004—Butterfly skate
Bathyraja matsubarai (Ishiyama, 1952)—Dusky-purple skate
Bathyraja meridionalis Stehmann, 1987—Darkbelly skate
Bathyraja microtrachys (Osburn and Nichols, 1916)—Fine-spine skate
Bathyraja minispinosa Ishiyama and Ishihara, 1977—Smallthorn skate
Bathyraja notoroensis Ishiyama and Ishihara, 1977–Notoro skate
?*Bathyraja ogilbyi* (Whitley, 1939)—Great skate (='Raja' whitleyi?)
Bathyraja pallida (Forster, 1967)—Pale skate
Bathyraja papilionifera Stehmann, 1985—Butterfly skate
Bathyraja parmifera (Bean, 1881)—Alaska skate
Bathyraja peruana McEachran and Miyake, 1984—Peruvian skate
Bathyraja richardsoni (Garrick, 1961)—Richardson's skate
Bathyraja scaphiops (Norman, 1937)—Cuphead skate
Bathyraja schroederi (Krefft, 1968)—Whitemouth skate
Bathyraja shuntovi Dolganov, 1985—Narrownose skate
Bathyraja simoterus (Ishiyama, 1967)—Hokkaido skate
Bathyraja smirnovi (Soldatov and Pavlenko, 1915)—Golden skate
Bathyraja smithii (Müller and Henle, 1841)—African softnose skate
Bathyraja spinicauda (Jensen, 1914)—Spinetail skate
Bathyraja spinosissima (Beebe and Tee-Van, 1941)—White skate
Bathyraja trachouros (Ishiyama, 1958)—Eremo skate
Bathyraja trachura (Gilbert, 1892)—Roughtail skate
Bathyraja tunae Stehmann, 2005—Cristina's skate
Bathyraja tzinovskii Dolganov, 1985—Creamback skate
Bathyraja violacea (Suvorov, 1935)—Okhotsk skate

Genus *Brochiraja* Last and McEachran, 2006—Thornsnout skates. An endemic genus found in the deepseas surrounding New Zealand. Six described species.

Brochiraja aenigma Last and McEachran, 2006—Enigma skate
Brochiraja albilabiata Last and McEachran, 2006—Whitelipped skate
Brochiraja asperula (Garrick and Paul, 1974)—Prickly deepsea skate
Brochiraja leviveneta Last and McEachran, 2006—Smooth blue skate
Brochiraja microspinifera Last and McEachran, 2006—Small prickly skate
Brochiraja spinifera (Garrick and Paul, 1974)—Spiny deepsea skate

Genus *Irolita* Whitley, 1931—Round skates. Endemic to Australia, found on the outer shelves and upper slopes. One described and one undescribed species.

Irolita waitii (McCulloch, 1911)—Southern round skate

Genus *Notoraja* Ishiyama, 1958—Velvet skates. Western Pacific, found mostly on continental and insular slopes. Four described and one undescribed species.

Notoraja laxipella (Yearsley and Last, 1992)—Eastern looseskin skate
Notoraja ochroderma McEachran and Last, 1994—Yellow skate
Notoraja subtilispinosa Stehmann, 1989—Deepwater skate
Notoraja tobitukai (Hiyama, 1940)—Leadhued skate

Genus *Pavoraja* Whitley 1939—Peacock skates. Western South Pacific, mainly on outer continental shelves and upper slopes. Two described species with at least four or more undescribed species.

Pavoraja alleni McEachran and Fechhelm, 1982—Allen's skate
Pavoraja nitida (Günther, 1880)—Peacock skate

Genus *Psammobatis* Günther, 1870—Sandskates. Endemic to the eastern South Pacific and western South Atlantic coasts of South America, mostly found on continental shelves. Eight described species.

Psammobatis bergi Marini, 1932—Blotched sandskate
Psammobatis extenta Garman, 1913—Zipper skate
Psammobatis lentiginosa (Bigelow and Schroeder, 1951)—Freckled sandskate
Psammobatis normani McEachran, 1983—Shortfin sandskate
Psammobatis parvacauda McEachran 1983—Smalltail sandskate
Psammobatis rudis Günther, 1870—Smallthorn sandskate
Psammobatis rutrum Jordan, 1890—Spade sandskate
Psammobatis scobina (Philippi, 1857)—Raspthorn sandskate

Genus *Pseudoraja* Bigelow and Schroeder, 1954—Fanfin skates. A single endemic species to the western North Atlantic, found on continental slopes.

Pseudoraja fischeri Bigelow and Schroeder, 1954—Fanfin skate

Genus *Rhinoraja* Ishiyama, 1952—Jointnose skates. A wide-ranging, but patchily distributed, group found in the North Pacific, western South Atlantic coast of South America, and Antarctica, found on continental and insular shelves and slopes. This genus differs from *Bathyraja* primarily in having a basal joint in the rostral cartilage and may not be separable from that genus. Twelve species.

Rhinoraja albomaculata (Norman, 1937)—Whitedotted skate
Rhinoraja interrupta (Gill and Townsend, 1897)—Bering skate
Rhinoraja kujiensis (Tanaka, 1916)—Dapple-bellied softnose skate
Rhinoraja longicauda Ishiyama, 1952—Whitebellied softnose skate
Rhinoraja macloviana (Norman, 1937)—Patagonian skate
Rhinoraja magellanica (Philippi, 1902)—Magellan skate
Rhinoraja multispinis (Norman, 1937)—Multispine skate
Rhinoraja murrayi (Günther, 1880)—Murray's skate
?*Rhinoraja obtusa* (Gill and Townsend, 1897)—Blunt skate.
Rhinoraja odai Ishiyama, 1958—Oda's skate
?*Rhinoraja rosispinis* (Gill and Townsend, 1897)—Flathead skate
Rhinoraja taranetzi Dolganov, 1985—Mud skate

Genus *Rioraja* Whitley, 1939—Rio skates. Endemic to the western South Atlantic coast of South America, found on continental shelves. One species.

Rioraja agassizi (Müller and Henle, 1841)—Rio Skate

Genus *Sympterygia* Müller and Henle, 1837—Fanskates. Endemic to the Atlantic and Pacific coasts of South America, found primarily on continental shelves. Four valid species.

Sympterygia acuta Garman, 1877—Bignose fanskate
Sympterygia bonapartei Müller and Henle, 1841—Smallnose fanskate
Sympterygia brevicaudata Cope, 1877—Shorttail fanskate
Sympterygia lima (Poeppig, 1835)—Filetail skate

Family Rajidae Blainville, 1816—Hardnose Skates

The Rajidae, in some arrangements (e.g. McEachran and Dunn 1998; McEachran and Aschliman 2004), are considered monophyletic, with all species occurring within this family due largely to their similarity in appearance. In the present arrangement this family has 13 genera, approximately 138 valid species, and at least 45 undescribed and dubious species world-wide.

Genus *Amblyraja* Malm, 1877—Stout skates. A wide-ranging, circumglobal, genus mostly found at higher latitudes and in deep tropical waters, mostly on continental and insular shelves and slopes, and around seamounts. Ten valid species and two undescribed species.

Amblyraja badia (Garman, 1899)—Broad skate
Amblyraja doellojuradoi (Pozzi, 1935)—Southern thorny skate
Amblyraja frerichsi (Krefft, 1968)—Thickbody skate
Amblyraja georgiana (Norman, 1938)—Antarctic starry skate
Amblyraja hyperborea (Collette, 1879)—Arctic skate

Amblyraja jenseni (Bigelow and Schroeder, 1950)—Jensen's skate

Amblyraja radiata (Donovan, 1808)—Thorny skate

Amblyraja reversa (Lloyd, 1906)—Reversed skate

Amblyraja robertsi (Hulley, 1970)—Bigmouth skate

Amblyraja taaf (Meisner, 1987)—Whiteleg skate

Genus *Breviraja* Bigelow and Schroeder, 1948—Lightnose skates. Found in the eastern and western Atlantic, mostly on continental slopes. Six described and possibly one undescribed species.

Breviraja claramaculata McEachran and Matheson, 1985—Bright spotted skate

Breviraja colesi Bigelow and Schroeder, 1948—Lightnose skate

Breviraja marklei McEachran and Miyake, 1987—Nove Scotia skate

Breviraja mouldi McEachran and Matheson, 1995—Blacknose skate

Breviraja nigriventralis McEachran and Matheson, 1985—Blackbelly skate

Breviraja spinosa Bigelow and Schroeder, 1950—Spiny skate

Genus *Dactylobatus* Bean and Weed, 1909—Skilletskates. Endemic to the western North Atlantic, primarily found on continental slopes. Two valid species.

Dactylobatus armatus Bean and Weed, 1909—Skillet skate

Dactylobatus clarki (Bigelow and Schroeder, 1958)—Hook skate

Genus *Dipturus* Rafinesque, 1810—Longnosed skates. A wide-ranging genus in cool temperate to tropical waters, found mostly on continental shelves and slopes. At least 31 species with 15 or more undescribed species; several species within this genus are questionably placed here.

Dipturus batis (Linnaeus, 1758)—Gray skate

Dipturus bullisi (Bigelow and Schroeder 1962)—Tortugas skate

Dipturus campbelli (Wallace, 1967)—Blackspot skate

Dipturus chilensis (Guichenot, 1848)—South American Yellownose skate

Dipturus crosnieri (Seret, 1989)—Madagascar skate

Dipturus diehli Soto and Mincarone, 2001—Thorny tail skate

Dipturus doutrei (Cadenat, 1960)—Javelin skate

Dipturus ecuadoriensis (Beebe and Tee-Van, 1941)—Ecuador skate

Dipturus garricki (Bigelow and Schroeder, 1958)—San Blas skate

Dipturus gigas (Ishiyama, 1958)—Giant skate

Dipturus gudgeri (Whitley, 1940)—Bight skate

Dipturus innominata (Garrick and Paul, 1974)—New Zealand smoothskate

Dipturus johannisdavisi (Alcock, 1899)—Travancore skate

Dipturus kwangtungensis (Chu, 1960)—Kwangtung skate

Dipturus laevis (Mitchell, 1817)—Barndoor skate

Dipturus lanceorostrata (Wallace, 1967)—Rattail skate

Dipturus leptocauda (Krefft and Stehmann, 1975)—Thintail skate

Dipturus linteus (Fries, 1838)—Sailray

Dipturus macrocauda (Ishiyama, 1955)—Japanese bigtail skate

Dipturus mennii Gomes and Parago, 200—South Brazilian skate

Dipturus nasutus (Banks *in* Müller and Henle 1841)—New Zealand longnose skate

Dipturus nidarosiensis Collett, 1880—Norwegian skate

Dipturus olseni (Bigelow and Schroeder, 1951)—Spreadfin skate

Dipturus oregoni (Bigelow and Schroeder, 1958)—Hooktail skate

Dipturus oxyrinchus (Linnaeus, 1758)—Longnosed skate

Dipturus pullopunctata (Smith, 1964)—Slime skate

Dipturus springeri (Wallace, 1967)—Roughbelly skate

Dipturus stenorhynchus (Wallace, 1967)—Prownose skate

Dipturus teevani (Bigelow and Schroeder, 1951)—Florida skate

Dipturus tengu (Jordan and Fowler, 1903)—Japanese acutenose skate

Dipturus trachyderma Krefft and Stehmann, (1975)—Roughskin skate

Genus *Fenestraja* McEachran and Compagno, 1982—Pluto skates. A widely scattered group with representatives in the western North Atlantic, western Indian Ocean, and western Pacific, found mostly on continental and insular slopes. Eight species.

Fenestraja atripinna (Bigelow and Schroeder, 1950)—Blackfin skate
Fenestraja cubensis (Bigelow and Schroeder, 1950)—Cuban skate
Fenestraja ishiyamai (Bigelow and Schroeder, 1962)—Plain pygmy skate
Fenestraja maceachrani (Seret, 1989)—Madagascar skate
Fenestraja mamillidens (Alcock, 1889)—Prickly skate
Fenestraja plutonia (Garman, 1881)—Pluto skate
Fenestraja sibogae (Weber, 1913)—Siboga pygmy skate
Fenestraja sinusmexicanus (Bigelow and Schroeder, 1950)—Gulf of Mexico pygmy skate

Genus *Gurgesiella* de Buen, 1959—Finless pygmy skates. Western South Atlantic and eastern South Pacific, these skates occur on continental and insular slopes. Three described species.

Gurgesiella atlantica (Bigelow and Schroeder, 1962)—Atlantic skate
Gurgesiella dorsalifera McEachran and Compagno, 1980—Onefin skate
Gurgesiella furvescens (de Buen, 1959)—Dusky finless skate

Genus *Leucoraja* Malm, 1877—Rough skates. This genus is most diverse in the North Atlantic and Mediterranean Sea, but includes a few outlying endemic species off southern Africa and Australia, mostly found on continental shelves and slopes. Twelve described and three undescribed species.

Leucoraja circularis (Couch, 1838)—Sandy skate
Leucoraja compagnoi (Stehmann, 1995)—Tigertail skate
Leucoraja erinacea (Mitchell, 1825)—Little skate
Leucoraja fullonica (Linnaeus, 1758)—Shagreen skate
Leucoraja garmani (Whitley, 1939)—Rosette skate
Leucoraja lentiginosa (Bigelow and Schroeder, 1951)—Freckled skate
Leucoraja leucosticta (Stehmann, 1971)—White-dappled skate
Leucoraja melitensis (Clark, 1926)—Maltese skate
Leucoraja naevus (Müller and Henle, 1841)—Cuckoo skate
Leucoraja ocellata (Mitchell, 1815)—Winter skate
Leucoraja wallacei (Hulley, 1970)—Yellowspot skate
Leucoraja yucatanensis (Bigelow and Schroeder, 1950)—Yucatan skate

Genus *Malacoraja* Stehmann, 1970—Soft skates. Restricted to the Atlantic Ocean, except for one undescribed species from the southwestern Indian Ocean, these are mostly deep-slope skates. Four described species and one possibly undescribed species.

Malacoraja kreffti (Stehmann, 1978)—Krefft's skate
Malacoraja obscura de Carvalho, Gomes, and Gadig, 2005—Dusky skate
Malacoraja senta (Garman, 1885)—Smooth skate
Malacoraja spinacidermis (Barnard, 1923)—Roughskin skate

Genus *Neoraja* McEachran and Compagno, 1982—Pygmy skates. Western North Atlantic and eastern Atlantic, mostly found on continental slopes. Four species.

Neoraja africana Stehmann and Seret, 1983—African skate
Neoraja caerulea (Stehmann, 1976)—Blue skate
Neoraja carolinensis McEachran and Stehmann, 1984—Carolina skate
Neoraja stehmanni (Hulley, 1972)—African dwarf skate

Genus *Okamejei* Ishiyama, 1958—Spiny rasp skates. This genus is most diverse in the western Pacific, but with outlier species in the Indian Ocean, occurs on continental shelves and slopes. Fourteen described and at least two undescribed species.

Okamejei acutispina (Ishiyama, 1958)—Japanese sharpspine skate
Okamejei australis (Macleay, 1884)—Common skate

Okamejei boesemani (Ishihara, 1987)—Black sand skate
Okamejei cerva (Whitley, 1939)—Whitespotted skate
Okamejei heemstrai (McEachran and Fechhlem, 1982)—East African skate
Okamejei hollandi (Jordan and Richardson, 1909)—Yellowspot skate
Okamejei kenojei (Müller and Henle, 1841)—Ocellate spot skate
Okamejei koreana (Jeong and Nakabo, 1997)—Korean skate
Okamejei lemprieri (Richardson, 1845)—Thornback skate
Okamejei meerdervoorti (Bleeker, 1860)—Bigeye skate
Okamejei philipi (Lloyd, 1906)—Aden ringed skate
Okamejei pita (Fricke and Al-Hussar, 1995)—Pita skate
Okamejei powelli Alcock, 1898—Indian ringed skate
Okamejei schmidti Ishiyama, 1958—Browneye skate

Genus *Raja* Linnaeus, 1758—Ocellate skates. Primarily found in the eastern North Atlantic, Mediterranean and Black seas, but with some species ranging into the southwestern Indian Ocean, usually found on continental and insular shelves and slopes. Thirteen valid species, two dubious species, and at least one undescribed species.

Raja asterias Delaroche, 1809—Starry skate
Raja brachyura Lafont, 1873—Blonde skate
Raja clavata Linnaeus, 1758—Thornback skate
Raja herwigi Krefft, 1965—Cape Verde skate
Raja maderensis Lowe, 1838—Madera skate
Raja microocellata Montagu, 1818—Small-eyed skate
Raja miraletus Linnaeus, 1758—Twineyed skate
Raja montagui Fowler, 1910—Spotted skate
Raja polystigma Regan, 1923—Speckled skate
Raja radula Delaroche, 1809—Rough skate
Raja rondeleti Bougis, 1959—Rondelet's skate
Raja straeleni Poll, 1951—False thornback skate or Biscuit skate
Raja undulata Lacepede, 1802—Undulate skate

Genus *Rajella* Stehmann, 1970—Gray skates. This genus is most diverse in the Atlantic, but with outliers in the eastern South Pacific, Indian Ocean, and Australia, mostly on continental and insular shelves and slopes. Fifteen valid species and at least four undescribed species, plus two dubious species.

Rajella annandalei (Weber, 1913)—Indonesian round skate
Rajella barnardi (Norman, 1935)—Bigthorn skate
Rajella bathyphila (Holt and Byrne, 1908)—Deepwater skate
Rajella bigelowi (Stehmann, 1978)—Bigelow's or chocolate skate
Rajella caudispinosa (Von Bonde and Swart, 1923)—Munchkin skate
Rajella dissimilis (Hulley, 1970)—Ghost skate
Rajella eisenhardti Long and McCosker, 1999—Eisenhardt's skate
Rajella fuliginea (Bigelow and Schroeder, 1954)—Sooty skate
Rajella fyllae (Lutken, 1887)—Round skate
Rajella kukujevi (Dolganov, 1985)—Mid-Atlantic skate
Rajella leopardus (von Bonde and Swart, 1923)—Leopard skate
Rajella nigerrima (de Buen, 1960)—Blackish skate
Rajella purpuriventralis (Bigelow and Schroeder, 1962)—Purplebelly skate
Rajella ravidula (Hulley, 1970)—Smoothback skate
Rajella sadowskii (Krefft and Stehmann, 1974)—Brazilian skate

Genus *Rostroraja* Hulley, 1972—Spearnose skates. A single species ranging from the eastern Atlantic to the southwestern Indian Ocean, mostly found on continental shelves and uppermost slopes.

Rostroraja alba (Lacepede, 1803)—Spearnose skate

Species nominally retained in *Raja* representing undescribed genus-group taxa

Genus "A" for North Pacific *Raja* Assemblage McEachran and Dunn, 1998. Six species.

Raja binoculata Girard, 1854—Big skate
Raja cortezensis McEachran and Miyake, 1988—Sea of Cortez skate
Raja inornata Jordan and Gilbert, 1881—California skate
Raja pulchra Liu, 1932–Mottled skate
Raja rhina Jordan and Gilbert, 1880—Longnose skate
Raja stellulata Jordan and Gilbert, 1880—Starry skate

Genus "B" for Amphi-American *Raja* Assemblage McEachran and Dunn, 1998. Seven species.

Raja ackleyi Garman, 1881—Ocellate skate
Raja bahamensis Bigelow and Schroeder, 1965—Bahama skate
Raja cervigoni Bigelow and Schroeder, 1964—Venezuela skate
Raja eglanteria Bosc, 1802—Clearnose skate
Raja equatorialis Jordan and Bollman, 1890—Equatorial skate
Raja texana Chandler, 1921—Roundel skate
Raja velezi Chirichigno, 1973—Rasptail skate

Genus "C" for western South Pacific species with provisional generic and familiar placement. Two described and at least three undescribed species.

Raja polyommata Ogilby, 1910—Argus skate
Raja whitleyi Iredale, 1938—Melbourne skate

Family Anacanthobatidae von Bonde and Swart, 1924—Legskates

McEachran and Dunn (1998) reduced this group to a sister taxon within the subfamily Rajinae. A small group, only two genera, these skates have distinctive leglike anterior pelvic fin lobes, and includes 18 valid species plus at least five undescribed species.

Genus *Anacanthobatis* von Bonde and Swart, 1924—Smooth legskates. A wide ranging, but sporadically distributed group found in the western North and Central Atlantic, Indian Ocean, and the western Central Pacific, on continental and insular slopes. Ten described and at least five undescribed species.

Anacanthobatis americanus Bigelow and Schroeder 1962—American legskate
Anacanthobatis borneensis Chan, 1965—Borneo skate
Anacanthobatis donghaiensis (Deng, Xiong and Zhan, 1983)—East China legskate
Anacanthobatis folirostris (Bigelow and Schroeder, 1951)—Leafnose legskate
Anacanthobatis longirostris Bigelow and Schroeder 1962—Longnose legskate
Anacanthobatis marmoratus (von Bonde and Swart, 1924)—Spotted legskate
Anacanthobatis melanosomus (Chan, 1965)—Blackbellied legskate
Anacanthobatis nanhaiensis (Meng and Li, 1981)—South China legskate
Anacanthobatis ori (Wallace, 1967)—Black legskate
Anacanthobatis stenosomus (Li and Hu, 1982)—Narrow legskate

Genus *Cruriraja* Bigelow and Schroeder 1948—Legskates. Western North and Central Atlantic, eastern South Atlantic, and southwestern Indian Ocean, found on continental and insular shelves and slopes. Eight valid species.

Cruriraja andamanica (Lloyd, 1909)—Andaman legskate
Cruriraja atlantis Bigelow and Schroeder, 1948—Atlantic skate
Cruriraja cadenati Bigelow and Schroeder 1962—Broadfoot legskate
Cruriraja durbanesis (von Bonde and Swart, 1924)—Smoothnose legskate
Cruriraja 'parcomaculata' Smith, 1964 (not von Bonde and Swart, 1924)—Roughnose skate
Cruriraja poeyi Bigelow and Schroeder, 1948—Cuban legskate
Cruriraja rugosa Bigelow and Schroeder 1958—Rough legskate
Cruriraja triangularis Smith, 1964 (=*C. parcomaculata* von Bonde and Swart, 1924)—Triangular skate

References

Bigelow HB, Schroeder WC (1948) New genera and species of batoid fishes. J Mar Res 7:543–566
Bigelow HB, Schroeder WC (1950) New and little known cartilaginous fishes from the Atlantic. Bull Mus Comp Zool Harvard 103:385–408

Bigelow HB, Schroeder WC (1951a) A new genus and species of anacanthobatid skate from the Gulf of Mexico. J Wash Acad Sci 41:110–113

Bigelow HB, Schroeder WC (1951b) Three new skates and a new chimaerid fish from the Gulf of Mexico. J Wash Acad Sci 41:383–392

Bigelow HB, Schroeder WC (1953) Fishes of the western North Atlantic. Mem Sears Found Mar Res 1(pt 2):588

Bigelow HB, Schroeder WC (1954) A new family, a new genus, and two new species of batoid fishes from the Gulf of Mexico. Breviora, Mus Comp Zool 24:1–16

Bigelow HB, Schroeder WC (1958) Four new rajids from the Gulf of Mexico. Bull Mus Comp Zool 119:201–233

Bigelow HB, Schroeder WC (1962) New and little known batoid fishes from the western Atlantic. Bull Mus Comp Zool 128:159–244

Bigelow HB, Schroeder WC (1964) A new skate *Raja cervigoni*, from Venezuela and the Guianas. Breviora Mus Comp Zool 209:1–5

Bigelow HB, Schroeder WC (1965) A further account of batoid fishes from the western Atlantic. Bull Mus Comp Zool 132:446–475

Carvalho MR de (1996) Higher-level elasmobranch phylogeny, basal squaleans, and paraphyly. In: Stiassny MLJ, Parenti LR, Johnson GD (eds) Interrelationships of fishes. Academic Press, San Diego, California, USA, pp 35–62

Carvalho MR de, Gomes UL, Gadig OBF (2005) Description of a new species of skate of the genus *Malacoraja* Stehmann, 1970: the first species from the southwestern Atlantic Ocean with notes on generic monophyly and composition (Chondrichthyes: Rajidae). Neotrop Ichthy 3:239–258

Chen CT, Joung SJ (1989) Fishes of the genus *Raja* (Rajiformes: Rajidae) from Taiwan. J Taiwan Mus 42:1–12

Compagno LJV (1999) Checklist of living elasmobranchs. In: Hamlett WC (ed) Sharks, skates, and rays: the biology of elasmobranch fishes. The Johns Hopkins University Press, Baltimore, MD, pp 471–498

Compagno LJV (2001) Sharks of the world. An annotated and illustrated catalogue of shark species known to date. Vol. 2. Bullhead, mackerel, and carpet sharks (Heterodontiformes, Lamniformes, and Orectolobiformes). FAO Species Catalogue for Fishery Purposes. No. 1, vol 2. Rome, FAO, 2001, 269 p

Compagno LJVC (2005) Checklist of living chondrichthyes. In: Hamlett WC (ed) Reproductive biology and phylogeny of chondrichthyes: sharks, batoids, and chimaeras. Science Publishers, Inc, Enfield, New Hampshire, USA, pp 501–548

Compagno LJV, Ebert DA (2007) Southern African skate biodiversity. Environ Biol Fish (this volume)

Compagno LJV, Ebert DA, Cowley PD (1991) Distribution of offshore demersal cartilaginous fishes (Class Chondrichthyes) off the west coast of southern Africa, with notes on their systematics. S Afr J Mar Sci 11:43–139

Compagno LJV, Ebert DA, Smale MJ (1989) Guide to the sharks and rays of southern Africa. Struik Pub, Cape Town, 160 pp

Cousseau MB, Figueroa DE, Diaz de Astarloa JM (2000) Clave de identificacion de las rayas del litoral maritime de Argentina y Uruguay (Chondrichthyes, Familia Rajidae). Publicaciones especiales, INIDEP, Mar del Plata, 35 pp

Dolganov VN (1983) Preliminary review of skates Rajidae family (Pacific coast of North America). Izv Tikhookean Nauchno-Issled Inst Rybn Khoz Okeanogr 107:56–72

Dolganov VN (1985) New aspects of the ray family Rajidae. Russian version of ''New species of skates of the family Rajidae from the northwestern Pacific Ocean''. J Ichthyol 25:121–132

Ebert DA (2003) The sharks, rays and chimaeras of California. University California Press, Berkeley, California, 284 pp

Eschmeyer WN, Herald ES, Hammond H (1983) A field guide to Pacific coast fishes of North America, vol 28. Houghton Mifflin Co Field Guide, 336 pp

Fowler HW (1941) The fishes of the groups Elasmobranchii, Holocephali, Isospondyli, and Ostariophysi obtained by United States Bureau of Fisheries Steamer Albatross in 1907 to 1910, chiefly in the Philippine Islands and adjacent seas. Bull US Natn Mus (100) 13:i–x, 1–879

Gomes UL (2002) Revisão taxonômica da família Rajidae no Brazil (Chondrichthyes, Elasmobranchii, Rajiformes). Unpublished doctoral thesis, Universidade Federal do Rio de Janeiro, 286 pp

Hulley PA (1970) An investigation of the Rajidae of the west and south coasts of southern Africa. Ann S Afr Mus 55:151–220

Hulley PA (1972a) The origin, interrelationships and distribution of Southern African Rajidae (Chondrichthyes, Batoidei). Ann S Afr Mus 60:1–103

Hulley PA (1972b) A new species of southern African Brevirajid skate (Chondrichthyes, Batoidei, Rajidae). Ann S Afr Mus 60:253–263

Ishihara H (1987) Revision of the Western North Pacific species of the genus *Raja*. Japan J Ichthyol 34:241–285

Ishihara H, Ishiyama R (1985) Two new North Pacific skates (Rajidae) and a revised key to *Bathyraja* in the area. Japan J Ichthyol 32:143–179

Ishihara H, Ishiyama R (1986) Systematics and distribution of the skates of the North Pacific (Chondrichthyes, Rajoidei). Indo-Pacific Fish Biology: Proc 2nd Int Conf Indo-Pacific Fishes 269–280

Ishiyama R (1952) Studies on the rays and skates belonging to the family Rajidae, found in Japan and adjacent regions. 4. A revision of three genera of Japanese rajids, with descriptions of one new genus and four new species mostly occurred in northern Japan. J Shimonoseki Coll Fish 1:1–34

Ishiyama R (1955) Studies on the rays and skates belonging to the family Rajidae, found in Japan and adjacent regions. 6. *Raja macrocauda*, a new skate. J Shimonoseki Coll Fish 4:43–51

Ishiyama R (1958) Studies on the Rajid fishes (Rajidae) found in the waters around Japan. J Shimonoseki Coll Fish 7:193–394

Ishiyama R (1967) Fauna Japonica. Rajidae (Pisces). Biogeogr Soc Japan, 84 pp

Ishiyama R, Ishihara H (1977) Five new species of skates in the genus *Bathyraja* from the western North Pacific, with reference to their interspecific relationships. Japan J Ichthy 24:71–90

Kuhnz LA, Bizzarro JJ, Chaney L, Ebert DA (2006) *In situ* video observations of deep-living skates and rays in the central and eastern North Pacific. 22nd Annual Meeting of the American Elasmobranch Society, abstract

Lamilla J, Sáez S (2003) Clave taxonomica para el reconocimiento de especies de rayas chilenas (Chondrichthyes, Batoidei). Invest Mar, Valparaiso 31:3–16

Last PR, McEachran JD (2006) New softnose skate genus *Brochiraja* from New Zealand (Rajidae: Arhynchobatinae) with description of four new species. NZ J Mar Freshwater Res 40:65–90

Last PR, Stevens JD (1994) Sharks and rays of Australia. CSIRO Division of Fisheries, Melbourne, Australia, 605 pp

Last PR, Yearsley GK (2002) Zoogeography and relationships of Australasian skates (Chondrichthyes: Rajidae). J Biogeogr 29:1627–1641

Leigh-Sharpe WH (1920) The comparative morphology of the secondary sexual characters of elasmobranch fishes. Memoir I J Morph 34:245–265

Leigh-Sharpe WH (1921) The comparative morphology of the secondary sexual characters of elasmobranch fishes. Memoir II J Morph 35:359–380

Leigh-Sharpe WH (1922a) The comparative morphology of the secondary sexual characters of elasmobranch fishes. Memoir III J Morph 36:191–198

Leigh-Sharpe WH (1922b) The comparative morphology of the secondary sexual characters of Holocephali and elasmobranch fishes. Memoir IV J Morph 36:199–220

Leigh-Sharpe WH (1922c) The comparative morphology of the secondary sexual characters of elasmobranch fishes. Memoir V J Morph 39:221–243

Leigh-Sharpe WH (1924a) The comparative morphology of the secondary sexual characters of elasmobranch fishes. Memoir VI J Morph 39:553–566

Leigh-Sharpe WH (1924b) The comparative morphology of the secondary sexual characters of elasmobranch fishes. Memoir VII J Morph 42:567–577

Leigh-Sharpe WH (1926a) The comparative morphology of the secondary sexual characters of elasmobranch fishes. Memoir VIII J Morph 42:307–320

Leigh-Sharpe WH (1926b) The comparative morphology of the secondary sexual characters of elasmobranch fishes. Memoir IX J Morph 42:321–334

Leigh-Sharpe WH (1926c) The comparative morphology of the secondary sexual characters of elasmobranch fishes. Memoir X J Morph 42:335–348

Leigh-Sharpe WH (1926d) The comparative morphology of the secondary sexual characters of elasmobranch fishes. Memoir XI J Morph 42:349–358

Malm AW (1877) Goteborgs och Bohusläns fauna, Ryggradjuren. Göteborg. Göteborgs Bohusläns Fauna, 674 pp

McEachran JD (1982) Revision of the South American skate genus *Sympterygia* (Elasmobranchii: Rajiformes). Copeia 4:867–890

McEachran JD (1983) Results of the research cruises of FRV "Walther Herwig" to South America. LXI. Revision of the South American skate genus *Psammobatis* Günther, 1870 (Elasmobranchii: Rajiformes, Rajidae). Arch FischWiss 34:23–80

McEachran JD, Aschliman N (2004) Phylogeny of Batoidea. In: Carrier JC, Musick JA, Heithaus MR (eds) Biology of sharks and their relatives. CRC Press, Boca Raton, FL, pp 79–114

McEachran JD, Compagno LJV (1982) Interrelationships of and within *Breviraja* based on anatomical structures (Pisces: Rajoidei). Bull Mar Sci 33:399–425

McEachran JD, Dunn KA (1998) Phylogenetic analysis of skates, a morphologically conservative clade of elasmobranchs (Chondrichthyes: Rajidae). Copeia 2:271–290

Menni RC, Stehmann MFW (2000) Distribution, environment, and biology of batoid fishes off Argentina, Uruguay, and Brazil. A review Rev Mus Argentino Cienc Nat 2:69–109

Seret B (1986) Deep water skates from Madagascar. Part 1. Anacanthobatidae (Pisces, Chondrichthyes, Batoidea), second record of the skate *Anacanthobatis ori* (Wallace, 1967) from off Madagascar. Cybium 10:307–326

Seret B (1989a) Deep water skates of Madagascar. Part 2. Rajidae. (Pisces, Chondrichthyes, Batoidea) *Gurgesiella (Fenestraja) maceachrani* sp. n. Cybium 13:55–64

Seret B (1989b) Deep water skates of Madagascar. Part 3. Rajidae. (Pisces, Chondrichthyes, Batoidea) *Raja (Dipturus) crosnieri* sp. n. Cybium 13:115–130

Stehmann M (1970) A taxonomic rearrangement of the northeastern Atlantic Rajidae (Chondrichthyes, Batoidea) based on comparative morphological and anatomical studies. Arch FischWiss 21:73–164

Stehmann M (1995) First and new records of skates (Chondrichthyes, Rajiformes, Rajidae) from the West African contintental slope (Morocco to South Africa), with descriptions of two new species. Arch Fish Mar Res 43:1–119

Stehmann M, Bürkel DL (1984) Rajidae. In: Whitehead PJP, Bauchot M-L, Hureau JC, Tortonese E (eds) Fishes of the Northeastern Atlantic and Mediterranean. UNESCO, Paris, pp 163–196

Stevenson DE (2004) Identification of skates, sculpins, and smelts by observers in North Pacific groundfish fisheries (2002–2003). NOAA Tech Memo NMFS-AFSC-142

Stevenson DE, Orr JW (2005) Records of two deepwater skate species from the eastern Bering Sea. Northwest Nat 86:71–81

Wallace JH (1967) The batoid fishes of the east coast of southern Africa. Part III: skates and electric rays. S Afr Assoc Mar Biol Res, Oceanogr Res Inst, Invest Rep 17:1–62

Whitley GP (1931) Studies in Ichthyology No 4. Rec Aust Mus v 18:96–133

Whitley GP (1939) Taxonomic notes on sharks and rays. Aust Zool 9:227–262

Zorzi GD, Anderson ME (1988) Records of the deep-sea skates, *Raja (Amblyraja) badia* Garman, 1899 and *Bathyraja abyssicola* (Gilbert, 1896) in the Eastern North Pacific, with a new key to California skates. Calif Fish Game 74:87–105

Southern African skate biodiversity and distribution

Leonard J. V. Compagno · David A. Ebert

Originally published in the journal Environmental Biology of Fishes, Volume 80, Nos 2–3, 125–145.
DOI 10.1007/s10641-007-9243-4 © Springer Science+Business Media B.V. 2007

Abstract The skates (Family Rajidae) have 12 genera and possibly 28 species off southern Africa (southern Angola, Namibia, South Africa and Mozambique). The geographic and bathymetric distribution and the taxonomic composition of the southern African skate fauna are analysed and the distribution mapped. The southern African skate fauna is best known off the temperate west coast of South Africa from the intertidal to approximately 1,200 meters, but poorly known below 1,200 m and sketchily known in warm-temperate and tropical parts of the area. Southern African skates of the temperate continental shelves above 100 m are not diverse and regularly include one species of the genus *Dipturus*, one species of *Leucoraja*, two species of *Raja* (including *R. straeleni*, the most abundant skate in southern African waters) and the giant skate *Rostroraja alba*. All of these skates are 'shelf overlap' species that range onto the outer shelves and uppermost slopes, and none are confined to inshore environments. Skate diversity increases on the outer shelves and upper slopes. At least half of the skate species are endemic to the southern African region; other species also occur off East or West Africa, a few extend to European waters, and records of one species, *Amblyraja taaf*, appear to be of strays from nearby sub-Antarctic seas. The genus *Bathyraja* and softnose skate group (Arhynchobatinae) are surprisingly limited (a single species) in deep-water off southern Africa (unlike other regions including the Antarctic), and almost all of southern African skates are members of the Rajinae. Amongst rajines, the tribes Amblyrajini (*Amblyraja*, two species, *Leucoraja*, two species, and *Rajella*, five species) Rajini (*Dipturus*, six species, *Okamejei*, one species, *Raja*, two species, and *Rostroraja*, one species), and Anacanthobatini (*Anacanthobatis*, two species, and *Cruriraja*, three species) predominate, while Gurgesiellini has a species of *Neoraja* and possibly two of *Malacoraja*.

Keywords Rajidae · Bathymetric and geographic distribution · Biodiversity · Systematics · Zoogeography · Southern Africa

L. J. V. Compagno (✉)
Shark Research Center, Iziko – Museums of Cape Town, Cape Town 8000, South Africa
e-mail: lcompagno@iziko.org.za

D. A. Ebert
Pacific Shark Research Center, Moss Landing Marine Laboratories, Moss Landing, CA, USA
e-mail: debert@mlml.calstate.edu

Introduction

This paper is an extension of our previous and ongoing work on the biodiversity and general biology of the cartilaginous fish fauna (Class

Chondrichthyes) of southern Africa (Compagno et al. 1989; Compagno et al. 1991; Compagno 1999; Compagno and Ebert in preperation). It focuses on the systematics, zoogeography and bathymetry of skates (Rajidae) as a major component of the southern African chondrichthyan fauna. Skates comprise approximately 12.3–12.6% of known chondrichthyan species from off southern Africa. Skate taxa discussed below are from a checklist in Compagno (2005) and a taxonomic database on cartilaginous fishes maintained by the senior author. It does not include lists of coordinate data for the station records comprising the study, but presents them as digital maps.

The zoogeographic boundaries of southern Africa for cartilaginous fishes used here are from Compagno et al. (1989), and include the coasts of Namibia, South Africa and Mozambique. Smith (1949, 1965) and Smith and Heemstra (1986) place the boundaries of southern Africa for systematic ichthyology at 20° S on both coasts. The boundaries are somewhat arbitrary in both cases, and for chondrichthyans it can be argued that on the west coast the temperate South African chondrichthyan fauna extends northwards to southern Angola, where it meets the tropical Atlantic fauna of West Africa. The warm-temperate and tropical chondrichthyan fauna off kwaZulu-Natal and Mozambique in turn blends with that off East Africa and Madagascar, with southern African taxa extending to East Africa and to Madagascar. Madagascar does have several endemic species not found in southern Africa (Seret 1986a) and a distinctive spectrum of species to that of southern Africa.

Southern Africa has long been recognized as a chondrichthyan faunal province with considerable endemism (Engelhardt 1913; Reif and Saure 1987; Compagno 1988, 1999) and this is true of the skates as part of that fauna. Over 50% of southern African skate species are endemics (Fig. 2C). The following account of the biodiversity of the southern African rajoid fauna extends that of the southern African chondrichthyan fauna by Compagno (1999). A checklist of southern African skates is presented in Table 1.

Southern Africa has one of the most diverse faunas of cartilaginous fishes (Class Chondrichthyes) and of skates in the world despite its relatively short coastline (Compagno 1999) and compares favourably in chondrichthyan and skate diversity to adjacent areas with coastlines three to 16 times as long and with relatively well-known chondrichthyan faunas. For comparison we list the skate diversity and coastline lengths of the CLOFNAM area (Eastern North Atlantic, Mediterranean and Black Seas), the United States (all coasts), and Australia (Table 2).

Skates are conspicuous by their absence from fresh-water lakes and rivers with only a single species present in Tasmanian fresh-water and brackish estuaries (Last and Yearsley 2002). Skates are bottom-dwellers that are known from the intertidal to the mid-slope, with few on the deep slopes and rises (below 1,500 m) and apparently with no pelagic taxa. Skates are generally found in deeper water towards the Equator (tropical submergence) and in shallow water in higher latitudes. Skates are generally absent from inshore environments in tropical seas where they are largely replaced by stingrays (Myliobatoidei).

Materials and methods

The South African Marine and Coastal Management (MCM, formerly Sea Fisheries Research Institute) research vessel *Africana* based in Cape Town, from 1985 to present has and is conducting yearly demersal cruises aimed at stock assessment of the Cape and deepwater hakes (*Merluccius capensis* and *M. paradoxus*), which are dominant bottom predators, are prey for many other demersal and pelagic species, and are the subject of a major bottom fishery along the west and southeast coast of southern Africa (Payne et al. 1987). Usually three separate month-long hake biomass cruises take place each year, initially with one on the southeast coast and a summer (usually in January) and winter (usually in July) west coast cruise. This schedule has varied over the years with ship availability and switches to two cruises a year or two southeast coast cruises and one west coast cruise.

The current *Africana* is the third and most capable MCM vessel of that name, being a 78 m, 2452 gross ton research vessel rigged as a stern trawler and capable of operating a 60 m German

Table 1 Checklist of southern African skates

Family Rajidae. Skates.
  Subfamily Arhynchobatinae, softnose skates.
Bathyraja smithii (Müller and Henle, 1841). African softnose skate.
  Subfamily Rajini, hardnose skates.
    Tribe Amblyrajini, rough skates.
Amblyraja robertsi (Hulley 1970). Bigmouth skate.
Amblyraja taaf (Meisner 1987). Whiteleg skate.
Leucoraja compagnoi (Stehmann 1995). Tigertail skate.
Leucoraja wallacei (Hulley 1970). Yellowspotted skate.
Rajella barnardi (Norman, 1935). Bigthorn skate.
Rajella caudaspinosa (von Bonde and Swart 1923). Munchkin skate.
Rajella dissimilis (Hulley 1970). Ghost skate.
Rajella leopardus (von Bonde and Swart 1923). Leopard skate
Rajella ravidula (Hulley 1970). Smoothback skate.
    Tribe Gurgesiellini, dwarf skates.
Malacoraja spinacidermis (Barnard, 1923). Roughskin skate.
Malacoraja sp. kwaZulu-Natal roughskin skate.
Neoraja stehmanni (Hulley, 1972). African pygmy skate.
    Tribe Rajini, typical skates.
Dipturus campbelli (Wallace 1967). Blackspot skate.
Dipturus doutrei (Cadenat, 1960). Javelin skate.
Dipturus lanceorostratus (Wallace 1967). Rattail skate.
Dipturus pullopunctatus (Smith 1964). Graybelly or slime skate.
Dipturus springeri (Wallace 1967). Roughbelly skate.
Dipturus stenorhynchus (Wallace 1967). Prownose skate.
Okamejei heemstrai (McEachran and Fechhelm, 1982). Narrow skate.
Raja miraletus Linnaeus, 1758. Twineyed skate
Raja straeleni Poll, 1951. Biscuit skate.
Rostroraja alba (Lacepede, 1803). Spearnose skate.
    Tribe Anacanthobatini, legskates.
Anacanthobatis marmoratus (von Bonde and Swart 1923). Smooth legskate.
Anacanthobatis ori (Wallace 1967). ORI legskate.
Cruriraja durbanensis (von Bonde and Swart 1923). Smoothnose legskate.
Cruriraja 'parcomaculata' Smith, 1964 (*not* von Bonde and Swart 1923). Roughnose legskate.
Cruriraja parcomaculata (von Bonde and Swart 1923) = *C. triangularis* Smith 1964. Triangular legskate.

bottom trawl down to 1,000–2,000 m as well as a variety of other collecting and sampling gear. We informally term her *Africana III* in this paper to distinguish her from *Africana I* and *II*, the two previous vessels which also contributed data to the present survey. The *Africana III* program has been invaluable in exploring the biodiversity and general natural history of the chondrichthyan fauna of southern Africa (Compagno et al. 1991; Compagno 1999). Thousands of specimens were examined for biological information, and extensive collections of cartilaginous fishes from *Africana* cruises were and are being preserved intact or skeletonized and are deposited at the South African Museum, Cape Town (now Iziko – South African Museum), and the South African Institute for Aquatic Biodiversity (formerly the J.L.B. Smith Institute of Ichthyology), Grahamstown. Skates were routinely collected on *Africana III's* bottom trawl stations over the past few decades. For details of the sampling methodology of the *Africana* program see Payne et al. (1984, 1985) and Compagno et al. (1991).

Bottom trawl stations were conducted by *Africana III* between Walvis Bay (23° S, 14° E), Namibia and the Agulhas Bank west of Cape Agulhas (36° S, 20° E) to Port Alfred, South Africa (34° S, 27° E), with skates collected during 31 cruises and 1,154 bottom trawl stations from 1985 to 1996 with a depth range of 17–1,150 m on the continental shelf and upper slope. After 1990 west coast cruises extended only to the Orange River mouth, South Africa and not into Namibian waters. Trawl stations were hour-long bottom

Table 2 Coastline lengths and diversity data for four chondrichthyan faunal areas including southern Africa. Coastline lengths from Microsoft Encarta electronic database; generic and species numbers from LJVC spreadsheet checklists

Area	Coastline length Km	Coastline length × SA	Species Chondrichthyes	Genera Skates	Species Skates
Southern Africa	6400	1	223	12	28
CLOFNAM	101866	15.9	141	10	32
United States	19929	3.1	213	14	53
Australia	36735	5.7	300	9	41
World			978–1,190	28	232–294

hauls and coordinates for the stations utilized here were taken at the end of each trawl haul. Skates comprised 2237 *Africana III* station records (species per station) and averaged 1.9 species taken per skate station (range 1–7 skate species per station).

Species record data of skates and other taxa for cruises we participated in were logged aboard *Africana III* with dBase IV and later combined and converted to FoxPro for DOS (dbf format) and stored in a single large database. We combined *Africana III* station record data for skates including our published data on west coast skates (Compagno et al. 1991) and additional *Africana III* east coast skate data with available data from other research vessels and expeditions that have collected skates off southern Africa, including predecessors (*Pickle, Africana I* and *II*) and contemporaries (*Algoa, Benguela, Sardinops*) of *Africana III* in research service for South Africa as well as South African commercial trawlers that did research sampling. Particularly important was the extensive fisheries survey work of Spanish scientists at the Instituto del Mar, Barcelona (Allue et al. 1984; Lleonart and Rucabado 1984; Lloris 1986; Turon et al. 1986; Mas-Reira and MacPherson 1989) with various research vessels working off Namibia which has greatly elucidated the distribution of skates and other chondrichthyans. Station and locality records were utilized from various published works listed under individual species.

Each database record includes fields for species name, family code, ordinal code, weight of catch, coast (east or west), station, cruise number, block data, date, depth, and latitude and longitude (including decimal coordinates). Additional coordinates for our database was added from hard copy or other databases and analyzed and initially mapped with Quattro Pro (Compagno et al. 1991) but eventually analyzed with Excel. Maps for the present paper were plotted with the shareware program Versamap using ASCII print data files for species coordinates and a CIA-derived high-resolution mapping database of Africa, with 200 and 1,000 m isobaths digitized from a bathymetric chart of southern Africa (Dingle et al. 1987).

A map of southern Africa with all the skate station coordinates plotted as hollow circles (Fig. 1) shows the extent and limits of our survey to date. Note that stations on the continental slope below 1,000 m are few and mostly confined to the west coast of South Africa. Coverage off Namibia is mostly on the upper slope between 200 and 1,000 m with the shelf fauna poorly known; it is mostly based on Spanish *Benguela* and *Valdivia* data and *Africana III* stations. The west coast of South Africa from the Orange River to Cape Agulhas is relatively well covered between 30 and 1,200 m and mostly from *Africana III* data; with gaps in data close inshore and deeper stations down to 1,350 m due in part to a cruise of the commerical trawler *Iris*. The southeast coast of South Africa from Cape Agulhas is again relatively well covered thanks to *Africana III*, including better inshore coverage than the west coast due to coastal surveys for sole (Soleidae); however deep stations below 1,000 m are very few and mostly off the Eastern Cape coast of South Africa from Mossel Bay to Port Elizabeth. A major gap in skate stations is present from East London to the southern border of kwaZulu-Natal, which has a very narrow shelf and is difficult to trawl because of the strong Agulhas current and precipitous slope. The kwaZulu-Natal coast is sketchily covered except around Durban (Wallace 1967), and the deep skate fauna below 1,000 m is essentially unknown off

Fig. 1 Map of southern Africa from northern Namibia to northern Mozambique, with 2676 skate station records plotted with hollow circles. 2237 stations are from the South African Marine and Coastal Managements research trawler *Africana III* (in silhouette). The map includes codes for distribution areas used in Table 3

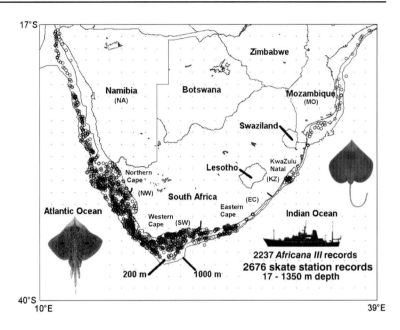

kwaZulu-Natal. Unfortunately *Africana III* has never had a regular collecting program off kwaZulu-Natal. Coverage of the skate fauna off Mozambique is sketchy, as the deep slope fauna is little-known (Wallace 1967) and a single cruise of the *Algoa* in 1994 off Mozambique supplied most of our station records for skates there on the outer shelf and upper slope at depths of 37–500 m.

Depth distribution of southern African skates

Comparison of bathymetric ranges for our sample of skates (Fig. 2A) reveals the skate fauna of southern Africa to be primarily an offshore and deep water fauna, with most diversity on the upper slope and outer shelf. As with the chondrichthyan fauna off the west coast of southern Africa (Compagno et al. 1991) the rajid fauna of southern Africa is stratified by depth, so that a different spectrum of species occurred on the continental shelves and upper slopes at 700–1,000 m compared to the shelf at 0–200 m although with much overlap of different species.

Depth zonation for southern African skates in our sample is less clear than in west coast cartilaginous fishes in general (Compagno et al. 1991) because of poor and erratic sampling on the east coast and because individual skate species often have a broad depth range and overlap with other species. Figure 2A indicates that there are no inshore species of skates presently known off southern Africa that are confined to the continental shelf from 0 to 100 m depth. There are some inshore overlap species that occur inshore at depths of 50 m or less and may enter shallow bays and the intertidal zone, including *Rostroraja alba*, *Raja miraletus*, and *Raja straeleni*, but these range onto the upper slopes to varying depths. Shelf overlap species occur on the outer shelf below 50 m, and generally range down the slope to 500 m or more. Many skates fit this category, including *Dipturus campbelli*, *D. pullopunctatus*, and *D. springeri*, *Leucoraja wallacei*, *Neoraja stehmanni*, *Rajella barnardi*, *R. caudaspinosa*, *R. leopardus*, *Cruriraja 'parcomaculata'*, *C. 'triangularis'* and possibly *Anacanthobatis marmoratus*. Upper slope species have upper limits within 200–300 m depth on the slopes and may range down to 1,000 m or more; southern African species include *Bathyraja smithii*, *Dipturus doutrei*, and *D. stenorhynchus*. Intermediate slope species includes skates that occur at a minimal depth of 350–500 m as part of a broad transitional zone of slope inhabitants, many of which range below 700 m depth. These include *Rajella dissimilis* (young) and *Rajella ravidula*. Finally there are deep slope species which are poorly known and with a minimum depth of 600–1,000 m or more,

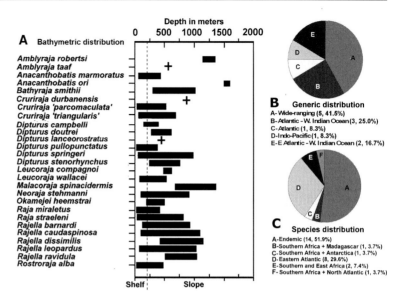

Fig. 2 Skate bathymetric and geographic records. (A) Graph of southern African skate species in alphabetical order, showing known depth ranges. Crosses represent single station records. Depth data for possible new *Malacoraja* from kwaZulu-Natal not available. (B–C) Skate geographic distribution types in southern Africa for B. Southern African genera. C. Southern African species. Note that there are no endemic genera, but over half the species are endemic

which includes *Malacoraja spinacidermis*, *Amblyraja robertsi*, and presumably *Anacanthobatis ori*. Skates do not include any eurybathic species with wide bathymetric ranges from close inshore to the deep slope, although *R. straeleni* comes closest to this.

Geographic distribution of southern African skates

The genera of southern African rajids include several wide-ranging taxa such as *Bathyraja*, *Amblyraja*, *Anacanthobatis* and *Dipturus*, while most of the remainder range into the northern Atlantic and western Indian Ocean (Fig. 2B). The southern African spectrum of skate genera has a decided similarity to those of the Eastern Atlantic with the exception of the presence of anacanthobatines which are most diverse in the western North Atlantic but are absent in the eastern Atlantic north of southern Africa. A single genus (*Okamejei*), however, is a primarily Indo-Pacific taxon with an endemic species (*O. heemstrai*) in the western Indian Ocean.

Southern African rajid species (Fig. 2C) are over half endemics, but include four species ranging to Madagascar, East Africa and Antarctica and a major component (nine species) that range to the Eastern Atlantic or have a bipolar distribution in the Eastern North Atlantic and southern Africa. Within southern Africa, skate species show zonation by latitude and longitude, with a fauna influenced by the cold Benguela current system and upwelling off Namibia and the west coast of South Africa down to Cape Agulhas. The cold skate fauna grades into the faunas influenced by the warm Agulhas Current including the skates in the warm-temperate area of the south-east coast from Cape Agulhas to the Eastern Cape and the subtropical to tropical skate faunas of kwaZulu-Natal and Mozambique. Table 3 shows the southern African rajifauna coded and sorted by geographic areas into distribution types including one or more of the areas. The coded areas include (see Fig. 1 and Table 1), from northwest to northeast, the Namibian coast (NA), the Atlantic coast of South Africa including the Northern Cape province and the Western Cape province to Cape Agulhas (NW), the south-western Indian Ocean coast of the Western Cape province (SW), the coast of the Eastern Cape province (EC), the coast of kwaZulu-Natal (KZ), and the coast of Mozambique (MO). Seven species occur in only one area (Table 3) but several extend to one or more adjacent areas and some range the entire coast of southern Africa. Distribution types that include two or more areas include EC to MO (1 species), KZ and MO (3 species), NA and NW (1 species), NA and NW, EC (4 species), NA and

Table 3 Southern African skates coded and sorted by distribution types from east to west. Abbreviations for geographic areas are listed from west to east: NA = Namibia. NW = Northern Cape province and western part of Western Cape provinces, Namibian border to Cape Agulhas. SW = Southwestern part of Western Cape province, Cape Agulhas to Plettenberg Bay. EC = Eastern Cape province, South Africa; KZ = kwaZulu-Natal, South Africa. MO = Mozambique. 1 = present, 0 = doubtful record

Species-area	NA	NW	SW	EC	KZ	MO	Distribution type
Rajella ravidula	1	1					NA and NW
Bathyraja smithii	1	1		1			NA and NW, EC
Dipturus doutrei	1	1		1			NA and NW, EC
Rajella barnardi	1	1		1			NA and NW, EC
Rajella dissimilis	1	1	0	1			NA and NW, EC
Cruriraja 'parcomaculata'	1	1	1	1			NA to EC
Dipturus pullopunctatus	1	1	1	1	0		NA to EC
Raja miraletus	1	1	1	1	1		NA to KZ
Leucoraja wallacei	1	1	1	1	1	1	NA to MO
Raja straeleni	1	1	1	1	0	1	NA to MO
Rostroraja alba	1	1	1	1	1	1	NA to MO
Rajella leopardus	1	1		1	1		NA and NW, EC and KZ
Dipturus springeri	1	1		1	1	1	NA and NW, EC to MO
Amblyraja robertsi		1					NW
Amblyraja taaf		1					NW
Cruriraja durbanensis		1					NW
Malacoraja spinacidermis	0	1					NW
Neoraja stehmanni		1		1			NW, EC
Rajella caudaspinosa	0	1	0	1			NW, EC
Leucoraja compagnoi		1		1	1		NW, EC and KZ
Dipturus stenorhynchus			0	1	1	1	EC to MO
Malacoraja sp.					1		KZ
Anacanthobatis marmoratus			0		1	1	KZ and MO
Cruriraja 'triangularis'					1	1	KZ and MO
Dipturus campbelli					1	1	KZ and MO
Anacanthobatis ori						1	MO
Dipturus lanceorostratus						1	MO
Okamejei heemstrai						1	MO
Area sums	13	20	6	16	11	11	

NW, EC and KZ (1 species), NA and NW, EZ to MO (1 species), NA to EC (2 species), NA to KZ (1 species), NA to MO (3 species, including the entire study area), NW and EC (2 species), and NW, EC and KZN (1 species).

The cold-water areas NA and NW have high skate diversity (13 and 20 species each), as do the transitional temperate EC (16 species) and the tropical–subtropical KZ and MO (11 species each), while SW has the least diversity (6 species). Southern Africa skate distribution recalls Ishiyama's (1967) division of Japanese skates into arhynchobatine species from cold northern areas (*Bathyraja* and *Rhinoraja*), and rajines from the southeast coast (*Dipturus* and *Okamejei*) that is warmed by the Kuroshio Current. Southern Africa is not as clear cut and more complicated than Japan, in part due to the different current systems and also the broad intermediate area between the Benguela system and the Agulhas Current as well as different and more complicated patterns of taxonomic diversity (more genera and higher groups than occur off Japan, with overlap in various areas). *Amblyraja* is confined to NW, *Bathyraja* mostly to the NA and NW areas with EC records, *Cruriraja*, *Dipturus*, *Leucoraja*, *Raja*, *Rostroraja* and possibly *Rajella* are wide-ranging, *Anacanthobatis* and *Okamejei* are confined to KZ and MO, *Neoraja* is in NW and EC, and *Malacoraja* is in NW and possibly KZ.

Family Rajidae Blainville, 1816

The classification of skates utilized here follows the cladistic analysis and classification of McEachran and Dunn (1998), which refined previous work by McEachran (1984) and McEachran and Miyake (1990a, 1990b). Their cladogram of skate genera (McEachran and Dunn 1998, Fig. 1) is followed here with additions from Last and Yearsley (2002) and Last and McEachran (2006) to show the relationships of southern African skate genera to other skates (Fig. 3A) and with the legskates placed in a separate tribe (Anacanthobatini).

The southern African skate fauna is noteworthy in having no endemic genera, unlike North and Central America, South America, and the Australian subregion. The predominance of genera of Rajinae and the paucity of Arhynchobatinae in the southern African fauna is shown in Fig. 3A, while the importance of species of *Dipturus* and *Rajella* and the tribe Anacanthobatini (legskates) in the fauna is indicated by Fig. 3B.

The southern African skate fauna is relatively 'new' in terms of description dates for the known species (Fig. 3C), with almost two-thirds of the species described from 1950 to the present. Earlier workers, including von Bonde and Swart (1923), Barnard (1925), and Smith (1949, 1965) had problems with the systematics and nomenclature of southern African skates, but this stabilized from the 1960s to the present with the work of Wallace (1967), Hulley (1966, 1969, 1970,

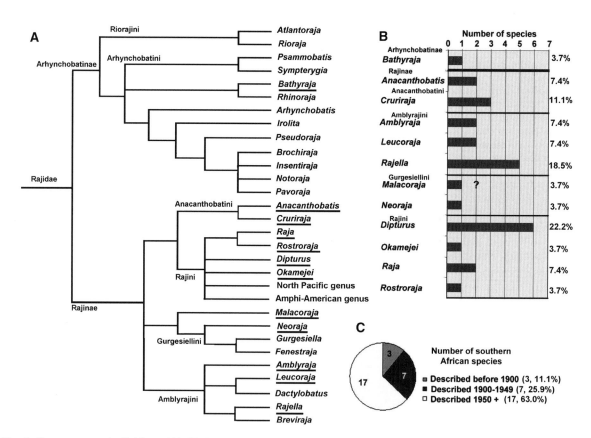

Fig. 3 Skate taxonomic divisions. (**A**) Cladogram of skate genera and higher taxa, modified from McEachran and Dunn (1998). Genera occuring in southern Africa are underlined. (**B**) Number of species in southern African skate genera. Note relative abundance of anacanthobatine, *Dipturus*, and *Rajella* species. (**C**) Number of southern African skate species described before 1900, between 1900 and 1950, and 1950 to present. Note that nearly two-thirds of the species were described after 1950

1972a, 1972b, 1973, 1986), Hulley and Stehmann (1977), Stehmann (1970, 1971, 1973, 1976, 1981, 1990, 1995), and our own work (Compagno et al. 1989, 1991; Compagno 1999). One suspects that additional rajid taxa will be described from southern Africa in the future, particularly from poorly sampled areas and below 1,300 m depth, and possibly there will be range extensions of species from adjacent areas.

Bathyraja smithii (Müller and Henle, 1841) (Fig. 4A)

Our sample of *B. smithii* includes 60 station records mostly on the west coast of South Africa and from 295 to 1,040 m depth. Distributional data is mostly from *Africana III* stations, with additional data from Hulley (1970), Turon et al. (1986), Stehmann (1995), and the commercial trawler *Iris* (R. Leslie, Marine and Coastal Management, personal communcation). The African softnose skate is a southern African deep-slope endemic of the west and southeast coasts, with a range from north-central Namibia to Cape Agulhas and a single record from the Eastern Cape. Compagno et al. (1991) noted that a nominal locality for one of the syntypes from the Bosphorus, between the Black Sea and Sea of Marmara in Turkey, is probably erroneous. *Bathyraja eatoni* from Kerguelen Island has been synonymized with this species but is distinct (Stehmann and Bürkel 1990; Compagno et al. 1991; Stehmann 1995). Most records of *B. smithii* are from below 600 m.

Amblyraja robertsi (Hulley 1970) (Fig. 4B)

This large, distinctive deepwater skate was described by Hulley (1970) from a single specimen from west of Cape Town. Compagno et al. (1991) noted that *A. robertsi* has not been recorded since its original description, but subsequently *Africana III* collected additional material from two stations near the type locality in 1,141–1,150 m. *Amblyraja robertsi* is apparently a deepwater South African endemic, close to but distinct from *A. hyperborea* and similar deep-dwelling *Amblyraja* skates (M. Endicott, University of London, personal communication).

Amblyraja taaf (Meisner 1987) (Fig. 4C)

This skate was reported by Hulley (1970) from two specimens taken by a commercial trawler off Cape Town in 548–640 m. Hulley identified it as *Raja radiata* Donovan 1808, = *Amblyraja radiata*, a common North Atlantic offshore species (Stehmann and Bürkel 1984), with a depth range of 55–604 m. In view of the abundance, wide depth and geographic range, and distinctiveness of *A. radiata* Compagno et al. (1991) found it surprising that it did not show up in *Africana III* trawl stations off southern Africa. However, the two specimens proved to belong to a sub-Antarctic species, described by Meisner (1987) as *A. taaf* (M. Endicott, personal communication, and confirmed by us and by P.A. Hulley personal communication). *Amblyraja taaf* is otherwise known from just south of the Crozet and Kerguelen Islands and well south of South Africa and Madagascar (Stehmann and Bürkel 1990). The South African records of *A. taaf* may prove to be of strays from its known range in sub-Antarctic seas.

Leucoraja compagnoi (Stehmann 1995) (Fig. 4D)

This rare skate was described by Stehmann (1995) from a single specimen collected by the Soviet research vessel *Poltava* off the west coast of South Africa northwest of Strandfontein. While his work was in progress we recorded additional specimens from four trawl stations by *Africana III* and *Benguela* from the upper slopes of South Africa near the type locality, in the Eastern Cape off Port Elizabeth and off kwaZulu-Natal. It is likely that this species can be confused with the abundant *L. wallacei* and could be overlooked by fisheries biologists that are not familiar with it.

Leucoraja wallacei (Hulley 1970) (Fig. 4E)

This common, distinctive skate is an endemic of southern Africa, with a range from southern Namibia to southern Mozambique at 73–517 m. Most of our 322 station records were from *Africana III* trawls but included here are a few additional records from the west and east coasts

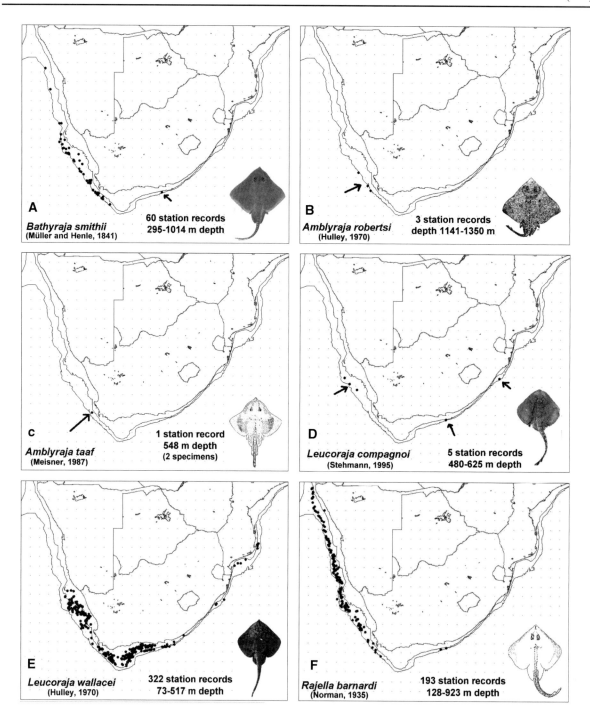

Fig. 4 Maps of southern Africa with records of skates (Rajidae) of the subfamily Arhynchobatinae (**A**) and subfamily Rajinae, tribe Amblyrajini (**B–F**) indicated by solid circles. A. *Bathyraja smithii*. B. *Amblyraja robertsi*. C. *Amblyraja taaf*. D. *Leucoraja compagnoi*. E. *Leucoraja wallacei*. F. *Rajella barnardi*

of South Africa by von Bonde and Swart (1923), Wallace (1967), Hulley (1970), and *Algoa* stations off southern Mozambique. Most records are from between 150 and 300 m. *Leucoraja wallacei* shows a distinctly bimodal geographic distribution off the west and southeast coasts of South Africa, with most of the records concentrated in an area between the Orange River and Cape Columbine, and a second area between Cape Point and Cape Agulhas to Algoa Bay.

Rajella barnardi (Norman, 1935) (Fig. 4F)

This distinctive common deepwater skate was named *Raja confundens* by Hulley (1970) who relegated *R. barnardi* to synonymy of *R. leopardus*, but Stehmann (1995) revived *R. barnardi* and synonymized *R. confundens* with it after comparing the holotypes of both species. *Rajella barnardi* occurs off tropical West Africa and the west coast of southern Africa from Mauritania to South Africa. Most of our 193 station records are from *Africana III* stations off the west coast of southern Africa and from Turon et al. (1986) from off Namibia. We also have a few station records of *R. barnardi* from *Africana III* off Jeffreys Bay and Port Elizabeth, Eastern Cape, from the commercial trawler *Iris* from the west coast of South Africa (R. Leslie, personal communcation), and data from Hulley (1970), Mas-Riera and MacPherson (1989), and Stehmann (1995) from off Namibia and the west coast of South Africa. Our depth range of 128–923 m mostly includes records below 200 m. Stehmann (1995) had additional records of this species from Rio de Oro, Mauritania, Senegal, the Gulf of Guinea, Gabon, and Angola.

Rajella caudaspinosa (von Bonde and Swart 1923) (Fig. 5A)

This southern African endemic skate is the smallest member of the subgenus *Rajella* in the area. There is a kwaZulu-Natal record of *R. caudaspinosa* but this skate is otherwise known from the upper slopes of Namibia and South Africa from Lüderitz to Cape Point and east to Algoa Bay in water 102–1,098 m deep with almost all records below 200 m. Most of our 60 station records are from *Africana III* stations but it was also present in a few stations of the *Pickle* (von Bonde and Swart 1923), *Africana I* (von Bonde 1933), and the commercial trawler *Iris* (R. Leslie, personal communication). Our survey did not collect *R. caudaspinosa* in Namibian waters north of the Orange River.

Rajella dissimilis (Hulley 1970) (Fig. 5B)

This deep-slope skate was described by Hulley (1970) from deep water off Cape Town, South Africa. *Africana III* collected *R. dissimilis* along the deep slope from southwest of Walvis Bay and off Lüderitz in Namibian waters, South Africa from the Orange River to southwest of Cape Point (Compagno et al. 1991) and the Eastern Cape from off Plettenberg Bay east to St Francis Bay in water 420–1,150 m deep. Most of our records of *R. dissimilis* are from below 700 m. Most of our 38 station records are from *Africana III* stations but also several records from Hulley (1970), the commercial trawlers *Iris* and *Boulonnias*, and Stehmann (1995). Stehmann (1995) recorded *R. dissimilis* from off Rio de Oro, Western Sahara, in water 1,200–1,640 m deep, and the Norwegian research vessel *F. Nansen* recently collected some small skates off Angola (R. Leslie, personal communication), which may be this species.

Rajella leopardus (von Bonde and Swart 1923) (Fig. 5C)

This deepwater skate is the commonest upper slope skate in the area along with *R. barnardi*. Von Bonde and Swart (1923) described it from the west coast of South Africa northwest of Saldanha Bay, and off Durban, kwaZulu-Natal. Most of our 149 station records are from *Africana III* stations mostly off the west coast of southern Africa from Walvis Bay to Cape Agulhas, with a few records from the Eastern Cape off St. Francis Bay and Port Elizabeth. We also include records from Turon et al. (1986), Mas-Riera and MacPherson (1989), Stehmann (1995), and the commercial trawler *Iris* (R. Leslie, personal communication). Stehmann (1995) recorded this species from off Mauritania, Senegal and Guinea. The majority of records were between 200 and 600 m, but there were a few records on the outer shelf to 73 m and a scattering of deep slope records down to 1,026 m.

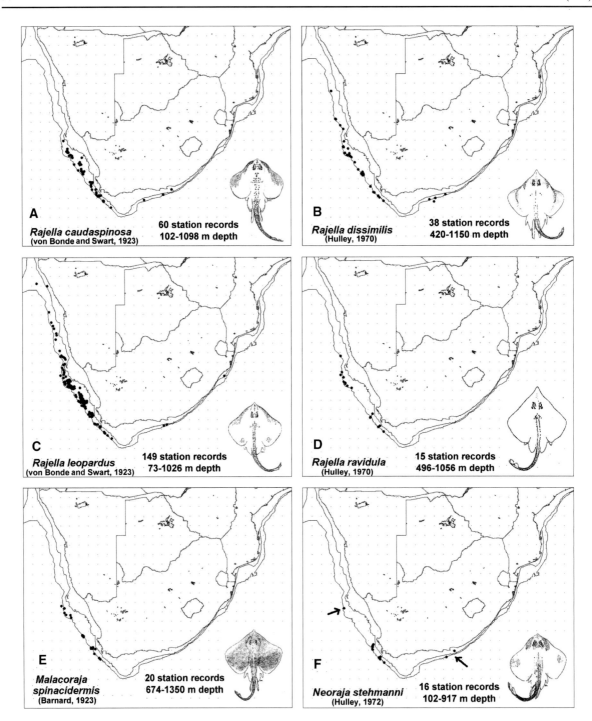

Fig. 5 Maps of southern Africa with records of skates (Rajidae, Rajinae) of the tribes Amblyrajini (**A–D**) and Gurgesiellinae (**E–F**) indicated by solid circles. A. *Rajella caudaspinosa*. B. *Rajella disimillis*. C. *Rajella leopardus*. D. *Rajella ravidula*. E. *Malacoraja spinacidermisi*. F. *Neoraja stehmanni*

Rajella ravidula (Hulley 1970) (Fig. 5D)

This deep-slope skate was described by Hulley (1970) off the west coast of South Africa on the deep slope off Cape Town, and we recorded it from 16 station records at depths of 496–1,056 m. Most of our records are from below 700 m depth. *Africana III* collected it southwest of Lüderitz off Namibia and in two areas from southwest of the Orange River to off Lambert's Bay, and from Saldanha Bay to southwest of Cape Point. The commercial trawler *Iris* also collected it off the Orange River mouth. Stehmann (1995) had a record of this species from deep water off Morocco (near the Canary Islands).

Malacoraja spinacidermis (Barnard, 1923) (Fig. 5E)

This is an uncommon, distinctive, moderate-sized, deepwater skate of the mid-slopes from southern Namibia to south of Cape Point, South Africa. It occurs in depths of 674–1,350 m with most records below 800 m. Of our 20 station records most are from *Africana III* collections, but we also include records from Hulley (1970) and Hulley and Stehmann (1977), including one station from the *Africana II*, and several from the commercial trawlers *Boulonnias* and *Iris*. Stehmann (1995) presents a record of this species from Rio de Oro, Western Sahara. *Malacoraja mollis* (Bigelow and Schroeder 1950) occurs in the North Atlantic (Stehmann and Bürkel 1984) and has been synonymized with *M. spinacidermis* (Stehmann 1970, 1973), but may be a separate species (Stehmann 1990, 1995).

Malacoraja sp.

Recently two hatchling skates were obtained from off kwaZulu-Natal, South Africa from the Oceanographic Research Institute, Durban, which are probably not identical with *M. spinacidermis* and may be an undescribed species of *Malacoraja*.

Neoraja stehmanni (Hulley, 1972) (Fig. 5F)

This deepwater skate, one of the smallest southern African rajids along with *Anacanthobatis marmoratus* and *Cruriraja 'triangularis'*, is an endemic which has been recorded from 16 station records from southwest of the Orange River mouth to south of Cape Point, South Africa, and off St. Francis Bay and Algoa Bay in the Eastern Cape. Most records are from *Africana III* and mostly are concentrated from off Saldanha Bay to south of Cape Point in water 102–917 m deep (usually below 600 m). Hulley (1972b) notes two station records (types) from *Africana II*. As presently known the distribution of *N. stehmanni* is unusually localized compared to other offshore southern African rajids, but this may change with further deep-slope exploration.

Dipturus campbelli (Wallace 1967) (Fig. 6A)

Dipturus campbelli is known from seven station records on the outer shelf and upper slope off kwaZulu-Natal, South Africa, and southern Mozambique in 137–403 m depth (four of these from Wallace 1967, with estimated coordinates). *Dipturus campbelli* has been synonymised with *D. pullopunctatus* (Hulley 1970) but is evidently distinct (Compagno et al. 1989). We examined the holotype of *D. campbelli* and obtained additional material from one station record by *Benguela* off kwaZulu-Natal and two by *Algoa* off Mozambique.

Dipturus doutrei (Cadenat, 1960) (Fig. 6B)

This Atlantic African skate has a broad range from Mauritania to Namibia and South Africa, with extremes including an *Africana III* record from south of Cape Agulhas and a second record from south of Port Elizabeth. Our 38 station records are mostly from north-central Namibia from the Cunene River to SW of Lüderitz, and include mostly records by Hulley (1970), Allue et al. (1984), Lleonart and Rucabado (1984), Turon et al. (1986), Stehmann (1995), and the South African *Benguela*. This species is apparently seldom recorded south of Namibia, and in

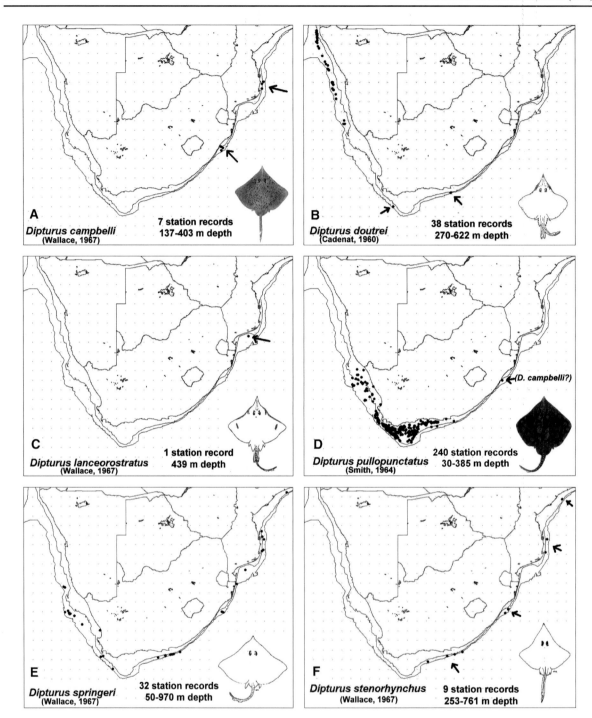

Fig. 6 Maps of southern Africa with records of skates (Rajidae, Rajinae, Rajini) indicated by solid circles. A. *Dipturus campbelli*. B. *Dipturus doutrei*. C. *Dipturus lanceorostratus*. D. *Dipturus pullopunctatus*. E. *Dipturus springeri*. F. *Dipturus stenorhynchus*

our station records ranges from 270 to 622 m with most below 300 m.

Dipturus lanceorostratus (Wallace 1967) (Fig. 6C)

This distinctive moderate-sized skate is known from three specimens (including the types) from off the Limpopo River mouth in Mozambique. We did not obtain additional records of this skate although the *Algoa* collected skates of other species in similar depths during an exploratory cruise off Mozambique in 1994.

Dipturus pullopunctatus (Smith 1964) (Fig. 6D)

Dipturus pullopunctatus was often confused with the European skate *Dipturus batis* (Linnaeus 1758) and the American *D. laevis* (Mitchill, 1817) prior to Smith's (1964) description. This is a common endemic southern African skate with a range from Lüderitz and possibly Walvis Bay, Namibia at least to Algoa Bay, South Africa and possibly to off Durban, South Africa. Our 240 station records are mostly from *Africana III* stations from the Orange River mouth to Algoa Bay at depths of 30–385 m. Additional specimens mentioned by Smith (1964) from 29° 59'S, 31° 07'E and 230 m depth off Natal need confirmation on identity as they might be *D. campbelli*.

Dipturus springeri (Wallace 1967) (Fig. 6E)

This uncommon giant deepwater skate (up to 1.9 m long) was recorded from kwaZulu-Natal, Mozambique (Wallace 1967), Madagascar (Compagno et al. 1989) and Kenya (Hulley 1986; Ochumba 1984, 1988), but collecting by *Africana III* extended its range to the west coast of southern Africa from Lüderitz, Namibia, to Cape Agulhas, South Africa, and the Eastern Cape from south of Cape St. Francis to southeast of Algoa Bay. Additional station records came from *Iris* off the Orange River, *Benguela* off kwaZulu-Natal, and *Algoa* off Mozambique. We have 32 station records from 50 to 970 m, with most on the upper slope between 200 and 600 m but with a few on the shelf at 50–194 m.

Dipturus stenorhynchus (Wallace 1967) (Fig. 6F)

This species was originally collected off Mozambique and known only from the holotype, but is currently known from scattered records off the southeastern coast of South Africa, southern Mozambique, and Kenya (Ochumba 1984, 1988). We include nine station records from 253 to 761 m depth, including Wallace's (1967) type locality, *Africana III* records from the southeastern Cape, South Africa off Plettenberg Bay to southeast of Algoa Bay, *Benguela* records off kwaZulu-Natal, and *Algoa* records from Mozambique.

Okamejei heemstrai (McEachran and Fechhelm, 1982) (Fig. 7A)

The most southwestern member of its genus, *O. heemstrai* was described from material collected by P.C. Heemstra from off Kenya. *Algoa* collected additional material of this distinctive species from three stations 200–500 m depth off central and southern Mozambique in 1994 and thus extended its range to southern Africa.

Raja miraletus (Linnaeus, 1758) (Fig. 7B)

This common, distinctive and attractve little skate is endemic to the Eastern Atlantic, Mediterranean and Western Indian Ocean, where it may form distinct subpopulations (McEachran et al. 1989). In southern Africa it occurs from Angola to southern Namibia in the vicinity of the Orange River, with an apparent gap in distribution along the Western Cape coast to off Cape Town and False Bay, where it reappears and ranges northeast to Natal, Mozambique, and Kenya (Ochumba 1984, 1988). Our 148 station records includes primarily *Africana III* stations from the Western Cape to Port Alfred, but also includes South African records from von Bonde (1933), Smith (1964), and Wallace (1967) and Namibian data from Allue et al. (1984) and Turon et al. (1986). Depth range for our records is 17–417 m with most records on the shelf at <200 m.

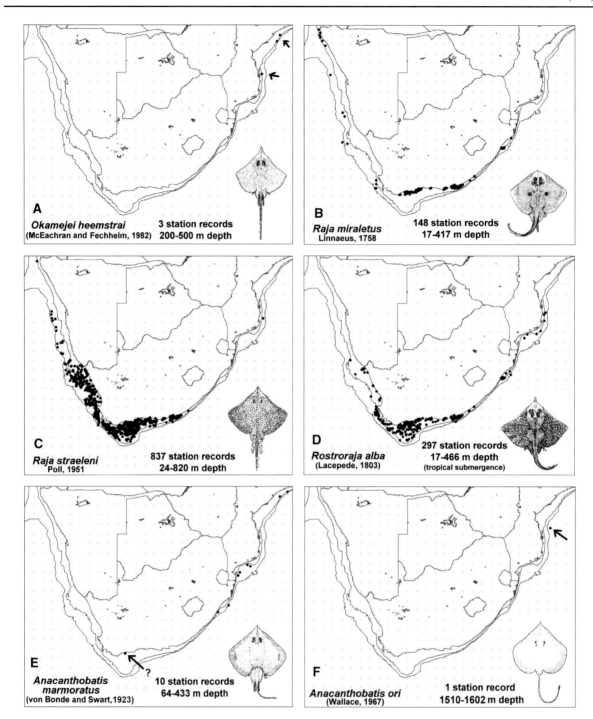

Fig. 7 Maps of southern Africa with records of skates (Rajidae, Rajinae) of the tribes Rajini (**A–D**) and Anacahthobatini (**E–F**) indicated by solid circles. A. *Okamejei heemstrai*. B. *Raja miraletus*. C. *Raja straeleni*. D. *Rostroraja alba*. E. *Anacanthobatis marmoratus*. F. *Anacanthobatis ori*

Raja straeleni (Poll, 1951) (Fig. 7C)

Classification and nomenclature of this skate have been problematical, with earlier literature often identifying it as the European thornback skate *Raja clavata* (Linnaeus, 1758) or considered a separate species, *R. rhizacanthus* (Regan, 1906) or *R. capensis* (Müller and Henle, 1841) (the latter a junior homonym, replaced by *R. bonaespaensis* (Fowler, 1910)). We use *R. straeleni* following Compagno et al. (1989, 1991). *Raja straeleni* is the commonest skate in southern African waters. We have 837 station records mostly from *Africana III* and at depths of 24–820 m, with most on the shelf between 24 and 200 m. Additional records are from von Bonde and Swart (1923), Wallace (1967), Smith and Smith (1966), *Algoa* (one station off Mozambique), *Sardinops* (longline, off Western Cape), and Lleonart and Rucabado (1984), Lloris (1986), and Turon et al. (1986) from Namibia.

Rostroraja alba (Lacepede, 1803) (Fig. 7D)

This is a giant skate (up to 2.5 m long) that is endemic to the eastern Atlantic, Mediterranean, and southwestern Indian Ocean. It ranges from the United Kingdom and the Mediterranean south to the Canary Islands, Mauritania and Senegal, with a gap in its known distribution south to central Namibia, where it ranges to Cape Agulhas, the Eastern Cape, kwaZulu-Natal, Mozambique, Madagascar (Seret 1986a) and Kenya (Ochumba 1984, 1988). Most of our 297 station records are from *Africana III* on the west coast from southern Namibia to Cape Agulhas and particularly the southeast coast from Cape Agulhas to Algoa Bay and Port Alfred. Additional records are from von Bonde (1933), Smith (1964), Wallace (1967), *Sardinops* (longline stations off South Africa), and *Algoa* (Mozambique). Our depth range is 17–466 m with most on the shelf above 200 m. Stations off kwaZulu-Natal and Mozambique average deeper (down to 466 m) than those on the eastern and western Cape (where it can occur in the intertidal).

Anacanthobatis marmoratus (von Bonde and Swart 1923) (Fig. 7E)

This is a dwarf skate of the outer shelf and uppermost slope that is known from a few records off kwaZulu-Natal and Mozambique and apparently is a southern African endemic. We have ten station records, including data from von Bonde and Swart (1923), Wallace (1967), and *Algoa* (Mozambique) with depths from 200 to 433 m. Two Western Cape records by *Africana I* (von Bonde 1933) collected near Mossel Bay in 64–66 m require confirmation as their area of capture has been intensively surveyed by *Africana III* without a single specimen being captured.

Anacanthobatis ori (Wallace 1967) (Fig. 7F)

This is a little-known deep slope skate collected from off Bazaruto Island, central Mozambique (Wallace 1967). It was subsequently collected from off Madagascar (Seret 1986a, b) at 1,000–1,725 m and is currently known from four juvenile specimens including the holotype and paratype. We examined the holotype but did not obtain additional material.

Cruriraja durbanensis (von Bonde and Swart 1923) (Fig. 8A)

This rare and distinctive skate differs from other southern African *Cruriraja* in lacking rostral spines and in having a dark brown dorsal surface and largely dark brown ventral surface (except for white areas around the mouth and on the abdomen). *Cruriraja durbanensis* was not recorded by any other research or commercial vessel since its collection by the *Pickle* off Hondeklip Bay (not kwa-Zulu Natal, despite its species name suggesting it was caught off Durban). We suspect that it may show up with further deep slope trawling off the west coast of South Africa, particularly in stations below 1,000 m depth.

Cruriraja 'parcomaculata' sensu (Smith 1964) (not von Bonde and Swart 1923) (Fig. 8B)

Smith (1964), followed by Wallace (1967), Hulley (1970), Hulley (1986) and Compagno et al. (1989,

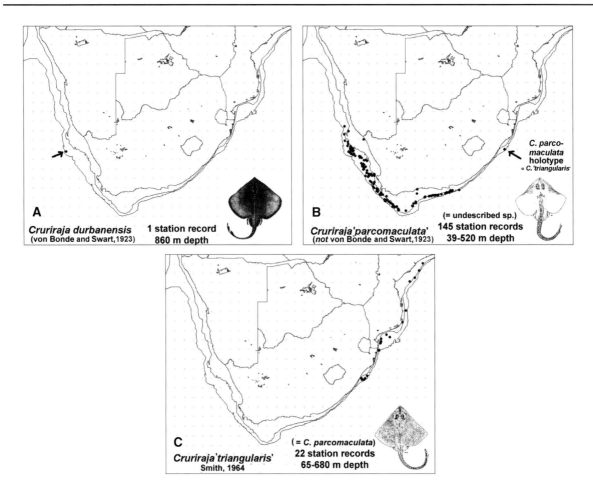

Fig. 8 Maps of southern Africa with records of skates (Rajidae, Rajinae, Anacanthobatini) indicated by solid circles. A. *Cruriraja durbanensis*. B. *Cruriraja 'parcomaculata'* of Smith, 1934 and subsequent authors, = an undescribed species. C. *Cruriraja 'triangularis'* Smith, 1964, = *C. parcomaculata* (von Bonde and Swart 1923)

1991), assumed that the *Cruriraja* present in the Western and Eastern Cape was the same as von Bonde and Swart's *R. parcomaculata* from kwaZulu-Natal. Smith redescribed *C. parcomaculata* apparently from Eastern Cape material but did not examine the holotype (or syntypes) in the BMNH collection. Smith's '*C. parcomaculata*' has a bluntly triangular disc and very numerous dark spots in the young. Smith further noted that two *Cruriraja* specimens collected from off Durban were distinct from his '*C. parcomaculata*' and named them *C. triangularis*. Our examination of a photo of the type of '*C. parcomaculata*' from the BMNH collection, the original descriptions and illustrations of *R. parcomaculata* and *C. triangularis*, a series of '*C. triangularis*' including spotted juveniles from kwaZulu-Natal and Mozambique, and a large collection of '*C. parcomaculata*' from the Western and Eastern Cape coasts of South Africa suggests that Smith erred in arranging these two species. *C. triangularis* Smith 1964 is apparently a junior synonym of *Raia parcomaculata* von Bonde and Swart 1923, while *C. parcomaculata* of Smith 1964 and subsequent authors apparently needs a replacement name.

'*C. parcomaculata*' *sensu* Smith (1964) is a moderately common skate and a southern African endemic. It occurs off Namibia and the west and Eastern Cape coasts of South Africa, from Lüderitz to Algoa Bay and possibly East London. Most of our 145 station records are from *Africana III* but also include records by Smith (1964),

Wallace (1967), and Turon et al. (1986). Depths were 39–545 m, with most station records between 200 and 500 m. Very few specimens were caught in any one trawl station, but the catch ranged from hatchlings to adults.

Cruriraja parcomaculata (von Bonde and Swart 1923) = *Cruriraja 'triangularis'* (Smith 1964) (Fig. 8C)

See '*C. parcomaculata*' sensu Smith 1964, above, for a discussion of the nomenclature of this species. This little endemic southern African skate is apparently smaller than '*C. parcomaculata*' *sensu* (Smith 1964), and may replace it in offshore waters of kwaZulu-Natal and Mozambique. Our records are from *Benguela* stations off kwaZulu-Natal, *Algoa* stations from Mozambique, von Bonde and Swart (1923; *Pickle* station off Durban), Smith (1964) and Wallace (1967). We have 22 station records from 65–680 m depth, with most stations below 200 m.

Acknowledgements We would like to thank A.I.L. Payne, C. J. Augustyn, B. Rose, G. Brill, and R. Leslie of Marine and Coastal Management, Cape Town, the late Capt. D. Krige and the officers and crew of RS *Africana*, P.D. Cowley of the South African Institute of Aquatic Biodiversity, M. Smale of the Port Elizabeth Museum, P. A. Hulley, the late M. A. Compagno Roeleveld, C. Goliath, M. Bougaardt, and E. Hoensen of the South African Museum, M. Endicott of the University of London, E. MacPherson of the Instituto de Ciencias del Mar, Barcelona, and the captain, officers and crew of the Spanish research vessel *Chicha Tousa* for much support on matters related to this paper. L.J.V. Compagno's research was supported by partial and comprehensive research grants and SANCOR grants from the South African Foundation for Research Development (FRD) and National Science Foundation (NRF) at the South African Museum and the former J.L.B. Smith Institute of Ichthyology, and by funds from the Fisheries Development Corporation and the J.L.B. Smith Institute of Ichthyology. D. A. Ebert's research was supported by a FRD predoctoral bursary, by Rhodes University, by the Shark Research Center, and by NOAA/NMFS to the National Shark Research Consortium, and the Pacific Shark Research Center.

References

Allue C, Borruel D, Lloris D, Rucabado JA (1984) Datos pesqueros de la campana "Benguela II". In: Rucabado JA, Bas C (eds) Resultados de las expediciones oceanografico-pesqueras "Benguela I" (1979) y "Benguela II" (1980) realizadas en el Atlantico sud-oriental (Namibia). Dat Inf Inst Inv Pesq Barcelona (9):85–190

Barnard KH (1925) A monograph of the marine fishes of South Africa. Part I (*Amphioxus*, Cyclostomata, Elasmobranchii, and Teleostei-Isospondyli to Heterosomata). Ann S Afr Mus 21(1):1–418

Compagno LJV (1988) Shark diversity in southern Africa. Naturalist 32(3):20–29

Compagno LJV (1999) An overview of chondrichthyan systematics and biodiversity in southern Africa. Trans Royal Soc S Afr 54(1):75–120

Compagno LJV (2005) Checklist of living chondrichthyes. In: Hamlett WC (ed) Reproductive biology and phylogeny of chondrichthyes: sharks, batoids, and chimaeras. Science Publishers, Inc., pp 501–548

Compagno LJV, Ebert DA, Cowley PD (1991) Distribution of offshore demersal cartilaginous fishes (Class Chondrichthyes) off the west coast of southern Africa, with notes on their systematics. S Afr J Mar Sci 11:43–139

Compagno LJV, Ebert DA, Smale MJ (1989) Guide to the sharks and rays of southern Africa. Struik Publishers, Cape Town, 160 pp

Dingle RV, Birch GF, Bremner JM, De Decker RH, Du Plessis A, Englebrecht JC, Fincham MJ, Fitton T, Flemming BW, Gentle RI, Goodlad SW, Martin AK, Mills EG, Moir GJ, Parker RJ, Robson SH, Rogers J, Salmon DA, Siesser WG, Simpson ESW, Summerhayes CP, Westall F, Winter A, Woodborne MW (1987) Deep-sea sedimentary environments around southern Africa (south-east Atlantic and south-west Indian Oceans). Ann S Afr Mus 98(1):1–27

Engelhardt R (1913) Monographie der Selachier der Münchener Zoologischen Staatssammlung (mit besonderer Berücksichtigung der Haifauna Japans). I. Teil: Tiergeographie de Selachier. Abh math-phys Klasse K Bayer Akad Wiss, Suppl, Beitr Naturg Ostasiens 4:p 110

Hulley PA (1966) The validity of *Raja rhizacanthus* Regan and *Raja pullopunctata* Smith, based on a study of the clasper. Ann S Afr Mus 48(20):497–514

Hulley PA (1969) The relationship between *Raja miraletus* Linnaeus and *Raja ocellifera* Regan based on a study of the clasper. Ann S Afr Mus 52(6):137–147

Hulley PA (1970) An investigation of the Rajidae of the west and south coasts of southern Africa. Ann S Afr Mus 55(4):151–220

Hulley PA (1972a) The origin, interrelationships and distribution of southern African Rajidae (Chondrichthyes, Batoidei). Ann S Afr Mus 60(1):1–103

Hulley PA (1972b) A new species of southern African brevirajid skate (Chondrichthyes, Batoidei, Rajidae). Ann S Afr Mus 60(9):253–263

Hulley PA (1973) Interrelationships within the Anacanthobatidae (Chondrichthyes, Rajoidea), with a description of the lectotype of *Anacanthobatis marmoratus* von Bonde and Swart, 1923. Ann S Afr Mus 62(4):131–158

Hulley PA (1986) Family Rajidae. In: Smith MC, Heemstra PC (eds) Smith's sea fishes. Macmillan, Johannesburg, pp 115–127

Hulley PA, Stehmann M (1977) The validity of *Malacoraja* Stehmann, 1970 (Chondrichthyes, Batoidei, Rajidae) and its phylogenetic significance. Ann S Afr Mus 72(12):227–237

Ishiyama R (1967) Fauna Japonica. Rajidae. Tokyo Electrical Engin. Coll. Press, Tokyo, 82 pp

Last PR, McEachran JD (2006) New softnose skate genus *Brochiraja* from New Zealand (Rajidae: Arhynchobatinae) with description of four new species. N Zeal J Mar Freshwater Res 40:65–90

Last PR, Yearsley GK (2002) Zoogeography and relationships of Australasian skates (Chondrichthyes: Rajidae). J Biogeogr 29:1627–1641

Lleonart J, Rucabado JA (1984) Datos pesqueros de la campana "Benguela I". In: Rucabado JA, Bas C (eds) Resultados de las expediciones oceanografico-pesqueras "Benguela I" (1979) y "Benguela II" (1980) realizadas en el Atlantico sudoriental (Namibia). Dat Inf Inst Inv Pesq Barcelona (9):11–93

Lloris D (1986) Ictiofauna demersal y aspectos biogeograficos de la costa sudoccidental de Africa (SWA/Namibia). Monogr Zool Mar, Barcelona 1:9–432

McEachran JD (1984) Anatomical investigations of the New Zealand skates *Bathyraja asperula* and *B. spinifera*, with an evaluation of their classification within the Rajoidei (Chondrichthyes). Copeia 1984(1):45–58

McEachran JD, Dunn KA (1998) Phylogenetic analysis of skates, a morphologically conservative clade of elasmobranchs (Chondrichthyes: Rajidae). Copeia 1998(2):271–290

McEachran JD, Miyake T (1990a) Phylogenetic interrelationships of skates: a working hypothesis (Chondrichthyes, Rajoidei). In: Pratt HL, Gruber SH, Taniuchi T (eds) Elasmobranchs as living resources: advances in the biology, ecology, systematics, and the status of the fisheries. NOAA Tech Rept 90, pp 285–304

McEachran JD, Miyake T (1990b) Zoogeography and bathymetry of skates (Chondrichthyes, Rajoidei). In: Pratt HL Jr, Gruber SH, Taniuchi T (eds) Elasmobranchs as living resources: advances in the biology, ecology, systematics, and the status of the fisheries. NOAA Tech Rept 90, pp 305–326

McEachran JD, Seret B, Miyake T (1989) Morphological variation within Raja miraletus and status of Raja ocellifera (Chondrichthyes, Rajoidei). Copeia 1989 (3):629–641

Mas-Riera J, Macpherson E (1989) Distribution and population structure of skates of the southern coast of Namibia. Int Comm SE Atl Fish, 10th Sess, Dec 1989:1–5

Meisner EE (1987) A new skate species (Rajidae Batoidei) from the Indian Ocean sector of Antarctica. Zool Zhurn 66(12):1840–1849

Ochumba PBO (1984) Notes on some skates and rays of the Kenya coast. E Afr Nat Hist Soc Bull 1984:46–51

Ochumba PBO (1988) The distribution of skates and rays along the Kenya coast. J E Afr Nat Hist Soc Natn Mus 78(192):25–45

Payne AIL, Leslie RW, Augustyn CJ (1984) Hake stock assessments in ICSEAF divisions 1.6 and 2.1/2.2. Coll Sci Pap Int Comm SE Atl Fish 11(2):23–33

Payne AIL, Augustyn CJ, Leslie RW (1985) Biomass index and catch of Cape hake from random stratified sampling cruises in division 1.6 during 1984. Coll Sci Pap Int commn SE Atl Fish 12(2):99–123

Payne AIL, Rose B, Leslie RW (1987) Feeding of hake and a first attempt at determining their trophic role in the South African west coast marine environment. S Afr J Mar Sci, 5:471–501

Reif W-E, Saure C (1987) Shark biogeography: vicariance is not even half the story. N Jb Geol Paläont Abh 175(1):1–17

Seret B (1986a) Deep water skates from Madagascar (Chondrichthyes, Rajoidei). Proc 2nd Int Conf Indo-Pacific Fishes 256–260

Seret B (1986b) Deep water skates of Madagascar. Part 1. Anacanthobatidae (Pisces, Chondrichthyes, Batoidea), second record of the skate *Anacanthobatis ori* (Wallace, 1967) from off Madagascar. Cybium 10(4):307–326

Smith JLB (1949) The sea fishes of southern Africa. Central News Agency Ltd, South Africa, p 550

Smith JLB (1964) Fishes collected by Dr. T. Mortenson off the coast of South Africa in 1929, with an account of the genus *Cruriraja* Bigelow and Schroeder, 1954 in South Africa. Vidensk. Medd fra Dansk Naturh Foren 126:283–300

Smith JLB (1965) The sea fishes of southern Africa, 5th edn. Central News Agency Ltd, South Africa, 580 pp

Smith JLB, Smith MM (1966) Fishes of the Tsitsikama National Park. Pub. of Nat. Parks Bd. Trust. Rep, South Africa, 161 pp

Smith MM, Heemstra PC (1986) Smith's sea fishes. Macmillian, South Africa, 1–1047

Stehmann M (1970) Vergleichend morphologische und anatomische Untersuchungen zur Neuordnung der Systematik der Nordostatlantischen Rajidae. Arch Fisch-Wiss Hamburg 21:73–164

Stehmann M (1971) Untersuchungen zur validät von *Raja maderensis* Lowe, 1839, zur geographischen variation von *Raja straeleni* Poll, 1951, und zum subgenerischen status beider arbeiten arten (Pisces, Batoidei, Rajidae). Arch Fischereiwiss 22(3):175–199

Stehmann M (1973) Rajidae. In: Hureau J-C, Monod T (eds) CLOFNAM. Check-list of the fishes of the north-eastern Atlantic and of the Mediterranean. UNESCO, Paris 1:58–69

Stehmann M (1976) Revision der Rajoiden-Arten des nördlichen Indischen Ozean und Indopazifik (Elasmobranchii, Batoidea, Rajiformes). Beaufortia 24(315):133–175

Stehmann M (1981) Batoid fishes. In: Fischer W, Bianchi G, Scott WB (eds) FAO species identification sheets for fisheries purposes. Eastern Central Atlantic, Fishing Areas 34, 47(part). Food and Agriculture Organization of the United Nations, Rome 6, 76 pp

Stehmann M (1990) Rajidae. In: Quero J-C, Hureau J-C, Karrer C, Post A, Sandanha L (eds) CLOFETA. Check-list of the fishes of the eastern tropical Atlantic. JNICT, Portugal, Union Européene d'Ichtyologie, Paris, UNESCO, Paris 1:29–50

Stehmann M (1995) First and new records of skates (Chondrichthyes, Rajiformes, Rajidae) from the West African continental slope (Morocco to South Africa), with descriptions of two new species. Arch FischWiss/Arch Fish Mar Res 43(1):1–119

Stehmann M, Bürkel DL (1984) Rajidae. In: Whitehead PJP, Bauchot M-L, Hureau JC, Tortonese E (eds) Fishes of the Northeastern Atlantic and Mediterranean. UNESCO, Paris, pp 163–196

Stehmann M, Bürkel DL (1990) Rajidae. In: Gon O, Heemstra PC (eds) Fishes of the Southern Ocean. J.L.B. Smith Institute of Ichthyology, Grahamstown, 86–97

Turon JM, Rucabado J, Lloris D, Macpherson E (1986) Datos pesqueros de las expediciones realizadas en aguas de Namibia durante los Anos 1981 a 1984 ("Benguela III" a "Benguela VII" y "Valdivia I"). Pp. 11–345 of E. MacPherson, E., *ed*. Resultados de las expediciones oceanografico-pesqueras "Benguela III" (1981) a "Benguela VII" (1984) y "Valdivia I" realizadas en el Atlantico Sudoriental (Namibia). Dat Inform Inst Cienc Mar, Barcelona (17):345 pp

Von Bonde C (1933) Report of Fisheries and Marine Biological Survey for 1932, including lists of fishes, etc., lists of stations of the R.S. *Africana*, and salinity results obtained during 1932. Un S Afr, Fish Mar Biol Surv, Rep (10), 1933:4–109

Von Bonde C, Swart DB (1923) The Platosomia (skates and rays) collected by the S. S. "Pickle". Un S Afr, Fish Mar Biol Surv, Rep (3), 1922, spec. Rep (5), 22 pp

Wallace JH (1967) The batoid fishes of the east coast of southern Africa. Part III: Skates and electric rays. S Afr Ass Mar Biol Res, Oceanogr Res Inst, Invest Rep (17):1–62

Food habits of the sandpaper skate, *Bathyraja kincaidii* (Garman, 1908) off central California: seasonal variation in diet linked to oceanographic conditions

Christopher S. Rinewalt · David A. Ebert · Gregor M. Cailliet

Originally published in the journal Environmental Biology of Fishes, Volume 80, Nos 2–3, 147–163.
DOI 10.1007/s10641-007-9218-5 © Springer Science+Business Media B.V. 2007

Abstract The stomachs of 130 sandpaper skates, *Bathyraja kincaidii* (Garman, 1908), were sampled from off central California to determine their diet composition. The overall diet was dominated by euphausiids, but shrimps, polychaetes and squids were also important secondary prey. A three-factor MANOVA demonstrated significant differences in the diet by sex, maturity status and oceanographic season using numeric and gravimetric measures of importance for the major prey categories. These three main factors explained more variation in diet than interactions between the factors, and season explained the most variance overall. A detailed analysis of the seasonal variation among the prey categories indicated that abundance changes in the most important prey, euphausiids, were coupled with seasonal changes in the importance of other prey. When upwelling occurred and productivity was great (Upwelling and Oceanic seasons), euphausiids were likely highly abundant in the study area and were the most important prey for *B. kincaidii*. As productivity declined (Davidson Current season), euphausiids appeared to decrease in abundance and *B. kincaidii* switched to secondary prey. At that time, gammarid amphipods and shrimps became the most important prey items and polychaetes, mysids and euphausiids were secondary.

Keywords Arhynchobatidae · MANOVA · Euphausiids · Gammarid amphipods

Introduction

The trophic ecology of a species, determined through diet analysis, gives an insight to its place in the food web, as well as that of its prey. This kind of study can also help to understand how a predator could influence its prey populations, and vice versa. Without this knowledge, problems could arise from changes to the food web when the abundance of one or more species is altered, such as those caused by overfishing.

Skates (Rajiformes) are common demersal fishes and the most speciose elasmobranch order, occurring in nearshore temperate environments and deep-water tropical and boreal regions (Compagno 1990). Skates are often taken as bycatch in important fisheries that target various gadoids, monkfish and shrimps, as well as in research trawls (Walmsley-Hart et al. 1999; Alonso et al. 2001; Brickle et al. 2003; Cedrola et al. 2005; Perez and Wahrlich 2005). Skates may also compete with commercial species by sharing

C. S. Rinewalt (✉) · D. A. Ebert · G. M. Cailliet
Pacific Shark Research Center, Moss Landing Marine Laboratories, 8272 Moss Landing Road, Moss Landing, CA 95039, USA
e-mail: crinewalt@mlml.calstate.edu

the same food resources (Berestovskiy 1990; Pedersen 1995; Orlov 1998a; Dolgov 2005). Smale and Cowley (1992) concluded that because of their wide breadth of diet and their biomass, skates are likely to have a significant influence on the benthos. These varied trophic interactions suggest that thorough dietary studies are needed (Stevens et al. 2000).

The unique biological attributes of elasmobranchs (see current volume), coupled with a lack of species-specific fishery data and unregulated bycatch could lead to overfishing in certain skate species (Holden 1977; Jennings et al. 1998; Dulvy et al. 2000; Musick et al. 2000; Zorzi et al. 2001). The commercial catch of skates has increased dramatically along the Pacific coast of the United States during the past decade (Camhi 1999). Though skates have been fished commercially off California since 1916, only recently have the fishery landings grown by an order of magnitude (Zorzi et al. 2001). From 1995 to 2003, annual skate landings, undifferentiated by species, in California ranged from 2 to 10 times the landings for each of the years from 1981 to 1994, and were often greater than the combined landings of all other elasmobranch species (PacFin Database 2006). This increase in landings indicates that skates have become an important component of commercial fisheries in the eastern North Pacific (ENP), yet these are some of the least studied elasmobranchs.

The sandpaper skate, *Bathyraja kincaidii* (Garman, 1908), is a deep-water elasmobranch endemic to the ENP. This species occurs between 55 m and 1,372 m (most commonly between 200 m and 500 m) from the Gulf of Alaska to northern Baja California (Miller and Lea 1972; Ebert 2003). *Bathyraja kincaidii* is the smallest skate along the ENP, growing to 635 mm total length (TL) with a longevity of at least 18 years (Perez 2005). Little research has been conducted on its life history, yet it is frequently caught in trawls off central California. Wakefield (1984) examined stomach contents from two individuals off the coast of northern Oregon and found seven prey taxa, including shrimp in the genus *Crangon*, *Citharichthys sordidus*, a pinnotherid crab and the mysid *Acanthomysis nephrophthalma*. Ebert (2003) reported anecdotal information on the diet of *B. kincaidii*, listing polychaetes, amphipods, crabs and shrimp. This study serves to increase the knowledge of an important aspect of the life history of *B. kincaidii* by identifying the prey items of this species and describing its place in the ENP food web. The diet of *B. kincaidii* is described and statistically tested for differences between sexes, maturities and among oceanographic seasons from central California.

Materials and methods

Bathyraja kincaidii were collected by approximately monthly trawl surveys along the central California coast from March 2002 to February 2005 by the National Marine Fisheries Service Santa Cruz Lab (Fig. 1). Specimens were collected from 24 hauls among four varying depth strata per cruise with average depths of 395 m (1), 285 m (2), 226 m (3) and 146 m (4). Skates were frozen onboard and later processed at which time the stomachs were removed. Stomach contents were sorted with a dissecting microscope and prey taxa were identified, counted and weighed to the nearest 0.001 g. Any prey item that did not register at least this was given a mass of 0.0005 g for use in calculations. Any material that was not identifiable to any taxonomic level was excluded. Prey taxa were grouped into nine higher taxonomic categories: polychaetes, cephalopods, small benthic crustaceans, shrimp-like crustaceans, crabs, unidentifiable crustaceans, teleosts, molluscs and echinoderms.

The importance of prey was described by their component indices: number, mass and frequency of occurrence. $\overline{\%N}$ is the mean percentage number of a given prey category (j) for the total number of all prey items, $\overline{\%M}$ is the mean percentage mass of a given prey category for the total mass of all prey items and %FO is the percentage frequency of occurrence of a given prey category from all stomachs. To estimate precision for %N and %M, each stomach was considered its own sample; the values reported here are the mean values on a stomach-by-stomach basis (Tirasin and Jørgensen 1999). Along with the component indices of importance,

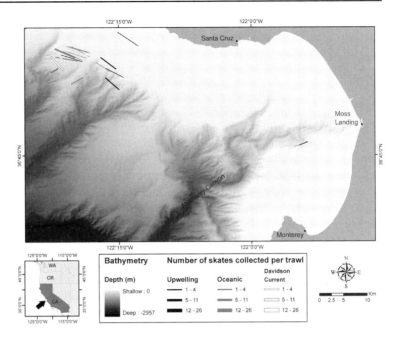

Fig. 1 Area map of central California indicating trawl locations and number of *Bathyraja kincaidii* captured whose stomachs were used in the study. Size of bar indicates length of trawl

a mean Index of Relative Importance ($\overline{\text{IRI}}$) was used to describe the diet of this skate (Pinkas et al. 1971; Hyslop 1980). This index was modified to incorporate percentage mass instead of percentage volume:

$$\overline{\text{IRI}}_j = (\overline{\% \ N_j} + \overline{\% \ M_j}) * \% \ FO_j$$

Mean percentage IRI, $\overline{\% \ \text{IRI}}$, was further calculated to provide the easiest measure to visualize the importance of any given prey:

$$\overline{\% \ \text{IRI}}_j = \left(\overline{\text{IRI}}_j / \sum \overline{\text{IRI}}\right) * 100$$

A randomized prey curve was generated using 100 resamplings (Ferry and Cailliet 1996), which plots the cumulative number of stomachs analyzed against the cumulative number of prey taxa encountered. A leveling of the curve and a reduction in variance indicates that enough stomachs have been examined to describe the taxonomic richness of the diet. An examination of lower taxa was conducted for prey that composed >5% of $\overline{\% \ N}$, $\overline{\% \ M}$ or $\overline{\% \ \text{IRI}}$.

The monthly samples were divided into three defined oceanographic seasons that characterize the study area, as described by Skogsberg (1936), Skogsberg and Phelps (1946) and Bolin and Abbott (1963). The Upwelling Season (UPS) (March–July) is characterized by the upwelling of cold, nutrient-rich water which can move far offshore due to strong southbound winds. This is followed by the Oceanic Season (OCS) (August–November), when the winds and upwelling weaken. During this weakening, oceanic water from the California Current moves close to shore. The Davidson Current Season (DCS) (December–February) is characterized by the continued weakening of the California Current, the development of an inshore northward current, a negligible thermocline and warm upper waters.

A three factor MANOVA was used to test the null hypothesis that there were no differences in the diet between sexes, maturity stages (mature versus immature) and among the three oceanographic seasons (Somerton 1991; Paukert and Wittig 2002). Compound indices should not be used because they can conceal the information of individual measurements, so both number and mass of the major prey categories were used separately in the analysis (Tirasin and Jørgensen 1999). The proportion of each category (not be confused with the percentage used in diet description) was arcsine transformed (Zar 1999, Eq. 13.8) to more closely meet the assumptions of homoscedasticity and normality. These assumptions were tested with Levene's Test (variance across groups for a single variable), Box's M test (covariance matrix across groups for all variables) and by examination of residual plots for each variable. Pillai's Trace was chosen as the reported test statistic as it is the most robust to violations of parametric assumptions (Olson 1974).

Six of the higher taxonomic categories (polychaetes, shrimp-like crustaceans, small benthic crustaceans, crabs, cephalopods and teleosts), accounting for the 12 variables (number and mass) were used in the statistical tests. Unidentifiable crustaceans and other taxa considered incidentally ingested (molluscs and echinoderms) were excluded as they contributed little to the diet. Additional randomized cumulative prey curves were created for each level of the three factors (e.g. immature females in the OCS, etc.) using these higher taxonomic categories.

A multivariate factor fit model using the three fixed factors (sex, maturity status and oceanographic season) was approximated from the univariate two factor fixed model provided by Graham and Edwards (2001). By describing how much of the observed variance is explained by these factors, the fit of these factors can often give more information about the model than their significance in the test itself. Because the data set was multivariate, the variance component for a single factor was calculated using the mean square and mean square error for each response variable as in the univariate model. These components were then averaged, with negative variances set to 0. The number of replicates per cell was not equal for this analysis, so the mean number of samples per cell was used for analysis. That averaged variance component for the factor was then used to calculate the magnitude of effects (ω^2), the percentage of the variance explained by that factor, which is analogous to the r^2 of regressions (Graham and Edwards 2001).

Results

In total, 138 *B. kincaidii* stomachs were collected, of which 8 (5.8%) were empty. The number of skates collected per trawl ranged from 1 to 26 individuals (mean 5.4 ± 6.5 SD) and ranged in size from 327 mm to 585 mm TL (mean 482 ± 53 SD mm TL). Examined by season, 21 skates with stomach contents were collected during the UPS, 78 from the OCS and 31 in the DCS. However, the majority of samples came from 3 months, January (22%), October (17%) and November (41%). The depth distribution of the specimens was also clumped, with most collected from the two deepest hauls at average depths of 395 m (46%) and 285 m (31%). The randomized cumulative prey curve revealed that enough stomachs had been collected to describe species richness accurately, averaging only three unique new prey taxa from the final 50 stomachs (Fig. 2).

Bathyraja kincaidii were found to prey on a wide variety of invertebrates and teleosts. The mean number, mass and IRI (68.4, 40.4, and 69.5%, respectively) revealed that the diet of *B. kincaidii* was dominated by shrimp-like crustaceans (euphausiids, mysids and shrimps), which were found in >98% of the stomachs examined (Fig. 3 and Table 1). Polychaetes were the second most important prey and were more important by mass than number and $\overline{IRI}$, with a large frequency of occurrence (11.6 $\overline{\% N}$, 21 $\overline{\% M}$, 65.4 %FO and 13.8 $\overline{\% IRI}$). Like polychaetes, cephalopods (5.3 $\overline{\% N}$, 19 $\overline{\% M}$ and 8.4 $\overline{\% IRI}$) and teleosts (4.6 $\overline{\% N}$, 12.3 $\overline{\% M}$ and 5.4 $\overline{\% IRI}$) were both most important by mass and were consumed by approximately half of the skates examined. Small benthic crustaceans (8.3 $\overline{\% N}$, 5.8 $\overline{\% M}$, 28.8 %FO and 2.7 $\overline{\% IRI}$) were not very important overall, but were more important by number than both cephalopods and teleosts. The remaining four major categories, crabs, unidentifiable crustaceans, molluscs and echinoderms, composed 0.2, 0.03, 0.01 and <0.01 $\overline{\% IRI}$, respectively, of the diet.

A qualitative examination of the importance of lower taxa revealed each categories major prey (Table 1). In terms of shrimp-like crustaceans, unidentifiable euphausiids (34.7 $\overline{\% N}$, 13.8 $\overline{\% M}$ and 50.1 $\overline{\% IRI}$) was by far the most important taxon, *Thysanoessa spinifera* (5.5 $\overline{\% N}$ and 5.7 $\overline{\% IRI}$) was the major identifiable species, and unidentifiable shrimps (5.2 $\overline{\% N}$) were of secondary importance. Onuphid (6 $\overline{\% M}$ and 5.8 $\overline{\% IRI}$) and nephtyid (8.4 $\overline{\% M}$) worms were the most important polychaete prey. Although a

Fig. 2 Cumulative prey curve for all prey items collected from *Bathyraja kincaidii* stomach samples. Error bars represent the standard deviation of the plotted mean generated from 100 resamplings

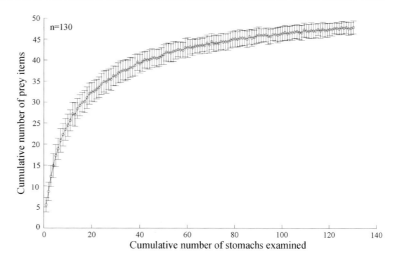

variety of cephalopod species were consumed, most of the remains were unidentifiable squids (7.5 $\overline{\% M}$ and 5.2 $\overline{\% IRI}$). Although not considered important, *Loligo opalescens* and *Octopus rubescens* comprised the largest percentage by mass of the identifiable cephalopods. Gammarid amphipods were also important (7.4 $\overline{\% N}$ and 6.5 $\overline{\% IRI}$) and ranked second by $\overline{\% IRI}$ and third by $\overline{\% N}$ in overall lower taxa importance. No species of teleost, crab, mollusc or echinoderm composed more than 5% of the diet by any of the three measures.

Analyses of the factorized random cumulative prey curves (Fig. 4) highlighted the low sample sizes of immature males and females in the UPS, so data from this season were excluded from the quantitative analysis, but were qualitatively compared to the other seasons. The randomized curves indicated enough stomachs were sampled to describe the richness of the eight other predator groupings (Fig. 4).

The assumptions for parametric tests were found to have been violated in this data set. Tests of the assumptions revealed that neither data series was homoscedastic. Levene's test was significant for polychaetes ($p = 0.004$) and Box's M was also significant ($p < 0.01$) for the mass data. Testing of the numeric data revealed only teleosts were not significant by Levene's test ($p = 0.189$) and Box's M was again significant ($p < 0.01$). An examination of the residuals indicated that both data sets were distributed normally.

The proportional mass data indicated significant differences in the diet by sex ($p = 0.028$,

Fig. 3 Graphical representation of the component indices of importance for major prey categories in the diet of *Bathyraja kincaidii*. Numbers in parentheses indicate $\overline{\% IRI}$ and error bars represent the standard error for their respective measurement. Crabs, unidentified crustaceans, molluscs and echinoderms are not included because of their low importance

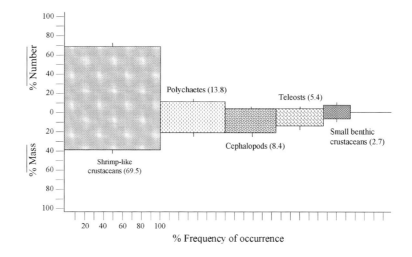

Table 1 The importance of prey items consumed by *Bathyraja kincaidii* at the major category (bold) and lowest identifiable taxa scales

Taxa	% N	% M	%FO	IRI	% IRI
Polychaetes	**11.61**	**20.99**	**65.38**	**2,131.57**	**13.83**
Onuphidae	4.65	5.98	26.92	286.28	5.83
Nephtys sp.	1.16	8.35	16.92	160.96	3.28
Polychaete A	1.22	0.86	10.77	22.48	0.46
Opheliidae	3.57	3.36	12.31	85.33	1.74
Polychaetes (unid)	1.01	2.43	20.00	68.71	1.40
Cephalopods	**5.34**	**18.97**	**53.08**	**1,290.18**	**8.37**
Octopus rubescens	0.98	3.94	16.15	79.47	1.62
Octopoda (unid)	0.02	0.04	0.77	0.05	0.00
Abraliopsis felis	0.42	0.77	3.85	4.58	0.09
Gonatus sp.	0.15	0.61	1.54	1.16	0.02
Loligo opalescens	0.65	4.66	6.92	36.70	0.75
Squids (unid)	1.93	7.50	26.92	253.93	5.17
Cephalopods (unid)	1.19	1.46	5.38	14.26	0.29
Small benthic crustaceans	**8.27**	**5.79**	**29.23**	**411.08**	**2.67**
Gammarid amphipods	7.39	4.20	27.69	320.99	6.54
Paraphronima sp.	0.19	0.06	0.77	0.19	0.00
Rocinela angustata	0.34	0.73	3.85	4.13	0.08
Isopods (unid)	0.35	0.80	1.54	1.76	0.04
Shrimp-like crustaceans	**68.44**	**40.41**	**98.46**	**10,717.33**	**69.52**
Euphausia pacifica	0.86	0.65	6.92	10.44	0.21
Thysanoessa raschi	0.04	0.01	0.77	0.04	0.00
Thysanoessa spinifera	5.49	4.02	29.23	278.14	5.67
Euphausiidae (unid)	34.65	13.76	50.77	2,458.01	50.09
Boreomysis californica	0.25	0.18	0.77	0.33	0.01
Holmsiella anomala	0.56	0.29	3.85	3.28	0.07
Mysidae (unid)	1.37	1.50	6.15	17.64	0.36
Euphausiid/mysid mix	2.22	1.40	4.62	16.71	0.34
Euphausiid/mysid/shrimp mix	1.17	0.84	2.31	4.64	0.09
Euphausiid/shrimp mix	9.54	5.39	17.69	264.11	5.38
Shrimp/Mysid mix	0.65	1.07	3.08	5.32	0.11
Neocrangon communis	0.39	0.90	3.85	4.96	0.10
Crangonidae (unid)	0.72	0.60	6.15	8.12	0.17
Heptacarpus sp.	0.38	0.40	0.77	0.60	0.01
Spirontocaris holmesi	0.21	0.33	3.08	1.65	0.03
Spirontocaris sica	1.18	1.42	3.85	9.99	0.20
Spirontocaris sp.	0.75	0.70	2.31	3.35	0.07
Hippolytidae (unid)	0.12	0.06	0.77	0.14	0.00
Pasiphaea pacifica	0.77	0.60	6.92	9.44	0.19
Sergestes similis	1.87	2.45	8.46	36.56	0.75
Shrimps (unid)	5.24	3.84	20.00	181.52	3.70
Crabs	**0.92**	**1.13**	**12.31**	**25.27**	**0.16**
Majoidea (unid)	0.18	0.24	1.54	0.65	0.01
Brachyuran (unid)	0.03	0.04	0.77	0.05	0.00
Crabs (unid)	0.47	0.56	9.23	9.50	0.19
Pagurus tanneri	0.23	0.23	0.77	0.35	0.01
Paguridae (unid)	0.02	0.05	0.77	0.06	0.00
Unidentifiable crustaceans	**0.57**	**0.09**	**6.15**	**4.08**	**0.03**
Unidentifiable crustaceans	0.57	0.09	6.15	4.08	0.08
Teleosts	**4.64**	**12.34**	**49.23**	**835.76**	**5.42**
Diaphus theta	0.31	0.26	2.31	1.32	0.03
Diaphus sp.	0.27	1.30	2.31	3.62	0.07
Stenobrachius leucopsarus	1.33	4.34	8.46	47.95	0.98
Myctophidae (unid)	0.19	0.54	4.62	3.36	0.07

Table 1 continued

Taxa	% N	% M	%FO	IRI	% IRI
Pleuronectidae (unid)	0.05	0.30	1.54	0.54	0.01
Sebastes sp.	0.50	1.55	8.46	17.33	0.35
Teleosts (unid)	1.99	4.05	26.92	162.68	3.31
Molluscs	**0.19**	**0.21**	**4.51**	**1.84**	**0.01**
Bivalve (unid)	0.01	0.00	0.75	0.01	0.00
Amphissa bicolor	0.07	0.20	1.50	0.40	0.01
Astyris gausapata	0.04	0.00	0.75	0.03	0.00
Gastropods (unid)	0.07	0.01	1.50	0.13	0.00
Echinoderms	**0.03**	**0.01**	**0.75**	**0.03**	**0.00**
Strongylocentrotus sp.	0.03	0.01	0.75	0.03	0.00

df = 6, 96), maturity ($p < 0.01$, df = 6, 96) and season ($p < 0.01$, df = 6, 96) with a significant interaction of maturity–season ($p = 0.014$, df = 6, 96). Pairwise comparisons revealed that males consumed a significantly larger proportion of shrimp-like crustaceans, small benthic crustaceans and crabs than females. Immature skates were found to have consumed more small benthic crustaceans and crabs than mature individuals. Greater proportions of small benthic crustaceans and crabs were consumed during the DCS than in the OCS. Though there was a significant maturity–season interaction, no prey category had significant interactions when examined by univariate tests (Fig. 5a). Most likely it was the combination of all prey categories that caused the interaction. Additionally there were significant sex–maturity interactions for the three crustacean prey categories ($p < 0.03$ for each) even though overall that interaction was not significant ($p = 0.13$, df = 6, 96) (Fig. 5b). In all cases (and for teleost prey, $p = 0.051$ for this interaction), immature males and females consumed nearly equal proportions of each category but mature males consumed significantly more of each than mature females.

Qualitatively, it appeared that there was no large difference in consumption by mass of certain prey during the UPS compared to the other two seasons. Predation on polychaetes was nearly equal among all three seasons, as was that of teleosts and cephalopods; the proportion ingested of the latter two categories during the UPS was less than the other seasons, but with a larger variance. However, shrimp-like crustaceans were consumed in a greater proportion during the UPS than the other two seasons. Consumption of small benthic crustaceans and crabs in the UPS was greater than that during the OCS; for the former prey category this was less than the proportion ingested in the DCS, while for the latter the UPS and DCS were similar.

Testing of the numeric data revealed significant differences in the diet by maturity ($p < 0.01$, df = 6, 96) and season ($p < 0.01$, df = 6, 96) with significant sex–season ($p = 0.037$, df = 6, 96) and maturity–season ($p < 0.01$, df = 6, 96) interactions. Shrimp-like crustaceans and teleosts were consumed in a greater proportion by mature skates, whereas immature individuals consumed more polychaetes and small benthic crustaceans. Between seasons, shrimp-like crustaceans were consumed more in the OCS, whereas polychaetes, teleosts, small benthic crustaceans and crabs were consumed in greater proportions in the DCS. The sex–season interaction was driven by shrimp-like crustaceans and polychaetes (Fig. 6a). Both sexes decreased their consumption of shrimp-like crustaceans from the OCS to DCS, but females displayed a greater decrease, having the greater consumption of the two sexes in OCS, but the lesser of the two in the DCS. Female skates greatly increased consumption of polychaetes from OCS to DCS, while males decreased slightly. The maturity–season interaction was caused by teleosts, small benthic crustaceans and cephalopods (Fig. 6b). Mature skates greatly increased their consumption of teleosts from the OCS to DCS, whereas immature skates increased only slightly. Predation on small

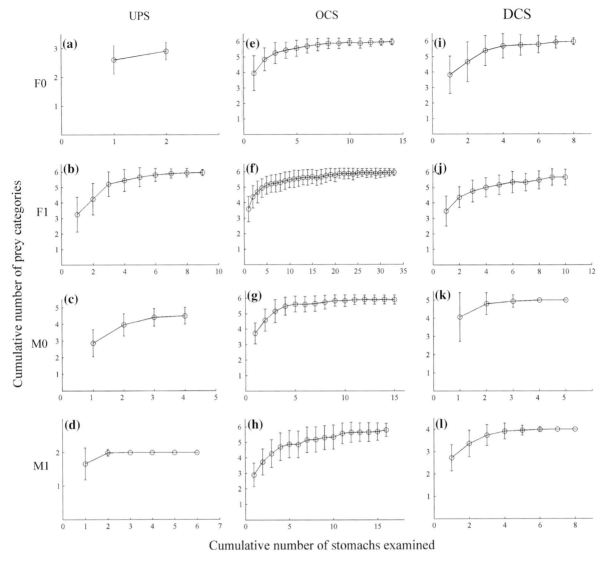

Fig. 4 Cumulative prey curves of the major prey categories for each combined factor grouping used in the analysis of the diet of *Bathyraja kincaidii*. Error bars represent the standard deviation of the plotted mean generated from 100 resamplings. For the following sex–maturity–season combinations, F = female, M = male, 0 = immature, 1 = mature, UPS = Upwelling season, OCS = Oceanic season, DCS = Davidson Current season

benthic crustaceans displayed a similar trend, but the maturity roles were reversed, immature skates having consumed more. Feeding on cephalopods by mature skates showed a marked increase from the OCS to DCS, though immature skates slightly decreased theirs.

In a qualitative examination of the numeric data, shrimp-like crustaceans were consumed in nearly equal proportion by skates during the UPS and the OCS; polychaetes and teleosts showed a similar pattern but the proportion ingested in the UPS was slightly less than the OCS but with a greater variance. The consumption of shrimp-like crustaceans in each of these two seasons was greater than the DCS, whereas predation on polychaetes and teleosts was less than during the DCS. Crab consumption during the UPS was similar to the other two seasons. The proportion of cephalopod prey taken by *B. kincaidii* was least in the UPS. There was a unique pattern in the proportion of small benthic crustacean prey, where consumption during the DCS was greater

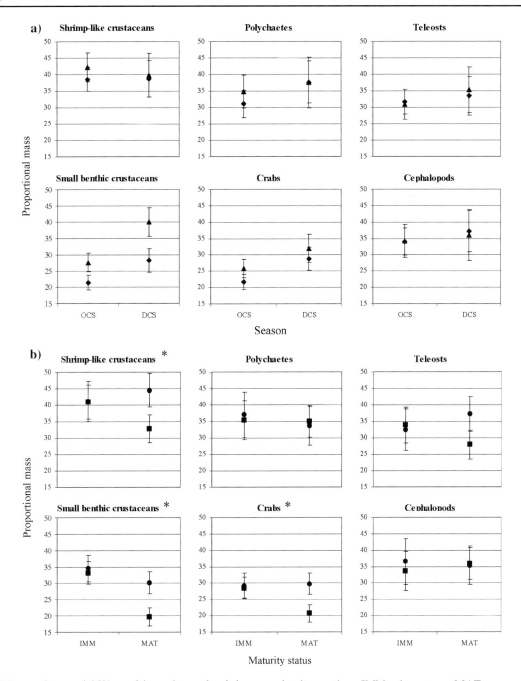

Fig. 5 Mean values and 95% confidence intervals of the gravimetric proportion of the six prey categories in the diet of *Bathyraja kincaidii*. (**a**) Maturity–season interaction, ▲ = immature skates, ♦ = mature skates. (**b**) Sex–maturity interaction, IMM = immature, MAT = mature, ■ = female skates, ● = male skates. *Significant ($p < 0.05$) for the interaction

than during the UPS which was in turn greater than the OCS.

Factor fit revealed that seasonal variation explained the most variance in both sets of *B. kincaidii* diet data (Table 2). By proportional number, season explained 17%, which was greater than the amount of variance explained by all variables in the mass model. Maturity stage was the second greatest factor by number and ranked third in importance by mass. This factor

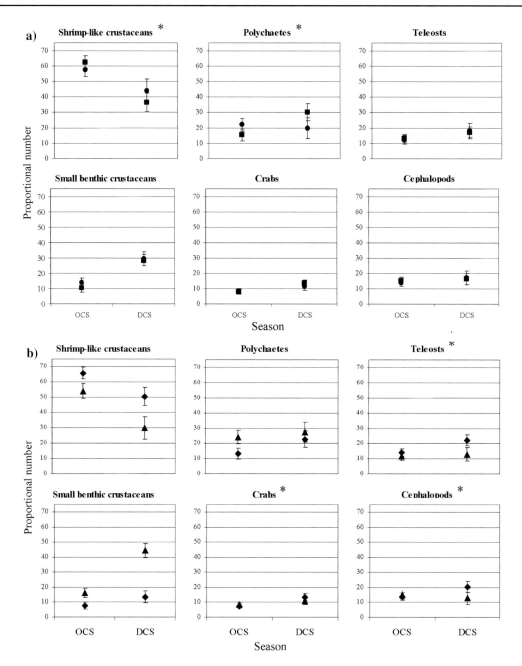

Fig. 6 Mean values and 95% confidence intervals of the numeric proportion of the six prey categories in the diet of *Bathyraja kincaidii*. (**a**) Sex–season interaction, ■ = female skates, ● = male skates. (**b**) Maturity–season, ▲ = immature skates, ♦ = mature skates. *Significant ($p < 0.05$) for the interaction

explained 14% of the variance by number, again more than the total explained in the mass model. Sex, which was a significant factor only for mass, explained the second most amount of variation in that model, but the second least amount numerically. Except for the maturity–season and sex–season interactions by number and the sex–maturity interaction by mass, the remaining interaction terms explained little of the variance in the diet.

Table 2 Factor fit (ω^2), Pillai's Trace p-value and the rank importance of both for the three main factors and their interactions resulting from the MANOVA performed on the gravimetric and numeric importance of the six main prey categories of *Bathyraja kincaidii*

Factor	ω^2	Rank	p-value	Rank
Gravimetric data				
Sex	2.05	2	0.03	4
Maturity	2.02	3	1.05×10^{-7}	1
Season	3.39	1	5.11×10^{-7}	2
Sex–maturity	1.85	4	0.13	5
Sex–season	0.21	5	0.23	6
Maturity–season	0.11	6	0.01	3
Sex–maturity–season	0.00	7	0.80	7
Error	90.37			
Numeric data				
Sex	0.04	6	0.53	6
Maturity	14.28	2	3.72×10^{-17}	1
Season	17.12	1	2.87×10^{-13}	2
Sex–maturity	0.01	7	0.86	7
Sex–season	2.31	4	0.04	4
Maturity–season	3.93	3	2.64×10^{-9}	3
Sex–maturity–season	0.61	5	0.29	5
Error	61.69			

Discussion

Crustaceans were by far the most important prey taxa to the overall diet of *B. kincaidii* from central California, comprising >72% of the prey by $\overline{\% \text{ N}}$ and $\overline{\% \text{ IRI}}$ and more than 47 $\overline{\% \text{ M}}$. This is a trait shared with many other nearshore and offshore, small bodied (<700 mm TL) benthic skates (McEachran et al. 1976; Berestovskiy 1990; Ebert et al. 1991; Smale and Cowley 1992; Pedersen 1995; Ellis et al. 1996; Orlov 1998a; Muto et al. 2001; Braccini and Perez 2005; Dolgov 2005; Mabragaña et al. 2005). The most important lower taxon of this category was euphausiids, which are known to be an important, and often primary, prey for cetaceans, birds and fishes (Schoenherr 1991; Brodeur and Pearcy 1992; Ainley et al. 1996; Croll et al. 1998, 2005; Yamamura et al. 1998). Small benthic crustaceans, mostly gammarid amphipods, played only a minor role overall but were seasonally important, becoming the most important lower prey taxon by all three measures during the DCS. Crabs were not an important food for these skates.

Polychaetes (e.g. Onuphidae and Nephtyidae) were the second most important prey category overall and their importance in the diets of skates has been well documented (McEachran et al. 1976; Templeman 1982; Berestovskiy 1990; Ellis et al. 1996; Brickle et al. 2003; Dolgov 2005; Mabragaña et al. 2005). The taxa of polychaeta consumed appeared to be related to skate maturity, with small-bodied worms (e.g. Opheliidae) more important to immature skates and larger nephtyid worms nearly absent from immature skate stomachs but important to mature skates.

The two remaining prey categories, cephalopods and teleosts, played only minor roles in the diet of *B. kincaidii*. The reduced importance of cephalopods and teleosts in the diet of smaller batoids is well established (McEachran et al. 1976; Templeman 1982; Berestovskiy 1990; Ebert et al. 1991; Smale and Cowley 1992; Pedersen 1995; Ellis et al. 1996; Walmsley-Hart et al. 1999; Muto et al. 2001; Dolgov 2005), and has been observed in other *Bathyraja* species (Orlov 1998a; Brickle et al. 2003). Both of these categories were dominated by unidentifiable prey.

Some items found in the stomachs of *B. kincaidii* during this study were considered incidentally ingested rather than prey. The gastropods *Amphissa bicolor* and *Astyris gausapata* along with a piece of an unidentifiable bivalve shell comprised the molluscs. Echinoderms, the other category, were represented by a single piece of *Strongylocentrotus* sp. Out of the 7 occurrences of these items, 6 were from stomachs that had benthic prey items in them, such as crabs, polychaetes and crangonid shrimp, suggesting that these items could have been ingested while feeding upon other prey.

The dietary importance of benthopelagic, vertically migrating prey such as euphausiids, myctophid fishes and the shrimp *Sergestes similis* raised the question of where these skates could be feeding. Though it is possible they could migrate into the water column to feed, a more likely explanation is the interaction of shoreward currents and the migration of their prey. It has been suggested that when these migrators are at their shallower nighttime depths, currents may advect them over shallower shelf waters, so that when they descend in the daytime they are near or in contact with the benthos (Isaacs and Schwartzlose 1965; Pereyra et al. 1969; Croll et al. 1998;

Ressler et al. 2005). Within Monterey Bay, it has also been suggested that the canyon walls can further serve to concentrate the prey (Croll et al. 2005); the area map (Fig. 1) indicates that many skates used in this study came from trawls near canyon edges. This interaction of currents, a nearshore shelf and steep canyon walls could allow *B. kincaidii* to feed on concentrations of these prey the skate otherwise might not encounter in its benthic habitat.

While sorting stomachs, it became apparent that certain prey items were biased in how they were considered important to the diet because of their differential digestion and degradation. Often with cephalopod and teleost prey there was little or no flesh remaining in the stomach, leaving only beaks and otoliths or other bones to be used, which underestimated the importance by mass of those prey. Similarly, polychaetes were often partially digested and at times counted by jaw parts rather than whole animals (except for opheliids which were almost exclusively whole). Even though these categories comprised the second, third and fourth greatest portions of the diet by mass, those values are considered to underestimate their importance to a certain degree. Because of this, the numeric abundance of these prey categories may more accurately estimate their importance to the diet. Significant differences for these numerically biased prey were only found when testing the numeric data. Shrimp-like and small benthic crustaceans showed the opposite relationship. These items, though not always whole, were rarely in an advanced state of digestion. However, their eyes, the characters used to enumerate them, were often degraded, somewhat underestimating their numeric importance. Similar to the other biased group above, these categories composed the first and third greatest percentages by number in the diet despite their bias. Despite this possible bias, these prey did indicate differences by number, but not always for the same factors or interactions as the mass data. It is possible that differences in the findings between the numeric and gravimetric data may have been influenced more by the digestion rates of certain prey rather than their importance in the diet. Analyses of both mass and number would be beneficial in such cases. A second possible source of bias could be from a low sample size (ten stomachs or less) in each sex–maturity grouping from the DCS. Though cumulative prey curves indicated those samples were enough to describe the richness of the diet, it could be argued they were inadequate for use in the statistical tests.

It is interesting that the data revealed differences in the diet between the sexes, though only by mass. Although not always analyzed, male and female diets frequently do not differ in elasmobranchs (Abdel-Aziz et al. 1993; Cortés et al. 1996; Alonso et al. 2001; Braccini and Perez 2005; Braccini et al. 2005), though sexual differences in diets have been observed in some species (King and Clark 1984; Gray et al. 1997; Orlov 1998b). In this study, consumption of the three major crustacean categories differed by sex. As previously stated, there were significant sex–maturity interactions which could explain these differences. Ingestion of these prey categories was not found to differ by sex when skates were immature but mature males consumed significantly more of each than mature females, which led to the result that males consumed more crustaceans than females.

Though the frequent significant interactions precluded the conclusion that differences in the diet could be due solely to main factors, fit revealed that season explained the most variance in the diet. Previous studies have also noted intra- and inter-annual changes in the diets of elasmobranchs (McEachran et al. 1976; Pedersen 1995; Cortés et al. 1996; Muto et al. 2001; Braccini and Perez 2005; Braccini et al. 2005). These results, coupled with previous studies on prey abundances, suggest that seasonal changes in the diet of *B. kincaidii* may be related to seasonal variation in the abundance of euphausiids, their most important prey. However, the majority of the variance remained unexplained in the gravimetric data, indicating additional factors are responsible for much of the variation in the diet by mass.

Euphausiid abundances have been found to vary intra-annually due to localized oceanographic changes, particularly upwelling, and inter-annually due to large scale El Niño/La Niña events (Brinton 1976; Ainley et al. 1996; Tanasichuk 1998a, b; Yamamura et al. 1998; Marinovic et al. 2002; Brinton and Townsend

2003). Though the timing can vary, when cool, nutrient rich water is upwelled, a predictable chain of events ensues (see Cushing 1971 for review) where phytoplankton increase in abundance followed by an increase in zooplankton, such as euphausiids. Because of variability and a lag between spawning and adulthood, peaks in euphausiid abundance can occur months after phytoplankton abundances begin to increase (Croll et al. 2005).

In the Monterey Bay area, upwelling most often occurs from late March/early April until late October/early November, peaking in June (Marinovic et al. 2002; Croll et al. 2005; Pacific Fisheries Environmental Laboratory 2006). This period encompasses the UPS and OCS used in this study, during which time euphausiids were most important to the diet of *B. kincaidii*. This is also the time of greatest euphausiid abundance in the study area (Marinovic et al. 2002; Croll et al. 2005). Upwelling decreases sharply starting in late July, and taking into account the 3- to 4-month time lag suggested by Croll et al. (2005), a decrease in the abundance of juvenile and adult euphausiids should first be seen in November, which corresponds to the start of the DCS. This decrease in abundance has been previously noted for euphausiids in Monterey Bay (Marinovic et al. 2002; Croll et al. 2005), *Euphausia pacifica* off southern California (Cailliet and Ebeling 1990) and for both *E. pacifica* and *Thysanoessa spinifera* off Vancouver Island (Brinton 1976; Tanasichuk 1998a, b).

The remaining shrimp-like crustaceans important in the diet of *B. kincaidii* included various shrimps and mysids, mostly unidentifiable. One of the more important identified shrimp species was *Sergestes similis*. In Monterey Bay, Barham (1957) found this species had a nearly constant abundance throughout the year due to two populations with a 6-month reproductive lag. There is no information currently available on the abundances of deep-water mysids in Monterey Bay, but Mauchline (1980) suggests that abundances of most species fluctuate seasonally with reproduction.

Myctophids were the most important identifiable teleosts in the diet of *B. kincaidii*. *Stenobrachius leucopsarus* abundance varies seasonally, peaking in winter and lowest from March to June (Neighbors and Wilson 2006). Barham (1957) noted that in Monterey Bay, *S. leucopsarus* was captured throughout the year, but was most abundant during the months of the DCS by a recalculated average. *Diaphus theta*, another myctophid consumed, was absent in all but one of the samples taken during the UPS, but like *S. leucopsarus* it was present in much greater numbers during the DCS (Barham 1957). These data suggest that myctophids are more abundant in the DCS than either of the other two seasons.

Little information is available on the seasonal abundance of *Loligo opalescens*, the most important identifiable squid species in the diet, aside from fishery dependant data. This is because of the difficulty in sampling this species with conventional gear such as trawls, which adults can easily evade or escape (Cailliet and Vaughan 1983). The fishery lands maximum catches from May to July (McInnis and Broenkow 1978; Hardwick and Spratt 1979; Cailliet and Vaughan 1983; Yaremko 2001). Assuming that catch was directly related to abundance (ignoring problems with fishing effort), October to March is the period of lowest *Loligo* abundance in Monterey Bay, suggesting that these squid were more abundant during the UPS and early OCS than in the DCS.

Data on the seasonality of deep-water small benthic crustaceans and polychaetes in the area is currently lacking. Slattery (1980) claimed that shallower amphipod species showed peaks in recruitment during spring and summer (UPS and early OCS), but in deeper water there was a reduced seasonality. Because there is no clear evidence, discussion of possible reasons for the fluctuation in importance of these prey in the diet is not discussed.

Synthesizing the abundances of the various prey from other studies, an explanation for the patterns observed in the diet of *B. kincaidii* is possible. Beginning in the UPS, euphausiids were likely highly abundant and remained so until approximately November. During this time they were the dominant prey of both males and females, but were more important to mature skates than immature skates. Also during this season, polychaetes were important prey to *B. kincaidii*, but were more important to immature skates. Squids and crabs were consumed but were not important to the diet. Though they did not contribute much to the diet, *Sergestes similis* was likely fairly abundant.

During the OCS euphausiids likely remained highly abundant and were the most important prey of *B. kincaidii*. However, there was a dramatic increase in the importance of myctophids and squid such as *Loligo opalescens*, which could be explained by an increased abundance in the area. These prey were exploited by both sexes, but were more important to mature than immature skates. Polychaetes remained secondarily important but the importance of gammarid amphipods declined.

Decreases in phytoplankton, likely associated with the DCS, led to decreased numbers of euphausiids. Presumably since their primary prey was no longer available in the same abundance, *B. kincaidii* began to prey more upon shrimps such as *S. similis*, which remained at roughly the same abundance all year. Mysids were also of greater importance to the diet in this season, more so than euphausiids. The importance of euphausiids in the diet decreased from an average of 47 $\overline{\%\ N}$, 21 $\overline{\%\ M}$ and 55 $\overline{\%\ IRI}$ in the UPS/OCS to 11, 6 and 6%, respectively, during the DCS. Though the overall importance of shrimp-like crustaceans declined somewhat during the same time period from an average of 77 $\overline{\%\ N}$, 45 $\overline{\%\ M}$ and 75 $\overline{\%\ IRI}$ to 43, 34 and 54%, respectively, it remained the most important prey category because of the increased importance of shrimps and mysids which masked the decline of euphausiids. *Bathyraja kincaidii* continued to prey on myctophids, which likely peaked in abundance during this season; they remained more important to mature skates than immature ones. Gammarid amphipods significantly increased in the diet and were much more important to immature skates, replacing teleosts and cephalopods that the mature skates fed upon. Polychaetes also increased in the diet, again more in immature skates. Squids remained important items in the diet, but not as much as during the OCS. This may reflect, but cannot be fully explained by, their likely minimal abundance during this season. Further sampling of the shelf-slope benthos should lead to a more complete understanding of any seasonal trends in abundance and to further discussion of the causes behind the observed seasonal dietary fluctuations of *B. kincaidii*.

Conclusion

Bathyraja kincaidii is a major predator of benthic and benthopelagic crustaceans. By mean number, mass and IRI the dominant prey were shrimp-like crustaceans, which were comprised primarily of euphausiids, but also contained shrimps and mysids. Although differences in the diet by season, maturity status and sex could not be ascribed solely to those factors because of frequent significant interactions, factor fit indicated that these main factors better explained the observed variance in the data than the interactions. The difference in findings between the numeric and gravimetric data may be related more to differences in digestion of certain prey categories than their importance by these measures. The seasonal variation in the diet is most likely attributable to the availability of euphausiids, the skate's primary prey. In the DCS when euphausiids are less abundant, *B. kincaidii* relies on secondary prey such as gammarid amphipods, shrimps, mysids, polychaetes and myctophids. Further research is needed to accurately assess the seasonal abundances of these prey. This is to determine whether the cause for their increased importance is related to a relative increase in their abundance compared to the lower euphausiid biomass or if they display comparatively similar or lower absolute abundances during this period and *B. kincaidii* actively chooses them.

Acknowledgements We thank the following people for assistance on various aspects of this project: Daniele Ardizzone, Joe Bizzarro, Aaron Carlisle, Chanté Davis, Colleena Perez, Heather Robinson, Wade Smith and Tonatiuh Trejo (Pacific Shark Research Center, Moss Landing Marine Laboratories), Don Pearson, John Fields, and E.J. Dick (NOAA Fisheries SWFSC, Santa Cruz Lab). Matt Levey created the area map and Josh Adams provided the computer program used to generate the randomized cumulative prey curves. We also thank Jim Ellis and an anonymous reviewer for their input. Specimens of *Bathyraja kincaidii* were collected under San Jose State University IACUC permit #801. Funding for this research was provided by NOAA/NMFS to the National Shark Research Consortium and Pacific Shark Research Center, and in part by the National Sea Grant College Program of the U.S. Department of Commerce's National Oceanic and Atmospheric Administration under NOAA Grant No. NA04OAR4170038, project number R/F-199, through the California Sea Grant College Program and in part by the California State Resources Agency. Partial funding was also provided by the American Museum of Natural History Lerner-Gray Grant for Marine Research, the Dr. Earl H. and Ethel M. Myers Oceanographic and Marine Biology Trust and the Packard Foundation.

References

Abdel-Aziz SH, Khalil AN, Abdel-Maguid SA (1993) Food and feeding habits of the common guitarfish, *Rhinobatos rhinobatos* in the Egyptian Mediterranean waters. Indian J Mar Sci 22:287–290

Ainley DG, Spear LB, Allen SG (1996) Variation in the diet of Cassin's auklet reveals spatial, seasonal and decadal occurrence patterns of euphausiids off California, USA. Mar Ecol Prog Ser 137:1–10

Alonso MK, Crespo EA, García NA, Pedraza SN, Mariotti PA, Vera BB, Mora NJ (2001) Food habits of *Dipturus chilensis* (Pisces: Rajidae) off Patagonia, Argentina. ICES J Mar Sci 58:288–297

Barham EG (1957) The ecology of sonic scattering layers in the Monterey Bay area, California. Doctoral Diss Ser Pub No 21,564. University Microfilms, Ann Arbor 182 pp

Berestovskiy EG (1990) Feeding in the skates, *Raja radiata* and *Raja fyllae*, in the Barents and Norwegian seas. J Ichthyol 29(8):88–96

Bolin RL, Abbott DP (1963) Studies on the marine climate and phytoplankton of the central coast area of California, 1954–1960. CalCOFI Rep 9:23–45

Braccini JM, Perez JE (2005) Feeding habits of the sandskate *Psammobatis extenta* (Garman, 1913): sources of variation in dietary composition. Mar Freshwater Res 56:395–403

Braccini JM, Gillanders BM, Walker TI (2005) Sources of variation in the feeding ecology of the piked spurdog (*Squalus megalops*): implications for inferring predator–prey interactions from overall dietary composition. ICES J Mar Sci 62:1076–1094

Brickle P, Laptikhovsky V, Pompert J, Bishop A (2003) Ontogenetic changes in the feeding habits and dietary overlap between three abundant rajid species on the Falkland Islands' shelf. J Mar Biol Ass UK 83:1119–1125

Brinton E (1976) Population biology of *Euphausia pacifica* off southern California. Fish Bull 74(4):733–762

Brinton E, Townsend A (2003) Decadal variability in abundances of the dominant euphausiid species in southern sectors of the California Current. Deep-Sea Res II 50:2449–2472

Brodeur RD, Pearcy WG (1992) Effects of environmental variability on trophic interactions and food web structure in a pelagic upwelling system. Mar Ecol Prog Ser 84:101–119

Cailliet GM, Ebeling AW (1990) The vertical distribution and feeding habits of two common midwater fishes (*Leuroglossus stilbius* and *Stenobrachius leucopsarus*) off Santa Barbara. CalCOFI Rep 31:106–123

Cailliet GM, Vaughan DL (1983) A review of the methods and problems of quantitative assessment of *Loligo opalescens*. Biol Oceanogr 2:379–400

Camhi M (1999) Sharks on the Line II: an analysis of Pacific state shark fisheries. Living Oceans Program, National Audubon Society, Islip, NY, 116 pp

Cedrola PV, González AM, Pettovello AD (2005) Bycatch of skates (Elasmobranchii: Arhynchobatidae, Rajidae) in the Patagonian red shrimp fishery. Fish Res 71:141–150

Compagno LJV (1990) Alternative life-history styles of cartilaginous fishes in time and space. Environ Biol Fish 28:33–75

Cortés E, Manire CA, Hueter RE (1996) Diet, feeding habits, and diel feeding chronology of the bonnethead shark, *Sphyrna tiburo*, in southwest Florida. Bull Mar Sci 58(2):353–367

Croll DA, Marinovic B, Benson S, Chavez FP, Black N, Ternullo R, Tershy BR (2005) From wind to whales: trophic links in a coastal upwelling system. Mar Ecol Prog Ser 289:117–130

Croll DA, Tershy BR, Hewitt RP, Demer DA, Fiedler PC, Smith SE, Armstrong W, Popp JW, Kiekhefer T, Lopez VR, Urban J, Gendron D (1998) An integrated approach to the foraging ecology of marine birds and mammals. Deep-Sea Res II 45:1353–1371

Cushing DH (1971) Upwelling and the production of fish. In: Russell FS, Yonge M (eds) Advances in marine biology, vol 9. Academic Press, New York, pp 255–334

Dolgov AV (2005) Feeding and food consumption by the Barents Sea skates. e- Journal of Northwest Atlantic Fishery Science v 35, art 34. http://journal.nafo.int/35/34-dolgov.html

Dulvy NK, Metcalfe JD, Glanville J, Pawson MG, Reynolds JD (2000) Fishery stability, local extinctions, and shifts in community structure in skates. Cons Biol 14(1):283–293

Ebert DA (2003) Sharks, rays and chimaeras of California. University of California Press, Berkeley, CA, 284 pp

Ebert DA, Cowley PD, Compagno LJV (1991) A preliminary investigation of the feeding ecology of skates (Batoidea: Rajidae) off the west coast of southern Africa. S Afr J Mar Sci 10:71–81

Ellis JR, Pawson MG, Shackley SE (1996) The comparative feeding ecology of six species of shark and four species of ray (Elasmobranchii) in the north-east Atlantic. J Mar Biol Ass UK 76:89–106

Ferry LA, Cailliet GM (1996) Sample size and data analysis: are we characterizing and comparing diet properly? In: MacKinley D, Shearer K (eds) Gutshop '96 feeding ecology and nutrition in fish: symposium proceedings. American Fisheries Society, pp 71–80

Graham MH, Edwards MS (2001) Statistical significance versus fit: estimating the importance of individual factors in ecological analysis of variance. Oikos 93(3):505–513

Gray AE, Mulligan TJ, Hannah RW (1997) Food habits, occurrence, and population structure of the bat ray, *Myliobatis californica*, in Humboldt Bay, California. Environ Biol Fish 49:227–238

Hardwick JE, Spratt JD (1979) Indices of the availability of market squid, *Loligo opalescens*, to the Monterey Bay fishery. CalCOFI Rep 20:35–39

Holden MJ (1977) Elasmobranchs. In: Gulland JA (ed) Fish population dynamics. John Wiley & Sons, London, pp 187–215

Hyslop EJ (1980) Stomach contents analysis – a review of methods and their application. J Fish Biol 17:411–429

Isaacs JD, Schwartzlose RA (1965) Migrant sound scatterers: interaction with the sea floor. Science 150:1810–1813

Jennings S, Reynolds JD, Mills SC (1998) Life history correlates of responses to fisheries exploitation. Proc R Soc Lond B 265(1393):333–339

King KJ, Clark MR (1984) The food of rig (*Mustelus lenticulatus*) and the relationship of feeding to reproduction and condition in Golden Bay. New Zeal J Mar Freshwater Res 18:29–42

Mabragaña E, Giberto DA, Bremec CS (2005) Feeding ecology of *Bathyraja macloviana* (Rajiformes:Arhynchobatidae): a polychaete-feeding skate from the south-west Atlantic. Sci Mar 69(3):405–413

Marinovic BB, Croll DA, Gong N, Benson SR, Chavez FP (2002) Effects of the 1997–1999 El Niño and La Niña events on zooplankton abundance and euphausiid community composition within the Monterey Bay coastal upwelling system. Prog Oceanogr 54: 265–277

Mauchline J (1980) The biology of mysids and euphausiids. In: Blaxter JHS, Russell FS, Yonge M (eds) Advances in marine biology, vol 18. Academic Press, New York, 681 pp

McEachran JD, Boesch DF, Musick JA (1976) Food division within two sympatric species-pairs of skates (Pisces:Rajidae). Mar Biol 35:301–317

McInnis RR, Broenkow WM (1978) Correlations between squid catches and oceanographic conditions in Monterey Bay, California. Ca Dept Fish Game, Fish Bull 169:161–170

Miller DJ, Lea RN (1972) Guide to the coastal marine fishes of California. Ca Dept Fish Game, Fish Bull 157, 249 pp

Musick JA, Berkeley SA, Cailliet GM, Camhi M, Hunstman G, Nammack M, Warren ML Jr (2000) Protection of marine fish stocks at risk of extinction. Fisheries 25(3):6–8

Muto EY, Soares LSH, Goitein R (2001) Food resource utilization of the skates *Rioraja agassizii* (Müller & Henle, 1841) and *Psammobatis extenta* (Garman, 1913) on the continental shelf off Ubatuba, southeastern Brazil. Rev Brasil Biol 61(2):217–238

Neighbors MA, Wilson RR Jr (2006) Chapter 13 Deep sea. In: Allen LG, Pondella DJ, Horn MH (eds) The ecology of marine fishes: California and adjacent waters. Univ. of California Press, Berkeley, pp 342–383

Olson CL (1974) Comparative robustness of six tests of multivariate analysis of variance. J Am Stat Assoc 69(348):894–908

Orlov AM (1998a) The diets and feeding habits of some deep-water benthic skates (Rajidae) in the Pacific waters off the northern Kuril Islands and southeastern Kamchatka. Alaska Fish Res Bull 5(1):1–17

Orlov AM (1998b) On feeding of mass species of deep-sea skates (*Bathyraja* spp., Rajidae) from the Pacific waters of the northern Kurils and southeastern Kamchatka. J Ichthyol 38(8):635–644

PacFin (2006) Washington, Oregon and California all species reports. http://www.psmfc.org/pacfin/woc.html (cited 10 Jan 2006)

Pacific Fisheries Environmental Laboratory (2006) PFEL coastal upwelling indices. http://www.pfeg.noaa.gov/products/PFEL/modeled/indices/upwelling/NA/upwell_menu_NA.html (cited 23 May 2006)

Paukert CP, Wittig TA (2002) Applications of multivariate statistical methods in fisheries. Fisheries 27(9):16–22

Pedersen SA (1995) Feeding habits of starry ray (*Raja radiata*) in west Greenland waters. ICES J Mar Sci 52:43–53

Pereyra WT, Pearcy WG, Carvey FE Jr (1969) *Sebastodes flavidus*, a shelf rockfish feeding on mesopelagic fauna, with consideration of the ecological implications. J Fish Res Board Can 26(8):2211–2215

Perez CR (2005) Age, growth and reproduction of the sandpaper skate, *Bathyraja kincaidii* (Garman, 1908) in the eastern north Pacific Ocean. MS Thesis California State University, Monterey Bay, 95 pp

Perez JAA, Wahrlich R (2005) A bycatch assessment of the gillnet monkfish *Lophius gastrophysus* fishery off southern Brazil. Fish Res 72:81–95

Pinkas L, Oliphant MS, Iverson ILK (1971) Food habits of albacore, bluefin tuna, and bonito in California waters. Ca Dept Fish Game, Fish Bull 152:1–105

Ressler PH, Brodeur RD, Peterson WT, Pierce SD, Vance PM, Røstad A, Barth JA (2005) The spatial distribution of euphausiid aggregations in the northern California Current during August 2000. Deep-Sea Res II 52:89–108

Schoenherr JR (1991) Blue whales feeding on high concentrations of euphausiids around Monterey Submarine Canyon. Can J Zool 69:583–594

Skogsberg T (1936) Hydrography of Monterey Bay, California. Thermal conditions, 1929–1933. Trans Am Phil Soc 29(1):1–152

Skogsberg T, Phelps A (1946) Hydrography of Monterey Bay, California. Thermal conditions, part II (1934–1937). Proc Am Phil Soc 90(5):350–386

Slattery PN (1980) Ecology and life histories of dominant infaunal crustaceans inhabiting the subtidal high energy beach at Moss Landing, California. MS Thesis, San Jose State University, 114 pp

Smale MJ, Cowley PD (1992) The feeding ecology of skates (Batoidea: Rajidae) off the cape south coast, South Africa. S Afr J Mar Sci 12:823–834

Somerton DA (1991) Detecting differences in fish diets. Fish Bull 89:167–169

Stevens JD, Bonfil R, Dulvy NK, Walker PA (2000) The effects of fishing on sharks, rays, and chimaeras (chondrichthyans), and the implications for marine ecosystems. ICES J Mar Sci 57:476–494

Tanasichuk RW (1998a) Interannual variations in the population biology and productivity of *Euphausia pacifica* in Barkley Sound, Canada, with special reference to the 1992 and 1993 warm ocean years. Mar Ecol Prog Ser 173:163–180

Tanasichuk RW (1998b) Interannual variations in the population biology and productivity of *Thysanoessa spinifera* in Barkley Sound, Canada, with special reference to the 1992 and 1993 warm ocean years. Mar Ecol Prog Ser 173:181–195

Templeman W (1982) Stomach contents of the thorny skate, *Raja radiata*, from the northwest Atlantic. J Northw Atl Fish Sci 3:123–126

Tirasin EM, Jørgensen T (1999) An evaluation of the precision of diet description. Mar Ecol Prog Ser 182:243–252

Wakefield WW (1984) Feeding relationships within assemblages of nearshore and mid-continental shelf benthic fishes off Oregon. MS Thesis, Oregon State University, 102 pp

Walmsley-Hart SA, Sauer WHH, Buxton CD (1999) The biology of the skates *Raja wallacei* and *R. pullopunctata* (Batiodea: Rajidae) on the Agulhas Bank, South Africa. S Afr J Mar Sci 21:165–179

Yamamura O, Inada T, Shimazaki K (1998) Predation on *Euphausia pacifica* by demersal fishes: predation impact and influence of physical variability. Mar Biol 132:195–208

Yaremko M (2001) California market squid. In: Leet WS, Dewees CM, Klingbiel R, Larson EJ (eds) California's living marine resources: a status report. The Resources Agency, California Department Fish and Game, pp 295–298

Zar JH (1999) Biostatistical analysis. Prentice-Hall, Upper Saddle River, NJ, 663 pp

Zorzi GD, Martin LK, Ugoretz J (2001) Skates and rays. In: Leet WS, Dewees CM, Klingbiel R, Larson EJ (eds) California's living marine resources: a status report. The Resources Agency, California Department Fish and Game, pp 257–261

Food habits of the longnose skate, *Raja rhina* (Jordan and Gilbert, 1880), in central California waters

Heather J. Robinson · Gregor M. Cailliet · David A. Ebert

Originally published in the journal Environmental Biology of Fishes, Volume 80, Nos 2–3, 165–179.
DOI 10.1007/s10641-007-9222-9 © Springer Science+Business Media B.V. 2007

Abstract Feeding studies can provide researchers with important insights towards understanding potential fishery impacts on marine systems. *Raja rhina* is one of the most common elasmobranch species landed in central and northern California demersal fisheries, yet life history information is extremely limited for this species and aspects of its diet are unknown. Specimens of *R. rhina* were collected between September, 2002 and August, 2003 from fisheries-independent trawl surveys. Percent Index of Relative Importance values indicated that the five most important prey items in 618 stomachs of *R. rhina* were unidentified teleosts (31.6% IRI), unidentified shrimps (19.6% IRI), unidentified euphausiids (10.9% IRI), Crangonidae (7.4% IRI), and *Neocrangon resima* (6.0% IRI). There were significant dietary shifts with increasing skate total length and with increasing depths. Smaller skates ate small crustaceans and larger skates ate larger fishes and cephalopods. With increasing depths, diet included bentho-pelagic teleosts and more cephalopods and euphausiids. The findings of this study are consistent with previous researchers that report similar diet shifts in skate species with size and depth.

Keywords Rajidae · Diet · Multivariate statistics · *Raja rhina*

Introduction

Information on diet provides researchers with important insight into the life history of a species. Knowing what a species eats can provide information about possible distribution and its position in food webs (Ebert et al. 1991; Barry et al. 1996; Ellis et al. 1996; Cortés 1999). Understanding a predator's trophic interactions can be crucial to developing sustainable management strategies.

Skates are almost one-fourth of the species of cartilaginous fishes living today (Nelson 2006). Together the three families (Anacanthobatidae, Arhynchobatidae and Rajidae) contain more than 280 species worldwide; making them the most diverse groups of elasmobranchs (Compagno 2005). Skates occur in all oceans of the world, yet are more common in temperate and polar waters. Despite the fact that skates are widespread and abundant, with new species being described each year, minimal research has been conducted to understand the life history traits of

H. J. Robinson (✉)
Moss Landing Marine Laboratories, Pacific Shark Research Center, 126 Second St. Winchendon, MA 01475, USA
e-mail: hrobinson@mlml.calstate.edu

G. M. Cailliet · D. A. Ebert
Moss Landing Marine Laboratories, Pacific Shark Research Center, 8272 Moss Landing Road, Moss Landing, CA 95039, USA

these fishes. Although the diets of several skates have been examined (e.g. McEachran et al. 1976; Berestovskiy 1989; Pedersen 1995; Morato et al. 2003), the majority of these studies were conducted in the Atlantic Ocean, and few researchers have analyzed populations of Pacific skate species (Orlov 1998, 2003; Yeon et al. 1999). The general lack of life history information along with poor fishery statistics, due to an inability to properly identify species, makes stock assessment and management decisions difficult.

Skates, while not specifically targeted, comprise a large portion of catches of bottom fishes off California. From 1916 to 1990, skate landings were 90% of all elasmobranch catch, and the most recent estimates indicate that these landings have increased tenfold for a net profit of greater than $340,000/year (Zorzi et al. 2001). In comparison with teleost fishes, skates, in general, are slow growing, late maturing, and are less fecund, making them more susceptible to overfishing (Zorzi et al. 2001). The limited amount of specific life history information for most eastern North Pacific Ocean skate species makes it difficult to determine what impacts these fisheries are having on these populations. Longnose skate, *Raja rhina*, big skate, *R. binoculata*, and California skate, *R. inornata* are the most commercially important skates landed in central and northern California fisheries (Zorzi et al. 2001).

Raja rhina (Jordan and Gilbert 1880) is one of eleven skate species living off California (Ebert 2003), and is easily distinguished from the other four members of the family Rajidae by its extremely long, acutely pointed snout. This skate ranges from the southeast Bering Sea southward to Cedros Island, Baja California and the Gulf of California (Mecklenburg et al. 2002). *Raja rhina* occurs in areas of mud-cobble bottoms with some vertical relief, nearshore to depths of 1,000 m (Ebert 2003). The only life history research on *R. rhina* is an age and growth study by Zeiner and Wolf (1993), however, age was not validated.

Other than anecdotal reports, the dietary composition of *R. rhina* is unknown. The objectives of this study were to characterize the overall diet of *R. rhina* and compare the diet of *R. rhina* between sexes, size classes, and among depth categories. In doing so, this study provides information that furthers the understanding of a highly understudied group of fishes.

Methods

Sample collection

Raja rhina were collected between September 2002 and August 2003 by the National Marine Fisheries Service, Southwest Fisheries Science Center (NMFS-SWFSC). Fishing vessels were contracted by NMFS-SWFSC to make a series of five otter trawls every month, weather permitting. Each of the five hauls was at a different depth, ranging from 15 m to 532 m, along the continental shelf and upper slope, over soft bottom habitats. The majority of hauls was off the coast of Davenport, California ($n = 27$), with the rest within Monterey Bay ($n = 17$).

Whole skates were frozen within 2–12 h from initial capture and at later dates (usually within several days after freezing) specimens were thawed and sorted by species. Skates were weighed to the nearest 0.1 kg. Measurements of total length (TL), disk length (DL) and disk width (DW) were made to the nearest millimeter (mm) for each specimen following Ebert (2003). Sex and maturity were determined following Ebert (2005). Stomachs of all skates were removed and re-frozen for later examination.

Overall diet characterization

For months when more than 100 *R. rhina* were collected a random sub-sample of 100 stomachs was chosen for processing. In months when fewer than 100 *R. rhina* were collected, all samples were processed. All sampled stomachs were thawed and contents sorted into prey categories over a 500 μm sieve. Once sorted, prey items were blotted on paper towels and enumerated. Wet-weight was recorded to the nearest 0.1 mg, and identifications were made to lowest possible taxa using a dissection microscope. When prey were greatly digested and only body fragments remained, the greatest number of individuals represented by those remains was recorded (Skjaeraasen and Bergstad 2000).

Cumulative prey curves were plotted to determine if enough stomach samples had been processed to assess the species richness of *R. rhina* diet. The curves were generated using a MATLAB computer program (Adams 2004), that randomized and resampled the data 100 times to provide mean and standard deviation (SD) values. This technique relies upon the fact that as sample size increases, variation in the estimate of species richness should decrease and the curve should reach an asymptote because new prey items are introduced only rarely (Ferry and Cailliet 1996; Cortés 1997).

Diet was characterized using measures discussed by Hyslop (1980). The percentage by number (%N) was calculated by dividing the number of individuals of one prey category by the total number of individuals from all prey categories in that stomach. The percentage by weight (%W) was determined in the same manner as %N. These two measures were generated for each stomach containing food and means were calculated to obtain average values for %N and %W (Cailliet et al. 1986). Frequency of occurrence (%FO) was obtained by dividing the total number of stomachs containing prey of one category by the total number of stomachs containing prey of any categories. The values of these three measures were combined into the Index of Relative Importance (IRI) to alleviate the biases of using any one measure alone, and to provide a more complete description of a prey item's importance in the diet of *R. rhina*. IRI was calculated by adding %N to %W and multiplying the sum by %FO (Pinkas et al. 1971). To facilitate comparisons with previous research, IRI values were standardized by converting them to percent IRIs (Cortés 1997, 1998; Hansson 1998). Diet was also described using the following resource indices: prey diversity, $H' = (\Sigma p_i * \ln p_i)$; prey evenness, $J = H'/H'_{max}$; and prey dominance, $D = \Sigma p_i^2$, where p_i = proportion of species i in diet (Cailliet et al. 1986; Krebs 1999).

Intraspecific dietary comparisons

Variations in the diet of *R. rhina* were assessed using the following intraspecific variables: sex, size class, and depth. Ontogenetic shifts in diet were determined by analyzing the following size classes: small (<60 cm TL) and large (>60 cm TL). Size categories were chosen based on reports that *R. rhina* reaches maturity at 62–74 cm TL for males and 70–100 cm TL for females (Zeiner and Wolf 1993; Ebert 2003). Depth categories were: shelf (<200 m), shallow slope (200–450 m), and deep slope (>450 m). These depth categories were chosen to reflect the local bathymetry of the sampling area (Wright and King 2002).

To facilitate the comparisons among intraspecific variables, all prey items were pooled into six higher taxonomic groupings: fishes (teleosts and elasmobranchs), shrimps, Euphausiidae, other crustaceans, Cephalopoda, and Gastropoda. Using %IRI values for these six groupings the diets were initially compared with five separate tests of Morisita's Simplified Index of Similarity (Krebs 1999). Morisita's Simplified Index of Similarity was calculated as follows:

$$C_\lambda = \frac{2(\sum p_{ij} p_{ik})}{\sum (p_{ij})^2 + \sum (p_{ik})^2}$$

where p_{ij} is the proportion that resource i is of the total resources used by species j; p_{ik} the proportion that resource i is of the total resources used by species k.

To further examine possible dietary patterns, and to provide support for the Morisita's tests, a principal components analysis (PCA) was conducted with the six prey groupings as dependant variables (McGarigal et al. 2000). Bi-plots were generated for each intraspecific variable with PC scores, and vector plots were created with eigenvectors. The resulting component variables from the PCA were then tested with a multivariate analysis of variance (MANOVA) to determine if the patterns indicated by the PCA were statistically significant (Paukert and Wittig 2002). Conducting the MANOVA on the component variables rather than on the original %IRI measures for the six prey groupings ensured the variables would not violate the assumptions of covariance and multivariate normality (Crow 1979). Within the MANOVA, significance of main effect was tested using Wilk's Lambda, Pillai's Trace Criterion, and Hotelling's Trace Criterion. Probability values were considered significant with a value less than α = 0.05. When multivariate tests were significant, post-hoc three-way analysis of variance (ANOVA) tests were conducted to determine which principal component (PC) was the cause for the difference in diets. The main factors (all fixed) for the ANOVA's were sex, size, and depth. Interactions were tested among all factors (sex*size, sex*depth, size*depth, and sex*size*depth), however only significant interaction terms among factors were considered in the results.

Results

Sample collection

A total of 527 females and 666 males was collected at depths ranging from 29 m to 532 m (Fig. 1). *Raja rhina* were absent from shallow waters <29 m. Due to cruise cancellations sampling was not carried out in April, May, or August.

Overall diet characterization

A total of 618 samples was randomly selected for stomach content analysis, of which 55 stomachs were empty, thereby making the total number of samples for this study 563. The cumulative species curve for all 563 samples (Fig. 2a) trended toward an asymptote near 450 samples, indicating that sufficient stomachs were processed to adequately describe the number of prey species in the diet.

Sixty-seven prey items were identified to lowest taxonomic level, containing at least 23 fish species, 10 shrimp species, four cephalopod species, one euphausiid species, three gastropod species, and six other crustacean species (Table 1). Those prey items having a percent Index of Relative Importance (%IRI) value greater than 5% were: teleost (unidentified), shrimp (unidentified), Euphausiidae (unidentified), Crangonidae,

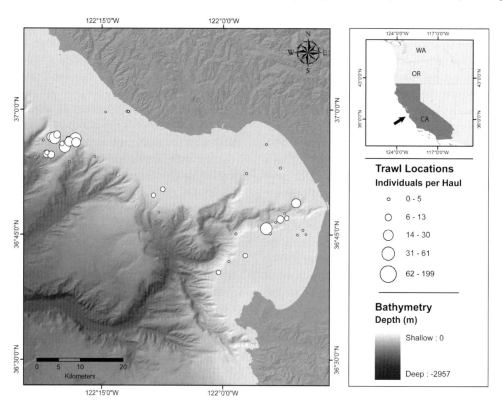

Fig. 1 Sample locations in central California where *R. rhina* were collected by National Marine Fisheries Service-Southwest Fisheries Science Center (NMFS-SWFSC) otter trawls. Each white circle is an individual trawl and size of white circle indicates the relative number of *R. rhina* collected in each trawl

Neocrangon resima, and *Sebastes* spp. (Fig. 3). The other 61 prey items collectively only had a %IRI value of 18.9. Results of the resource indices indicate a diverse diet: prey diversity, $H' = 3.18$; prey evenness, $J = 0.76$; prey dominance, $D = 0.06$.

Intraspecific dietary comparisons

Cumulative species curves for each category of all intraspecific variables (sex, size, and depth) trended toward asymptotes (male ~ 250, female ~ 225, small ~ 200, large ~ 200, shelf ~ 250, shallow slope ~ 100, and deep slope ~ 70), indicating that enough samples were processed from each category to adequately describe and compare the prey species in the diets (Fig. 2 b–h). Sixteen samples lacked any depth information so they were excluded from the Morisita's Index comparing depth classes. These samples were also excluded from PCA and MANOVA testing, thereby making the total number of samples in these two analyses 547.

The five separate Morisita's Index of Overlap tests demonstrated a low level of similarity between the size classes (Table 2). Diets of males and females almost completely overlapped (99.4%), as did the diets of skates caught at the two shallower depths (94.3%). When the deep slope diet was compared with either of the shallower depths (76.2% and 64.4%) overlap was considered high based upon standards set forth in Cailliet and Barry (1978), however, the values were much less than the comparison of the two shallow depths, indicating a possibility that different prey items were eaten at deeper depths.

Examination of the PCA bi-plots and vector plots revealed several dietary patterns among the intraspecific variables compared (Fig. 4). In combination, the first three PCs explained 71% of the total variation in *R. rhina* diet. PC 4 explained less than 10% of the total variance; therefore, only the first three PCs will be discussed further. Prey categories that loaded heavily on PC 1 were shrimps (positive) and fishes (negative). Cephalopods, euphausiids (both positive) and shrimps (negative) loaded heavily on PC 2. Other crustaceans and gastropods (both positive) loaded heavily on PC 3. When the data were coded for

Fig. 2 Cumulative prey species curve for all *R. rhina* samples processed (**a**) to lowest possible taxonomic level ($n = 67$), and for all intraspecific variable categories (**b–h**). Variable categories are male, female, small (<60 cm), large (>60 cm), shelf (<200 m), shallow slope (200–450 m), and deep slope (>450 m). The numbers of samples processed within each category are given with the corresponding curves

sex (Fig. 4a, e) no pattern was observed, indicating there was no difference in the diets of males and females. However, when the data were coded for size (Fig. 4b, f) a pattern emerged in which the large individuals were trending towards the lower left quadrant (Fig. 4b) and to the left of the *x*-axis (Fig. 4f), whereas the small individuals were interspersed with a slight trend to the lower right quadrant (Fig. 4b). When these patterns were examined in conjunction with the vector plots (Fig. 4d, h) it appeared that the diet of large individuals was dominated by fishes and the diet of small individuals was dominated by shrimps. When the data were coded for depth (Fig. 4c, g), the individuals caught on the deep slope trended towards the left hand side of the group of points (Fig. 4c), and below the *y*-axis (Fig. 4g). When examined in conjunction with the vector plots (Fig. 4d, h) this observed dietary pattern appeared to be driven by the presence of fishes, euphausiids, and cephalopods. There was no apparent pattern in diet between the two shallower depths (shelf and shallow slope).

MANOVA supported the dietary patterns observed in the PCA bi-plots. All three tests for main effect (Wilks' Lambda, Pillai's Trace, and Hotelling's Trace) provided consistent values, therefore, only the results for Wilks' Lambda will be presented and discussed further. A low value of Wilks' Lambda indicated no statistical significance between diets of male and female *R. rhina*. However, statistically significant results were detected when testing for effects of size and depth (Table 3). Size-based effects on diet were the strongest and were significant on all three PCs. Although depth-based effects were significant, they were less so than size-based effects and were only driven by PC 2. It is important to point out, however, that prey categories that loaded heavily on PC 2 were cephalopods and

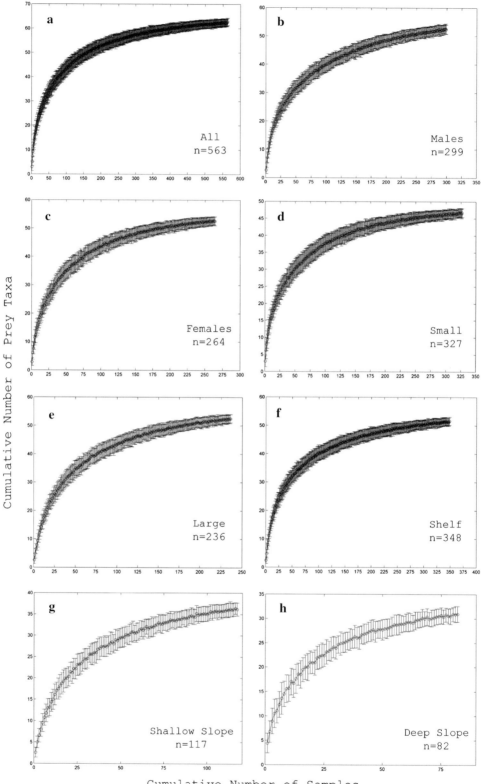

Table 1 Percent Index of Relative Importance (%IRI) values calculated for all prey items ($n = 67$) for all *R. rhina* stomach samples ($n = 563$) with sex, size and depth combined

Prey category		%N	%W	%FO	%IRI
Fishes		**41.13**	**51.75**	**71.05**	**63.47**
	Teleost (unidentified)	13.29	13.81	33.39	31.66
	Sebastes spp.	4.60	6.22	14.39	5.45
	Sebastes jordani	4.53	6.01	6.75	2.49
	Citharichthys sordidus	3.26	4.22	6.75	1.77
	Merluccius productus	2.63	3.79	6.22	1.40
	Chilara taylori	2.81	3.39	5.15	1.12
	Stenobrachius leucopsarus	1.44	2.50	5.51	0.76
	Glyptocephalus zachirus	1.22	2.14	2.66	0.31
	Lycodes diapterus	0.68	1.36	1.78	0.13
	Pleuronectidae	0.64	0.90	2.31	0.12
	Sebastes saxicola	0.91	1.27	1.42	0.11
	Genyonemus lineatus	1.01	1.01	1.24	0.09
	Diaphus theta	0.52	0.58	1.60	0.06
	Sebastes diploproa	0.75	0.81	0.89	0.05
	Parophrys vetulus	0.42	0.60	0.89	0.03
	Myctophidae	0.41	0.19	1.42	0.03
	Microstomus pacificus	0.43	0.46	0.89	0.03
	Lyopsetta exilis	0.28	0.50	0.89	0.02
	Zoarcidae	0.20	0.35	0.53	0.01
	Citharichthys spp.	0.22	0.23	0.53	0.01
	Icelinus spp.	0.10	0.19	0.36	0.00
	Citharichthys stigmaeus	0.09	0.20	0.36	0.00
	Porichthys notatus	0.12	0.09	0.36	0.00
	Tarletonbeania crenularis	0.02	0.17	0.36	0.00
	Lycodes cortezianus	0.18	0.18	0.18	0.00
	Xeneretmus spp.	0.09	0.18	0.18	0.00
	Engraulis mordax	0.09	0.16	0.18	0.00
	Zalembius rosaceus	0.09	0.16	0.18	0.00
	Apristurus brunneus	0.09	0.08	0.18	0.00
	Agonidae	0.01	0.01	0.18	0.00
Shrimps		**31.55**	**26.57**	**47.96**	**26.81**
	Shrimp (unidentified)	13.93	8.93	24.51	19.61
	Crangonidae	7.78	6.72	14.56	7.39
	Neocrangon resima	7.10	7.37	11.90	6.03
	Pasiphaea pacifica	1.05	0.89	3.02	0.20
	Pandalidae	0.34	0.89	1.24	0.05
	Spirontocaris spp.	0.35	0.36	0.89	0.02
	Spriontocaris sica	0.13	0.28	0.89	0.01
	Crangon alaskensis	0.23	0.24	0.71	0.01
	Spirontocaris holmesi	0.13	0.30	0.71	0.01
	Sergestes similis	0.24	0.16	0.53	0.01
	Spirontocaris snyderi	0.06	0.14	1.07	0.01
	Pandalopsis spp.	0.18	0.18	0.18	0.00
	Pandalus jordani	0.04	0.10	0.18	0.00
	Heptacarpus spp.	0.01	0.02	0.18	0.00
Cephalopoda		**6.94**	**11.05**	**27.00**	**4.67**
	Octopus rubescens	3.15	5.54	12.61	3.83
	Loligo opalescens	2.19	4.53	8.70	2.05
	Cephalopoda (unidentified)	1.43	0.84	6.75	0.53
	Octopoda	0.09	0.06	0.36	0.00
	Gonatus spp.	0.07	0.00	0.36	0.00
	Histioteuthis spp.	0.01	0.08	0.18	0.00
Euphausiidae		**11.59**	**5.61**	**18.83**	**3.11**
	Euphausiidae (unidentified)	11.43	5.48	18.47	10.93
	Thysanoessa spinifera	0.16	0.13	0.71	0.01

Table 1 continued

Prey category		%N	%W	%FO	%IRI
Other Crustaceans		**8.06**	**4.83**	**15.28**	**1.89**
	Eucarida	6.50	3.00	10.48	3.48
	Brachyura	0.33	0.40	1.07	0.03
	Schmittius politus	0.31	0.50	0.89	0.03
	Mursia gaudichaudii	0.34	0.46	0.89	0.02
	Crustacean (unidentified)	0.27	0.24	0.89	0.02
	Euphilomedes carcharodonta	0.15	0.01	1.07	0.01
	Chionoecetes tanneri	0.06	0.16	0.18	0.00
	Stomatopoda	0.04	0.02	0.18	0.00
	Chirostylidae	0.03	0.02	0.18	0.00
	Galathaidae	0.01	0.02	0.18	0.00
	Cumacea	0.03	0.00	0.18	0.00
Gastropoda		**0.72**	**0.20**	**4.09**	**0.04**
	Amphissa bicolor	0.30	0.11	1.95	0.03
	Gastropoda (unidentified)	0.29	0.08	1.24	0.02
	Astyris gausapata	0.07	0.01	0.71	0.00
	Amphissa spp.	0.06	0.00	0.53	0.00

Percent N and W are mean values. Values in bold are those for the six prey categories of higher taxonomic grouping

euphausiids that were more commonly observed in samples from the deep slope category.

The outcome of the post-hoc ANOVA tests indicated a significant interaction between size and depth for PC 1 and PC 2 (Table 3). The resulting plots of these interactions for each PC (Fig. 5) demonstrated that fish and shrimp prey that loaded heavily on PC 1 were consumed by small and large skates, respectively, such that size-effects were less extreme in skates from the deeper depths (Fig. 5 a–c). The opposite trend was observed for PC 2. Cephalopods and euphausiids, which loaded heavily on PC 2, were more abundant in small *R. rhina* from deeper depths. Size effects were not extreme in skates from shallower depths (Fig. 5d–f). There was no difference in the consumption of prey items that loaded heavily on PC 3 (other crustaceans and gastropods) for either size class among depths (Fig. 5g–i).

Table 2 Results of five separate Morisita's Index of Overlap tests that were calculated to compare the intra-specific variables of sex, size, and depth

Intraspecific	Morisita's
Variables compared	Index of Similarity
Females/males (264/299)	0.994
Small/large (327/236)	0.491*
Shelf/shallow slope (348/117)	0.943
Shelf/deep slope (348/82)	0.762
Shallow slope/deep slope (117/82)	0.646

Asterisks indicate values of low similarity when using a significance level of 0.53 (Cailliet and Barry 1978). Values in parentheses are the number of samples within each variable

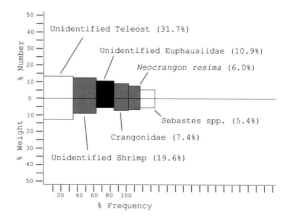

Fig. 3 *Raja rhina* prey items that have %IRI values (in parentheses) greater than 5%. Total number of prey items was 67 and total number of stomach samples in this analysis was 563

Discussion

Overall diet characterization

Intermittent feeding (bouts of feeding followed by longer periods of resting or non-feeding) is known for several shark species (Wetherbee et al. 1990; Joyce et al. 2002; Braccini et al. 2005), however, skates employ a more constant feeding strategy (Abd El-Aziz 1986; Ezzat et al. 1987; Muto et al. 2001). Within our study only 8.9% of stomachs examined were empty. This value is less than other rajid feeding studies that used similar collection methods (Pedersen 1995; Skjaeraasen and Bergstad 2000). State of digestion among all stomachs containing prey items was fairly consistent. Within each stomach, however, it was common to find similar prey items in various states of decomposition. The low occurrence of empty stomachs along with the lack of homogeneous digestion demonstrates that R. rhina have no periodicity of feeding. This study, therefore, supports the idea that skates exhibit continuous feeding activities.

Many researchers have classified skates as specialist feeders (Ebert et al. 1991; Orlov 1998; Braccini and Perez 2005; Mabragaña et al. 2005), whereas, several others have found skate species to be generalist feeders (Holden and Tucker 1974; Smale and Cowley 1992; Koen Alonso et al. 2001). The diet of R. rhina consisted of 67 prey items from a wide array of biological groups (i.e. fishes, shrimps, gastropods). Washington (1984) reported that in most biological communities the value of diversity (H') does not exceed 5.0. Therefore, the value of prey diversity measured for R. rhina, 3.18, can be considered quite high, thus making the probability of finding any one of the 67 prey items within this skate's diet highly uncertain. Although only six prey items had a %IRI value greater than 5%, the overall diet was not dominated by any one prey species, and the remaining 61 prey items caused the diet to remain fairly even. These factors: great number of prey items, great diversity of prey species, and an even diet, lead to the conclusion that R. rhina can be classified as a generalist feeder.

Although there are only two anecdotal reports of the diet of R. rhina (Wakefield 1984; Ebert 2003) the findings of this study support those of previous researchers. Ebert (2003) described its diet as mainly benthic crustaceans and bony fishes. Wakefield (1984) found mostly fishes and decapod crustaceans, but only had four stomachs to analyze. The overall diet of R. rhina from our current study was mostly teleosts (~23 species) and shrimps (~10 species). The major fish groups were the families Scorpaenidae, Myctophidae, and the order Pleuronectiformes, whereas the major shrimp groups were the families Crangonidae, Hippolytidae, and Pandalidae. Cephalopods were the third most important prey group. Several other skate species feed on cephalopods (McEachran et al. 1976; Pedersen 1995; Kabasakal 2002; Morato et al. 2003), however, this has not been previously reported for R. rhina. In Allen's (2006) recent book chapter, he mentions that R. rhina is an important member of the outer shelf (100–200 m) and mesobenthal (200–500 m) slope habitats from northern California through northern Baja California. Our study in central California supports these claims, indicating that R. rhina is an upper lever predator feeding on other important fish species in these habitats (pacific hake, *Merluccius productus*, spotted cusk-eel, *Chilara taylori*, rex sole, *Glyptocephalus zachirus*, splitnose, *Sebastes diploproa*, stripetail, *S. saxicola*, and shortbelly rockfishes, *S. jordani*).

The utilization of pelagic food sources by demersal elasmobranchs has been well documented by several authors (Holden and Tucker 1974; Mauchline and Gordon 1983; Smale and Cowley 1992; Koen Alonso et al. 2001; Braccini et al. 2005). Although the actual mechanism of how these prey items are preyed upon remains unknown there are two possible explanations. The first is simply that demersal predators are consuming fishery discards or animals that have died of natural causes. Berestovskiy (1989) argued that the relative body shape and

Fig. 4 Principal component analysis (PCA) bi-plots (**a–c, e–g**) and vector plots (**d, h**) for sex (triangles), size classes (squares) and depths (circles). Together, the three principal components explain 71% of the variance in the diet of R. rhina. The six bi-plots (**a–c, e–g**) depict patterns in the diet of R. rhina, while the vector plots (**d, h**) help explain which prey categories are driving the bi-plot patterns

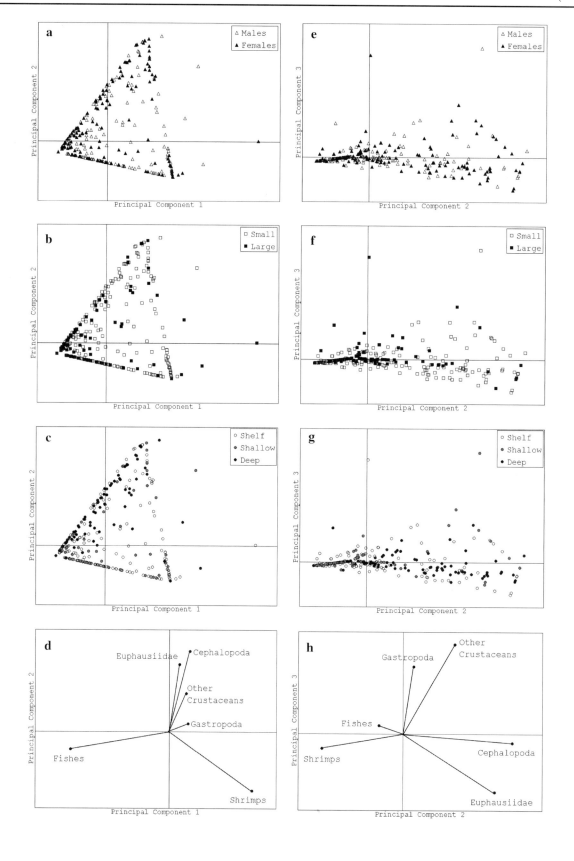

Fig. 5 Plots of analysis of variance (ANOVA) interaction terms between size and depth for all three principal components (PC). Prey categories that loaded heavily on each PC are presented on the *y*-axes. The rows of plots (**a–c**, **d–f**, and **g–i**) represent the different PCs, whereas the columns of plots (**a**, **d**, **g**; **b**, **e**, **h**; and **c**, **f**, **i**) represent the three depth categories

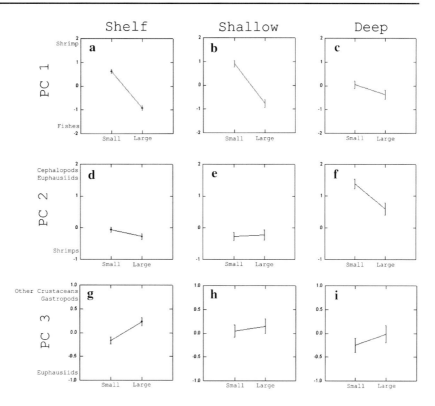

lie-and-wait feeding behavior of skates prevents them from successfully hunting large pelagic species, therefore, their presence in skate diets can only be explained by scavenging on dead fishes.

The second explanation is that demersal predators swim off the substrate to actively hunt pelagic species that have vertical migratory patterns. Kabasakal (2002) provided evidence that demersal predators were able to consume several semi-pelagic cephalopod species because during daylight hours the cephalopods live in close proximity to the bottom. Orlov (2003) speculated that due to the relative shape of the continental slope and the migratory patterns of mesopelagic fishes these prey species could be eaten by demersal skate species. Morato et al. (2003) also reported they thought active hunting was why pelagic species were in the diet of *Raja clavata*. The general trend of skates to eat pelagic species is also apparent in the diet of *R. rhina* with the inclusion of *Loligo*, *Thysanoessa*,

Table 3 Multivariate analysis of variance (MANOVA) results for all intraspecific variables tested (sex, size, and depth) using %IRI values

Variable tested	Multivariate test statistics			Univariate test statistics							
	Wilks' Lambda			Principal component 1			Principal component 2			Principal component 3		
	df	F	P	df	F	P	df	F	P	df	F	P
Sex	3,540	1.709	0.164	–	–	–	–	–	–	–	–	–
Size	3,540	77.473	<0.0001*	1	105.471	<0.0001*	1	8.156	0.004*	1	5.115	0.024*
Depth	6,1080	16.025	<0.0001*	2	1.836	0.160	2	40.260	<0.0001*	2	1.189	0.305
Size*depth	–	–	–	2	8.662	<0.0001*	2	3.639	0.027*	2	0.952	0.387

Wilks' Lambda values are presented. Univariate *F*-tests are presented for size and depth as the MANOVA was significant for these effects. Only size*depth yielded a significant interaction in the ANOVA tests and is included here

* Significant result at $\alpha = 0.05$

and several myctophid fishes in the list of prey items consumed by this skate. Although we did not investigate how these pelagic prey items were caught, the frequency with which they were eaten by *R. rhina*, relative to other minor prey items, indicates they were actively preyed upon rather than passively scavenged. These findings are consistent with the notion that some skates have a more benthopelagic feeding strategy.

Intraspecific dietary comparisons

Several investigators have compared the diet of skates between sexes (Orlov 1998; Koen Alonso et al. 2001; Morato et al. 2003; Braccini and Perez 2005; Dolgov 2005). In the majority of species examined males and females fed on similar prey items. Only Orlov (1998, 2003) found different diets between sexes of several species of bathyrajid skates; males consumed more crab and cephalopod species whereas females consumed more fish species. Orlov (1998) attributed these differences to size dimorphism of male and female skates. Although size dimorphism does occur in some rajid species, no other researcher has found that it plays a role in dietary segregation. Our results are consistent with previous papers, diets of male and female *R. rhina* overlapped by 99.4%, despite the fact that females may reach larger sizes than males.

Ontogenetic shifts in diet is a common phenomenon among rajid species. Many researchers have reported a general trend that younger smaller skates fed more on smaller prey groups such as gammarid amphipods and small shrimps, whereas older, larger skates fed more on larger shrimps, polychaetes, and fishes (Ajayi 1982; Yeon et al. 1999; Muto et al. 2001; Brickle et al. 2003). Lucifora et al. (2000) provided evidence that large and small *Dipturus chilensis* fed on the same prey species but the relative size of the fish consumed increased with larger predator size. These general patterns were clearly observed in the diet of *R. rhina*. Smaller *R. rhina* had a diet consisting of more shrimps, whereas larger *R. rhina* consumed a greater amount of fishes. Although there was some degree of overlap between the two size classes examined (due to similar prey species being consumed by both size classes) the relative sizes of those prey species increased with increasing size of *R. rhina*. It is believed that the reason for these patterns is morphological constraints in which smaller skates are unable to consume prey sizes available to larger skates because of gape width limitations and a weaker foraging ability (Smale and Cowley 1992; Lucifora et al. 2000; Braccini and Perez 2005). It is commonly thought that these shifts in diet are a mechanism to reduce intraspecific competition among members of a community. The central Californian population of *R. rhina* exhibited ontogenetic shifts in diet that are common within this group of fishes.

Variations in diet with increasing depth has been a highly understudied aspect of skate feeding, and only a few authors have included it into their research. Hacunda (1981) reported that in the Gulf of Maine demersal fishes selected prey from different depth strata as a means of food resource partitioning. Templeman (1982) determined that in the diet of *Raja radiata* certain fish species and cephalopods were more important in deeper water whereas crabs and different fishes were consumed more in shallower water. Ebert et al. (1991) concluded that off the west coast of southern Africa there were two distinct skate communities, one shallower and one deeper than 380 m. Although these two communities were determined to be distinct, both groups had members that filled similar niches (i.e. crustacean specialists). Even though all three of these studies concluded there were depth-based dietary differences not one of them performed any analyses to determine if these differences were statistically significant, therefore, these conclusions have to be accepted with caution. In the present study, the diet of *R. rhina* varied with increasing depths, with cephalopods, euphausiids, and certain fishes (mainly rockfishes) being more important in the diets of skates living deeper than 450 m. Unlike previous studies, these results were significant by means of multivariate analyses. Although we did not measure prey abundance, it seems likely that these variations with depth were a function of prey species availability in the deeper sections of *R. rhina* range. Therefore, it is possible that *R.*

rhina has incorporated these prey species into its diet as a way to further reduce intraspecific competition.

Importance of statistical methodology

Compound measures of importance (IRI, etc.), while being used more commonly, are still highly underutilized in diet and feeding studies. The importance of measuring more than one individual parameter (i.e. number, weight/volume, or frequency of occurrence) and incorporating these measures into one encompassing index has been discussed by several authors (Hyslop 1980; Ferry and Cailliet 1996; Cortés 1997; Braccini et al. 2005). Using any one measure alone will provide different results than if another measure was chosen for analysis. For instance, in the current study, if only data for %W were examined, *Sebastes* species would have had greater importance in *R. rhina* diet whereas, if only %FO was examined, the importance of cephalopod species in *R. rhina* diet would have been over accentuated. Only by combining all three measures into the %IRI was this study able to remove the bias of numerically abundant prey items, heavier prey items, and frequently consumed prey items to provide a complete comprehensive description of the overall diet of *R. rhina*. This study thereby provides evidence to support and to endorse the use of compound measures of importance in any future feeding studies.

There is an overwhelming lack of any statistical support for results presented in the majority of published literature of feeding studies. Most authors utilize common overlap or similarity measures to compare diets within and among species. Although there is nothing inherently wrong with using these measures, the problem arises in that researchers try to classify the resulting data as significant. Ferry and Cailliet (1996) point out that, in spite of an index's power, they are not probability-based statistics and cannot infer any significance. The only means for a researcher to make these claims is to follow a similarity index with some sort of parametric or non-parametric statistical test. Crow (1979) stated early on, that if testing for differences in two or more groups of fish with more than one species multivariate tests are mandatory, and although this was published more than 25 years ago, researchers continue to publish these types of comparisons without the proper statistics.

Another problem with simply using an index is that sometimes patterns in data are not fully realized until a more complex statistic is employed. The results of this current study make this point very clearly. If *R. rhina* diet was only compared with the Morisita's Index of Overlap, then it would have been thought that the diet of skates at any depth were highly overlapped, therefore similar. The diets of small and large skates also would have been considered only marginally dissimilar. However, by also using a PCA and a MANOVA, some significant patterns were observed between the size classes and among the depth categories. These are patterns that can be highly important for proper inclusion in any management models, yet they would have remained unnoticed without the use of parametric statistics. By clearly describing significant patterns in *R. rhina* diet we have supported previous reports that complex statistical methods are mandatory to completely characterize diet.

Conclusion

Despite obvious regional differences in species composition, the diet of *Raja rhina* is consistent with other skate species (Orlov 1998; Yeon et al. 1999; Lucifora et al. 2000; Koen Alonso et al. 2001). The main prey groups utilized by *R. rhina* are fishes, crustaceans, and cephalopods. No difference was detected between males and females. However, there was a significant dietary shift from small shrimps to large teleosts with increasing skate total length. Along the central California coast, *R. rhina* also displayed dietary variations in which cephalopods, euphausiids, and certain fishes became more important at deeper depths. These details of *R. rhina* feeding, along with other life history parameters yet to be examined, are crucial in understanding how this species fits into the demersal food web of this region and how this skate may be impacted by commercial fisheries. This life history information can further be used to develop important models for sustainable management of the skate community of central California.

Acknowledgments Funding for this research was provided by NOAA/NMFS to the National Shark Research Consortium and Pacific Shark Research Center, and in part by the National Sea Grant College Program of the U.S. Department of Commerce's National Oceanic and Atmospheric Administration under NOAA Grant no. NA04OAR4170038, project number R/F-199, through the California Sea Grant College Program, and in part by the American Elasmobranch Society and the Western Groundfish Conference. We are grateful for the help of Don Pearson of National Marine Fisheries Service, Southwest Fisheries Science Center for his aid in collection of skate specimens. We thank the many employees of PSRC for their help in skate dissections, Joe Bizzarro for his continuous input into this study, and Lara Ferry-Graham and James T. Harvey for contributing to this manuscript.

References

Abd El-Aziz SH (1986) Food and feeding habits of *Raja* species (Batoidea) in the Mediterranean waters off Alexandria. Bull Inst Oceanogr Fish 12:265–276

Adams J (2004) Foraging ecology and reproductive biology of Cassin's Auklet (*Ptychoramphus A aleuticus*) in the California Channel Islands. Thesis (M.S.). California State University, San Francisco. 119 pp

Ajayi TO (1982) Food and feeding habits of *Raja* species (Batoidei) in Carmarthen Bay, Bristol Channel. J Mar Biol Assoc UK 62:215–223

Allen MJ (2006) Chapter 7. Continental Shelf and Upper Slope. In: Allen LG, Pondella II DJ, Horn MH (eds) The ecology of marine fishes. California and adjacent waters. University of California Press, Berkeley, California, pp 167–202

Barry JP, Yoklavich MM, Cailliet GM, Ambrose DA, Antrim BS (1996) Trophic ecology of the dominant fishes in Elkhorn Slough, California, 1974–1980. Estuaries 19:115–138

Berestovskiy EG (1989) Feeding on the skates, *Raja radiata* and *Raja fyllae*, in the Barents and Norwegian Seas. Voprosy Ikhtiologii 6:994–1002

Braccini JM, Perez JE (2005) Feeding habits of the sandskate *Psammobatis extenta* (Garman, 1913): sources of variation in dietary composition. Mar Freshw Res 56:395–403

Braccini JM, Gillanders BM, Walker TI (2005) Sources of variation in the feeding ecology of the piked spurdog (*Squalus megalops*): implications for inferring predator–prey interactions from overall dietary composition. ICES J Mar Sci 62:1076–1094

Brickle P, Laptikhovsky V, Pompert J, Bishop A (2003) Ontogenetic changes in the feeding habits and dietary overlap between three abundant rajid species on the Falkland Islands' Shelf. J Mar Biol Assoc UK 83:1119–1125

Cailliet GM, Barry JP (1978) Comparison of food array overlap measures useful in fish feeding habit analysis. In: Lipovsky SJ, Simenstad CA (eds) Fish food habit studies. Proceedings of the second Pacific Northwest Technical Workshop, pp 67–79

Cailliet GM, Love MS, Ebeling AW (1986) Fishes: a field and laboratory manual on their structure, identification, and natural history. Waveland Press Inc., Prospect Heights, 186 pp

Compagno LJV (2005) Chapter 16: Checklist of living chondrichthyes. In: Hamlett WC (ed) Reproductive biology and phylogeny of chondrichthyes. Sharks, Batoids and Chimaeras. Science Publishers, Inc. Enfield, New Hampshire, pp 503–548

Cortés E (1997) A critical review of methods of studying fish feeding based on analysis of stomach contents: application to elasmobranch fishes. Can J Fish Aquat Sci 54:726–738

Cortés E (1998) Methods of studying fish feeding: reply. Can J Fish Aquat Sci 55:2708

Cortés E (1999) Standardized diet compositions and trophic levels of sharks. ICES J Mar Sci 56:707–717

Crow ME (1979) Multivariate statistical analysis of stomach contents. In: Lipovsky SJ, Simenstad CA (eds) Fish food habit studies: proceedings of the second Pacific northwest technical workshop. Washington Sea Grant Program, University of Washington, Seattle, WA, pp 87–96

Dolgov, AV (2005) Feeding and food consumption by the Barents Sea skates. e-Journal of Northwest Atlantic Fisheries Science 35: art. 34

Ebert DA (2003) Sharks, rays, and chimaeras of California. University of California Press, Berkeley, California, 284 pp

Ebert DA (2005) Reproductive biology of skates, *Bathyraja* (Ishiyama), along the eastern Bering Sea continental slope. J Fish Biol 66:618–649

Ebert DA, Cowley PD, Compagno JV (1991) A preliminary investigation on the feeding ecology of skates (Batoidea: Rajidae) off the west coast of Southern Africa. South African J Mar Sci 10:71–81

Ellis JR, Pawson MG, Shackley SE (1996) The comparative feeding ecology of six species of shark and four species of ray (Elasmobranchii) in the north-east Atlantic. J Mar Biol Assoc UK 76:89–106

Ezzat A, Abd El-Aziz SM, El-Gharabawy MM, Hussein MO (1987) The food of *Raja miraletus* Linnaeus, 1758 in Mediterranean waters off Alexandria. Bull Inst Oceanogr Fish 13:59–74

Ferry LA, Cailliet GM (1996) Sample size and data analysis: are we characterizing and comparing diet properly? In: MacKinlay D, Shearer K (eds) Gutshop '96. Feeding ecology and nutrition in fish symposium proceedings. San Francisco State University, pp 71–80

Hacunda JS (1981) Trophic relationships among demersal fishes in a costal area of the Gulf of Maine. Fish Bull 79:775–788

Hansson S (1998) Methods of studying fish feeding: a comment. Can J Fish Aquat Sci 55:2706–2707

Holden MJ, Tucker RN (1974) The food of *Raja clavata* Linnaeus 1758, *Raja montagui* Fowler 1910, *Raja naevus* Muller and Henle 1841 and *Raja brachyuran* Lafont 1873 in British waters. Journal du Conseil International pour l'Exploration de la Mer 35:189–193

Hyslop EJ (1980) Stomach content analysis – a review of methods and their application. J Fish Biol 17:411–429

Jordan DS, Gilbert CH (1880) Description of a new species of ray, RAIA RHINA, from the coast of California. Proceedings of the United States National Museum, pp 251–253

Joyce WN, Campana SE, Natanson LJ, Kohler NE, Pratt HL Jr, Jensen CF (2002) Analysis of stomach contents of the porbeagle shark (*Lamna nasus* Bonnaterre) in the northwest Atlantic. ICES J Mar Sci 59:1263–1269

Kabasakal K (2002) Cephalopods in the stomach contents of four Elasmobranch species from the northern Aegean Sea. Acta Adriatica 43:17–24

Koen Alonso M, Crespo EA, García NA, Pederaza SN, Mariotti PA, Berón Vera B, Mora NJ (2001) Food habits of *Dipturus chilensis* (Pisces: Rajidae) off Patagonia, Argentina. ICES J Mar Sci 58:288–297

Krebs CJ (1999) Ecological methodology, 2nd edn. Addison Wesley Longman, Inc., California, 620 pp

Lucifora LO, Valero JL, Bremec CS, Lasta ML (2000) Feeding habits and prey selection by the skate *Dipturus chilensis* (Elasmobranchii: Rajidae) from the south-western Atlantic. J Mar Biol Assoc UK 80:953–954

Mabragaña E, Giberto DA, Bremec CS (2005) Feeding ecology of *Bathyraja macloviana* (Rajiformes: Arhynchobatidae): a polychaete-feeding skate from the South-west Atlantic. Sci Mar 69:405–413

Mauchline J, Gordon JDM (1983) Diets of sharks and chimaeroids of the Rockall Trough, northeast Atlantic Ocean. Mar Biol 75:269–278

McEachran JD, Boesch DF, Musick JA (1976) Food division within two sympatric species-pairs of skates (Pisces: Rajidae) Mar Biol 35:301–317

McGarigal K, Cushman S, Stafford S (2000) Multivariate statistics for wildlife and ecology research. Springer Science + Buisness Media, Inc. New York, 283 pp

Mecklenburg CW, Mecklenburg TA, Thorsteinson LK (2002) Fishes of Alaska. American Fisheries Society, Bethesda, Maryland, 1037 pp

Morato T, Solá E, Grós MP, Menezes G (2003) Diets of thornback ray (*Raja clavata*) and tope shark (*Galeorhinus galeus*) in the bottom longline fishery of the Azores, northeastern Atlantic. Fish Bull 101:590–602

Muto EY, Soares LSH, Goitein R (2001) Food resource utilization of the skates *Rioraja agassizii* (Muller, Henle, 1841) and *Psammobatis extenta* (Garman, 1913) on the continental shelf off Ubatuba, southeastern Brazil. Revista Brasileira de Biologia 61:217–238

Nelson JS (2006) Fishes of the world, 4th edn. John Wiley, Sons, Inc, New Jersey, 601 pp

Orlov AM (1998) On feeding of mass species of deep-sea skates (*Bathyraja* spp., Rajidae) from the Pacific waters of the northern Kurils and southeastern Kamchatka. J Ichthyol 38:635–644

Orlov AM (2003) Diets, feeding habits, and trophic relations of six deep-benthic skates (Rajidae) in the western Bering Sea. J Ichthyol Aquat Biol 7:45–60

Paukert CP, Wittig TA (2002) Applications of multivariate statistical methods to fisheries. Fisheries 27:16–22

Pedersen SA (1995) Feeding habits of starry ray (*Raja radiata*) in West Greenland waters. ICES J Mar Sci 52:43–53

Pinkas L, Oliphant MS, Iverson ILK (1971) Food habits of Albacore, Bluefin Tuna, and Bonito in California Waters. Fish Bull 152:1–105

Skjaeraasen JE, Bergstad OA (2000) Distribution and feeding ecology of *Raja radiata* in the northeastern North Sea and Skagerrak (Norwegian Deep) ICES J Mar Sci 57:1249–1260

Smale MJ, Cowley PD (1992) The feeding ecology of skates (Batoidea: Rajidae) off the Cape South Coast, South Africa. South African J Mar Sci 12:823–834

Templeman W (1982) Stomach contents of the thorny skate, *Raja radiata*, from the northwest Atlantic. J Northwest Atlantic Fish Sci 3:123–126

Wakefield WW (1984) Feeding relationships within assemblages of nearshore and mid-continental shelf benthic fishes off Oregon. Oregon State University, Master's Thesis

Washington HG (1984) Diversity, Biotic and Similarity Indices: a review with special relevance to aquatic ecosystems. Water Res 18:653–694

Wetherbee BM, Gruber SH, Cortés E (1990) Diet, feeding habits, digestion, and consumption in sharks, with special reference to the lemon shark, *Negaprion brevirostris*. In: Pratt HL, Gruber SH, Taniuchi TT (eds) Elasmobranchs as living resources: advances in biology, ecology, systematics, and the status of the fisheries. NOAA Technical Report, NMFS 90, pp 29–47

Wright N, King C (2002) California bathymetry at 10 m contour intervals (Teale Albers/NAD 27) California Department of Fish and Game. Marine Region, GIS Laboratory, Monterey, CA

Yeon J, Hong SH, Cha HK, Kim ST (1999) Feeding habits of *Raja pulchra* in the Yellow Sea. Bull Natl Fish Res Dev Inst Korea 57:1–11

Zeiner SJ, Wolf P (1993) Growth characteristics and estimation of age at maturity of two species of skates (*Raja binoculata* and *Raja rhina*) from Monterey Bay, California. NOAA Technical Report National Marine Fisheries Service 115, pp 87–99

Zorzi GD, Martin LK, Ugoretz J (2001) Skate and Rays. pp. 257–261 In: Leet, WS, Dewes CM, Klingbeil R, Larson EJ (ed) California's living marine resources: a status report. California Fish and Game, Resources Agency

Dietary comparisons of six skate species (Rajidae) in south-eastern Australian waters

Michelle A. Treloar · Laurie J. B. Laurenson · John D. Stevens

Originally published in the journal Environmental Biology of Fishes, Volume 80, Nos 2–3, 181–196.
DOI 10.1007/s10641-007-9233-6 © Springer Science+Business Media B.V. 2007

Abstract The diet of six skate species caught as bycatch in south-eastern Australian waters was examined over a 2-year period. The skates were segregated into two regions (continental shelf and continental slope) based on prey species and depth of capture. The shelf group consisted of four species, *Dipturus* sp. A, *D. cerva*, *D. lemprieri* and *D. whitleyi*, while the slope group comprised two species, *Dipturus* sp. B and *D. gudgeri*. The two groups varied in feeding strategies with the shelf species generally occupying a broader feeding niche and preying on a larger diversity of prey including a variety of crustaceans (brachyurans, anomurans, achelates, carideans and dendobranchiates), cephalopods, elasmobranchs and teleosts. Within the slope group, *Dipturus* sp. B and *D. gudgeri* were more specialised. *Dipturus* sp. B preyed primarily on anomurans (galatheids) and bachyurans (homolids), whereas *D. gudgeri* preyed primarily on teleosts. A size related change in diet was evident for all species with the exception of *D. gudgeri* in which all sizes preyed predominantly on teleosts. Smaller representatives of the four shelf species all preyed on numerous amounts of caridean shrimps, in particular *Leptochela sydniensis*. In contrast, the continental slope species, *Dipturus* sp. B consumed anomurans when small, shifting to brachyurans with increasing size. Of the six skate species examined in this study, three were secondary consumers (trophic level <3) and the remaining three tertiary consumers (trophic level >4). Although ANOSIM found significant differences in dietary composition between species within groups, there was some overlap in prey species amongst co-existing skates, which suggests that there is some degree of resource partitioning amongst them.

Keywords Trophic level · *Dipturus* · Prey · Niche · Partitioning

M. A. Treloar · L. J. B. Laurenson
School of Life and Environmental Sciences, Deakin University, PO Box 423, Warrnambool, VIC 3280, Australia

M. A. Treloar (✉) · J. D. Stevens
CSIRO Marine and Atmospheric Research, GPO Box 1538, Hobart, Tasmania 7001, Australia
e-mail: michelle.treloar@csiro.au

Introduction

Skates (Rajiformes) are a frequent bycatch in the southeast Australian trawl (Wayte et al. 2006), drop-line (Daley et al. 2006) and gillnet fishery (Walker et al. 2006). They are regarded as a low value by-product and are mostly discarded. However, as a result of their abundance and wide distribution they constitute a substantial percentage of the benthic community (Graham et al. 2001). At least 12 species of skates occur in the benthic habitat of southeast Australia (Last and Stevens 1994). The benthic habitat where these skates reside is poorly understood with only limited ecological studies on the demersal fish communities of the area. Therefore, our understanding of trophic interactions and relationships in the region are equally poorly documented (Bulman et al. 2001), and especially so with the skates.

Despite skates (Rajiformes) containing about 245 described species world-wide (Ebert and Compagno 2007) and being caught regularly as bycatch of demersal fisheries (Bonfil 1994), they are poorly studied in comparison to other elasmobranchs (Orlov 1998). Few studies exist on the comparative feeding ecology of four or more sympatric skate species in the one location (Ebert et al. 1991; Ellis et al. 1996; Orlov 1998, 2003; Dolgov 2005; Fariasa et al. 2006) and less information on trophic levels and their placement within food webs (Morato et al. 2003; Braccini and Perez 2005). Results from these comparative feeding studies have shown that some competition can occur between skate species and other benthic organisms and that a high biomass of skates in an area can have a significant impact on the benthic fauna.

Total reported landings of skates (excluding the animals that are either misidentified or classified under the generic 'rays') from the southeast trawl fishery have increased from 30 tonnes in 1992 to 46 tonnes in 2002 (Treloar unpublished data). This data suggests that skates are being landed in increasing quantities and it is therefore necessary to determine trophic levels to gain an insight into the impact catches may have on the trophic ecology of the demersal community. Given the lack of dietary information on the southeast Australian skate assemblage this study provides the first information on the comparative feeding ecology and trophic levels of six skate species in south-eastern Australian waters. The species considered in this paper include *Dipturus* sp. A [sensu Last and Stevens 1994 = *Raja* sp. A], *Dipturus* sp. B [sensu Last and Stevens 1994 = *Raja* sp. B], *Dipturus cerva*, *Dipturus gudgeri*, *Dipturus* (*Dentiraja*) *lemprieri* [Last and Yearsley 2002] and *Dipturus* (*Spiniraja*) *whitleyi* [Last and Yearsley 2002].

Methods

Study area and sample collection

From March 2002 to June 2004, six skate species (*Dipturus* sp. A, *Dipturus* sp. B, *D. cerva*, *D. gudgeri*, *D. lemprieri*, and *D. whitleyi*) were collected from commercial and scientific vessels operating in south-eastern Australian waters (Fig. 1). Samples were collected from fishers operating (24 h operation) demersal trawls (18–700 m), Danish-seines (20–64 m), gill nets (5–20 m) and bottom drop-lines (300–585 m).

Skates were identified using Last and Stevens (1994), sexed, weighed (nearest gram) and the total length (TL) measured (nearest 1.0 mm). Stomachs were excised and their contents weighed (nearest gram) and preserved in 70% ethanol for subsequent examination. Stomach fullness was assessed using a visually estimated scale of 1–5 (1, empty; 2, 0–25% full; 3, 26–50% full; 4, 51–75% full; 5, 76–100% full). Level of digestion was assessed using a visual ranking scale where: 1 = prey items that were only slightly digested and easily distinguishable; 2 = medium level of digestion, prey items distinguishable; and 3 = an advanced level of digestion, prey items not easily distinguishable. Prey items assigned to a level 3 were largely unidentifiable and excluded from further analysis. Prey items were identified to the lowest possible taxon using various identification guides (Hale 1927–1929; Gomon et al. 1994; Jones and Morgan 1994; Edgar 2000; Poore 2004) and each individual item weighed (wet) using an analytical balance (±0.0001 g). A collection of otoliths extracted from known species was used to identify fish fragments that were otherwise unrecognisable (Furlani et al. 2007). Prey items such as hydroids, algae, salps and sponges were considered to be incidentally ingested and were not included in diet analysis. Parasitic nematodes were excluded from the analysis.

Dietary analyses

Prey items were sorted into 11 major categories to facilitate dietary comparisons: Dendobranchiates (prawns), carideans (shrimp), brachyurans (crabs), isopods, achelates (crayfish and pill-

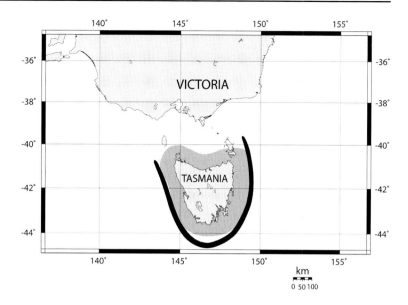

Fig. 1 Skate sample collection sites from south-eastern Australian waters. *Dipturus* sp. A, *D. cerva*, *D. lemprieri* and *D. whitleyi* () occurring in waters along the continental shelf and *Dipturus* sp. B and *D. gudgeri* (■) occurring along the continental slope

bugs), anomurans (squat lobsters), cephalopods (octopus and squid), molluscs (gastropods and bivalves), teleosts (fish excluding Anguilliformes), Anguilliformes (eels), and elasmobranchs (sharks, skates and rays). Where fewer than four higher taxa items were identified (in total) from all stomachs, these were excluded from further analysis (Moura et al. 2005). The importance of prey was assessed using several indices to avoid the bias inherent in each (Cortés 1997).

Percentage Frequency of Occurrence $\%F = \frac{N_i}{N_s} * 100$ where N_i is the number of individuals containing dietary category i and N_s is the total number of stomachs examined (Hyslop 1980).

Percentage Mass $\%M = \frac{W_i}{W_{sc}} * 100$ where W_i is the total weight of prey category i taken from all skates and W_{sc} is the total weight of all prey from all skates, and

Percentage Number $\%N = \frac{N_i}{N_{sc}} * 100$ where N_i is the total number of prey category i taken from all skates and N_{sc} is the total number of all prey from all skates (Hyslop 1980).

Comparisons among species were carried out using the Percentage Index of Relative importance (%IRI):

$$\%IRI = \frac{100 \, IRI}{\sum_{i=1}^{n} IRI} \quad \text{(Cortés 1997) where}$$

$IRI = (\%N + \%M) * \%F$ (Pinkas et al. 1971) and $\%N$ and $\%M$ are defined above.

Some authors (MacDonald and Green 1983; Hansson 1998) suggest that IRI values under represent lower taxonomic categories. We incorporate both IRI and the stomach fullness index, expressed as the product of the ratio of the mass/volume of a particular prey category in a stomach and the total mass/volume of all prey in that stomach (described later in the MDS analysis methods), to describe the difference in diet between species.

The mean %IRI contributions to the diet of skates of the top five prey categories (based on contribution by %IRI) were plotted against size class (Table 1) to provide a visual representation of ontogenetic changes in diet.

Preliminary multivariate analysis indicated that the dietary compositions of the females and males of all species were not statistically different ($P > 0.05$). In all subsequent analysis sexes have been pooled, this was also the case for spatial and temporal data.

To determine whether sufficient stomach samples were collected to characterise the diet for each species and size class, randomised cumulative prey curves were generated (Ferry and Cailliet 1996). Each skate was assigned a size class depending on species and sample size (Table 1). Prey species diversity was calculated using the pooled quadrat method based on the Brillouin Index of Diversity (Pielou 1966; Braccini

Table 1 Size classes (cm) of all skate species for data analysis

Dipturus sp. A		*Dipturus* sp. B		*D. cerva*		*D. gudgeri*		*D. lemprieri*		*D. whitleyi*	
<45	(47)	<82	(16)	<45	(40)	<75	(15)	<45	(21)	<75	(17)
45–55	(90)	>83	(17)	45–55	(78)	76–100	(34)	>46	(42)	76–100	(13)
>56	(109)			>56	(25)	>101	(20)			>101	(16)

Sample sizes in parentheses

et al. 2005). Stomach samples of each size class in each species were randomised (×100) and re-sampled with replacement to calculate cumulative mean diversity and standard deviation (Koen Alonso et al. 2001). When the curve reached an asymptote, the number of stomachs was considered sufficient for accurately describing the diet.

The trophic level of skates was calculated using the trophic index (TR):

$$TR = 1 + \left(\sum_{j=1}^{n} P_j * TR_j\right) \quad \text{(Cortés 1999)}$$

where TR_j is the trophic level of each prey taxa j and P_j is the proportion of each prey taxa in the diet of a skate species, based on %IRI values (Braccini and Perez 2005). A list of trophic levels of prey categories ($n = 8$) from the lowest taxonomic level was taken from various sources (Pauly et al. 1998a; Cortés 1999; University of Columbia 2005). Unidentified crustaceans were excluded from this analysis. Cluster analysis was performed in PRIMER v5.2.2 (Clarke and Gorley 2001) to determine similarities in trophic levels between the six skate species.

Niche breadth was quantified using Levins' Index (B): $B = 1/\sum p_j^2$ (Levins 1968) where p_j is the fraction by weight of prey items in the diet of food category $j (\sum p_j = 1)$. Niche breadth was calculated using 20 prey categories (see Table 3). Because few teleosts were identified to species or family level they were placed into one category. A lower index indicated a narrow niche breadth (Ellis et al. 1996).

Non-metric multi-dimensional scaling ordination (MDS) using the PRIMER v5.2.2 (Clarke and Gorley 2001) was used to examine changes in diet with respect to size (Table 1). Primer was also used for size comparisons across species with size classes standardised to <50 cm TL, 51–100 cm TL, and >101 cm TL. The data used for this analysis follows the methods of Linke et al. (2001) and White et al. (2004) in which the Stomach Fullness Index is expressed as the product of the ratio of the mass/volume of a particular prey category in a stomach and the total mass/volume of all prey in that stomach. In most instances, the mean of a random sub-sample (Table 2) of these values (based on the desired analysis categories e.g. size, etc) were square root transformed and subjected to MDS using a Bray–Curtis similarity matrix (White et al. 2004).

One-way analyses of similarities (ANOSIM) were used to determine if prey categories differed significantly amongst skate species or sizes and whether any paired relationships existed. Similarity percentages (SIMPER) were used to identify significantly important dietary categories. Multivariate dispersion (MVDISP) was used to determine the amount of dispersion within one group of points, i.e. species (White et al. 2004).

Results

Diet comparison

Depth segregation occurred among the six skate species, with two distinct groups being present. *Dipturus* sp. B and *D. gudgeri* (frequently caught by bottom drop-liners) occurred in depths of 300–

Table 2 The number by which each species was pooled for size comparisons

	Dipturus sp. A	*Dipturus* sp. B	*D. cerva*	*D. lemprieri*	*D. gudgeri*	*D. whitleyi*
Size	24	4	12	6	8	4

700 m, whereas *Dipturus* sp. A, *D. cerva*, *D. lemprieri* and *D. whitleyi* were caught in depths ranging from 1 to 280 m. Cumulative prey curves were constructed for each species by size class and showed either well-defined asymptotes or trends towards an asymptote (Fig. 2). This indicated that sufficient stomachs were examined to describe the diet for most size classes and species.

Thirty-six species, 37 genera, and 36 families of prey items were identified in the six skate species examined (Table 3). The most diverse taxa were crustaceans (containing 17 families and over 19 identified species); amongst which brachyurans were the most numerous infraorder, followed by carideans. Teleosts were of importance to two species with Anguilliformes being consumed by five of the six skate species and Platycephalidae (flathead) occurring in the diet of three skate species. Juvenile elasmobranchs occurred in three of the six skate species examined.

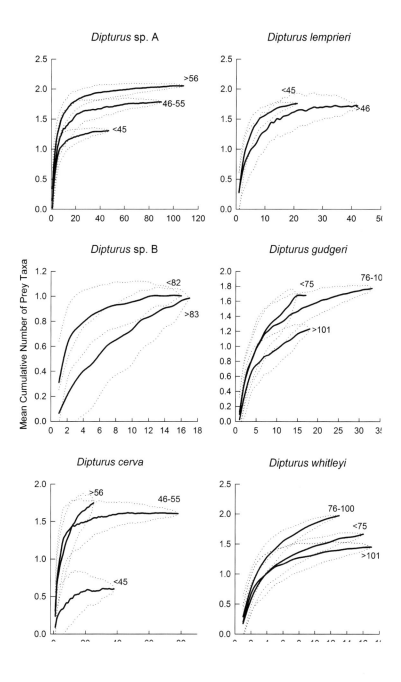

Fig. 2 Mean cumulative diversity of prey taxa plotted against cumulative number of stomach samples for all size classes (TL cm) in *Dipturus* sp. A, *Dipturus* sp. B, *D. cerva*, *D. gudgeri*, *D. lemprieri* and *D. whitleyi*. Dotted line indicates ± 1 SD

Table 3 Diet composition of six skate species in terms of frequency of occurrence (%F), contribution by number (%N), mass (%M) and index of relative importance on a percent basis (%IRI) for the major prey taxa and identifiable dietary categories

Major taxa	*Dipturus* sp. A %F	%N	%M	%IRI	*Dipturus* sp. B %F	%N	%M	%IRI	*D. cerva* %F	%N	%M	%IRI	*D. gudgeri* %F	%N	%M	%IRI	*D. lemprieri* %F	%N	%M	%IRI	*D. whitleyi* %F	%N	%M	%IRI
Mollusca		**10.2**	**26.3**	**23.9**						**2.4**	**1.8**	**0.9**		**2.3**	**0.2**	**0.1**		**3.2**	**5.0**	**0.0**		**8.4**	**27.0**	**27.7**
Cephlapoda[a]	20.8	10.0	26.1	23.9					13.3	2.4	1.8	0.9	3.2	2.3	0.2	0.1	13.2	3.2	5.0	0.0	23.6	7.9	26.9	27.7
Gastropoda[a]	0.5	0.2	0.2	0.0																	1.8	0.5	0.1	0.0
Crustacea		**78.0**	**27.2**	**56.8**		**96.4**	**90.2**	**98.9**		**95.3**	**86.5**	**97.4**		**58.7**	**21.8**	**19.7**		**91.0**	**75.7**	**89.7**		**73.3**	**29.5**	**32.0**
Callapidae[a]	15.7	25.0	2.5	13.7					48.8	59.7	12.6	65.7	3.2	9.0	0.7	0.5	18.4	16.2	1.7	9.3	10.9	27.2	0.2	11.2
Cancridae[a]	2.2	1.8	0.5	0.2					5.4	0.6	2.5	0.1					6.6	1.6	2.6	0.3				
Caridea[a]	0.5	0.4	0.1	0.0					3.0	1.6	1.2	0.2												
Dendobranchiata[aa]	3.4	3.9	1.1	0.5	34.1	55.5	15.7	37.8	24.1	12.9	39.0	20.4	19.4	16.9	5.2	6.9	17.1	13.3	30.7	21.1	3.6	2.0	0.1	0.3
Galatheidae[a]					2.3	0.7	0.3	0.0	3.6	0.5	0.4	0.1												
Goneplacidae[a]	0.3	0.1	0.3	0.0					1.2	0.1	1.1	0.0									3.6	1.0	5.0	0.8
Homolidae[a]	0.3	0.1	0.1	0.0																	1.8	3.5	2.5	0.4
Hymenosomatidae[a]	2.9	1.2	1.2	0.2					1.2	0.1	0.3	0.0					15.8	9.2	16.1	11.2	1.8	0.5	0.1	0.0
Isopoda[a]	1.7	0.8	0.8	0.1	9.1	10.7	2.5	1.9	5.5	1.1	2.5	0.4					2.6	4.3	2.8	0.5				
Leucosidae[a]					40.9	21.0	70.1	58.0					15.1	10.2	14.0	5.9					1.8	0.5	1.7	0.2
Majidae[a]	0.7	2.1	0.2	0.1					0.6	0.1	0.0	0.0					1.3	0.3	0.0	0.0				
Palinuridae[a]	12.0	10.7	4.9	6.0					0.6	0.1	1.3	0.0					11.8	7.2	3.4	3.5	3.6	1.5	0.2	0.1
Pilumnidae[a]	0.3	0.1	0.2	0.0	4.6	1.4	0.7	0.1													1.8	0.5	1.5	0.1
Portunidae[a]	0.3	0.1	0.1	0.0																				
Raninidae[a]	8.8	7.6	5.1	3.6					10.8	6.3	11.0	3.9					14.5	14.2	3.5	5.7	9.1	15.4	6.9	3.8
Scyllaridae[a]																					1.8	1.5	0.5	0.0
Unidentifiable	29.7	24.2	10.1	32.3	13.6	7.1	1.0	1.1	35.5	12.1	14.6	6.5	16.1	22.6	1.9	6.4	55.3	24.7	14.9	38.1	20.0	19.8	10.8	15.0
Elasmobranchs[a]		**0.1**	**0.2**	**0.0**						**0.5**	**2.3**	**0.1**						**0.3**	**0.6**	**0.0**		**0.5**	**1.1**	**0.1**
Scyliorhinidae	0.3	0.1	0.2	0.0					2.4	0.4	1.9	0.1												
Rajidae									0.6	0.1	0.2	0.0												
Urolophidae									0.6	0.1	0.3	0.0					1.3	0.3	0.6	0.0	1.8	0.5	1.1	0.1
Teleosts[a]		**11.7**	**46.3**	**19.3**		**3.5**	**9.8**	**1.0**		**2.0**	**9.1**	**1.6**		**39.1**	**78.0**	**80.3**		**5.6**	**18.4**	**10.3**		**17.8**	**42.4**	**40.2**
Anguillidae	6.9	3.8	4.3	1.8					6.0	1.0	1.9	0.4	3.2	4.5	1.4	0.3	1.3	0.3	0.0	0.0	9.1	3.5	4.3	2.7
Antennariidae	0.5	0.2	2.7	0.0																				
Bythitidae																					1.8	1.0	0.4	0.1
Callionymidae																					1.8	1.0	2.2	0.2
Gempylidae									0.6	0.1	0.0	0.0									1.8	1.0	5.9	0.5
Macrouridae					2.3	0.7	5.4	0.2					1.1	0.8	1.0	0.0								
Moridae																					1.8	0.5	4.0	0.3
Pegasidae	0.3	0.2	0.2	0.0																				
Platycephalidae	1.7	0.7	7.2	0.4													2.6	0.9	2.9	0.3	1.8	1.0	1.9	0.2
Pleuronectidae	0.3	0.1	0.6	0.0																				
Sebastidae																					1.8	0.5	1.2	0.1
Syngnathidae																					1.8	0.5	0.5	0.1
Triglidae	1.0	0.3	6.8	0.2													1.3	0.4	0.3	0.0				

Table 3 continued

Major taxa	Dipturus sp. A %F	%N	%M	%IRI	Dipturus sp. B %F	%N	%M	%IRI	D. cerva %F	%N	%M	%IRI	D. gudgeri %F	%N	%M	%IRI	D. lemprieri %F	%N	%M	%IRI	D. whitleyi %F	%N	%M	%IRI
Unidentifiable	17.2	6.5	24.5	16.9	6.8	2.8	4.4	0.8	7.2	0.9	7.2	1.2	45.2	33.8	75.6	79.9	18.4	4.0	15.2	9.9	30.9	8.9	22.1	36.0
Total no. of stomachs	410.0				113.0				166.0				263.0				77.0				60.0			
No. of stomachs containing food[b]	335.0				37.0				160.0				72.0				72.0				48.0			
Percentage of stomachs containing food	81.7				33.0				96.4				27.4				93.5				80.0			
Mean stomach fullness	2.7				1.9				2.9				1.9				2.9				2.8			
Niche breadth (B)	4.1				1.7				3.7				2.2				3.8				4.1			

[a] Prey categories for Niche Breadth analysis
[b] This excluded stomachs containing 100% unidentified digested matter

A total of 410 *Dipturus* sp. A stomachs were examined, 75 of which were empty (Table 3). A wide variety of prey was consumed with crustaceans representing the largest proportions (%N) and diversity (%O). Brachyurans contributed the most to this subphylum, consisting of at least eight families. Carideans were of secondary importance (14% IRI) in this subphylum with *Leptochela sydniensis* dominating (13% IRI). Cephalopods were also important (24% IRI), occurring in 22% of stomachs. Teleosts occurred in 21% of stomachs and contributed the most in mass (46%). Other prey categories such as Elasmobranchs, Dendobranchiata, Achelata, Anomura, Isopoda and Gastropoda were of lesser importance. The importance of each prey category changed with the increase of skate size and ANOSIM values indicated that dietary composition differed among size classes ($P < 0.001$; global R statistic 0.78) (Fig. 3). Dendobranchiata/carideans (50%) and brachyurans (44%) dominated the diet of small individuals (<45 cm TL), whereas teleosts (46%), brachyurans (30%) and cephalopods (24%) were of more importance in the larger size classes (>56 cm TL).

One hundred and thirteen *Dipturus* sp. B stomachs were examined, of which 76 were empty (Table 3). *Dipturus* sp. B had the least diverse diet of all six skate species. Crustaceans (99% IRI) were the most important taxa, with brachyurans predominantly *Dagnaudus petterdi* (Homolidae) dominating the diet. Galatheids (anomurans) were second most important, occurring in 34% of stomachs. Very few teleosts occurred in the stomachs examined (1% IRI). ANOSIM showed that there were significant differences in dietary composition among size classes ($P < 0.03$; global R statistic 0.79) (Fig. 3). Although brachyurans were the predominant prey found in *Diptrurus* sp. B, small skates (<82 cm TL) consumed large amounts of anomurans and small amounts of brachyurans and teleosts (Fig. 3). A change in diet occurred with the larger skates (>83 cm TL) consuming predominantly brachyurans.

A total of 166 *D. cerva* stomachs were examined, six of which were empty (Table 3). The diet of *D. cerva* consisted of predominantly crustaceans with carideans (66%), primarily *L. sydniensis* being of greatest importance. Brachyurans

Fig. 3 Percentage of IRI by size class for the six skate species

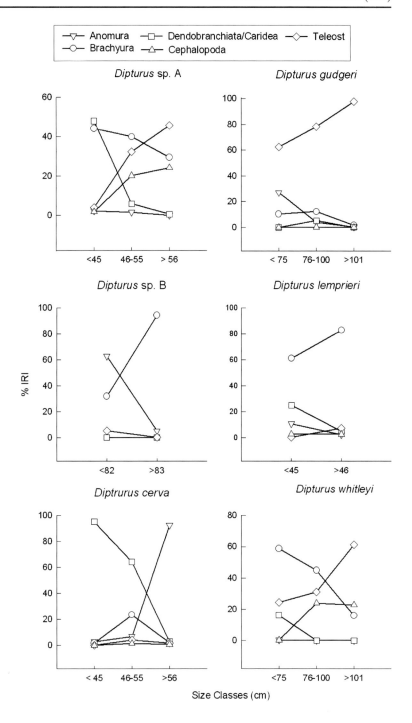

consisted of at least five families occurring in 29% of stomachs, whereas anomurans occurred in 24% of stomachs but contributed the most by mass (39%). Teleosts, elasmobranchs, dendobranchiates, achelates, cephalopods and isopods were of lesser importance to the diet of *D. cerva*.

Although elasmobranchs were of lesser importance, a diverse range of taxa occurred in the diet including *Asymbolus* spp. (including egg cases), rajids and urolophids (Table 3). All prey items varied in level of importance throughout size classes (Fig. 3), which was reflected in the ANO-

SIM analysis ($P < 0.001$; global R statistic 0.67) suggesting significant differences between size classes. Dendobranchiates/carideans were dominant in two size classes (<45 and 46–56 cm TL) but decreased in importance as the skate increased in size, whereas brachyurans and teleosts started to become of more importance in the 46–56 cm TL size class. Anomurans dominated the diet of individuals >56 cm TL.

A total of 263 *D. gudgeri* stomachs were analysed, of which 191 of these were empty (Table 3). *Dipturus gudgeri* consumed a narrow range of prey species with teleosts dominating the diet. Most teleosts could not be identified, with the exception of two families, Anguillidae and Macrouridae. Brachyurans, predominantly *D. petterdi* (6% IRI) and anomurans (7% IRI), were of less importance. ANOSIM showed no changes in diet across size classes ($P > 0.05$; R statistic 0.14) although anomurans were of more significance in small individuals (<75 cm TL) and teleosts increased in %IRI with size but still dominated all size classes (Fig. 3). Brachyurans (Homolidae) and dendobranchiates/carideans were less important in all size classes.

Stomachs were collected from 77 *D. lemprieri*, of which five were empty (Table 3). The diet of *D. lemprieri* consisted of a wide range of prey items with crustaceans dominating the diet. Brachyurans were the most dominant infraorder with five families identified in 74% of the stomachs, although the majority of the brachyurans were unidentified. Anomurans were important in the diet of *D. lemprieri* in terms of %IRI and contributed the most in mass. Carideans, cephalopods, isopods, elasmobranchs, and teleosts were of less importance. ANOSIM showed significant differences in diet among size classes ($P < 0.04$; global R statistic 0.22) (Fig. 3). Dendobranchiates/carideans were of some importance in small individuals (<45 cm TL), although brachyurans dominated this size class and increased in importance with increasing size (>46 cm TL). Cephalopods, teleosts and anomurans were of little importance.

Sixty *D. whitleyi* stomachs were examined, 12 of these were empty (Table 3). The diet of *D. whitleyi* was the most diverse out of all six skate species. Teleosts contributed the most in terms of %IRI and mass. *Dipturus whitleyi* consumed the greatest diversity of teleosts of all the skate species consisting of at least eight families, although 36% by IRI were still unidentified. The majority of these teleosts were benthic; however two benthopelagic species (Gemiphylidae spp.) were identified. Anguilliformes were only present in 9% of the stomachs, although they occurred most in the diet of this species. Cephalopods, primarily *Octopus* spp. occurred in 24% of stomachs contributing 28% (IRI), whereas brachyurans (including unidentified) occurred in 36% of stomachs contributing 19% in terms of percentage IRI. Carideans, anomurans, achelates, elasmobranchs and gastropods were of lesser importance. ANOSIM showed that diet differed amongst size classes ($P < 0.03$; global R statistic 0.28) with brachyurans dominating both the <75 cm TL and 76–100 cm TL size classes but becoming of less importance as the skate increased in size (Fig. 3). Cephalopods became more important in the 76–100 cm TL size class and teleosts became increasingly important from 76 cm TL onwards (Fig. 3). Dendobranchiates/carideans were of little importance throughout all size classes.

Skate trophic levels

Skate trophic levels ranged from 3.6 in *Dipturus* sp. B to 4.2 in *Dipturus* sp. A and *D. whitleyi*. Fish, eels, elasmobranchs, and octopus had the highest trophic levels and therefore contributed the most to species that consumed these prey groups (Table 4). All skate species that had a TR of <4.0, fed primarily on crustaceans, whereas the diet of *D. gudgeri* and *D. whitleyi* (TR >4.0) consisted primarily of teleosts, and several elasmobranchs, crustaceans, and octopus. *Dipturus* sp. A fed on a mixture of these.

Cluster analysis identified two distinct TR groupings within the six skate species (Fig. 4). The greatest similarity was found between *Dipturus* sp. A and *D. whitleyi*, which both largely preyed on fish and octopus. The second group included *Dipturus* sp. B and *D. lemprieri* which targeted primarily crustaceans (Fig. 4; Table 4).

Table 4 Standardised trophic compositions and trophic levels of six skate species

Species	Fish (3.2)	Eels (4.2)	Squid (3.2)	Octo (3.8)	Crabs (2.6)	Sh/Pr (2.7)	Crus (2.5)	Elasmo (4.1)	TR
Dipturus sp. A	0.22	0.02	–	0.30	0.27	0.18	0.01	0.00	4.2
Dipturus sp. B	0.00	–	–	–	0.60	0.00	0.39	–	3.6
D. cerva	0.01	0.00	0.00	0.01	0.07	0.71	0.22	0.00	3.7
D. gudgeri	0.85	0.00	0.00	0.00	0.06	0.00	0.07	–	4.1
D. lemprieri	0.12	0.00	–	0.02	0.40	0.13	0.25	0.08	3.8
D. whitleyi	0.37	0.03	–	0.28	0.19	0.11	0.01	0.00	4.2

Prey taxa are abbreviated as follows: (Octo = octopus; Sh/Pr = Shrimps/Prawns; Crus = Other Crustacea (i.e. anomurans, isopods, etc); and Elasmo = Elasmobranchs). Trophic levels for each prey group are in parentheses

Comparisons of feeding habits

Ordination of the %M contributions for the two deep-sea skate species (*Dipturus* sp. B and *D. gudgeri*) identified significant differences ($P = 0.001$; global R statistic 0.68, Fig. 5A). SIMPER demonstrated that brachyurans and teleosts were responsible for this difference. Niche breadth differed between the two species with *Dipturus* sp. B (1.7) having a lower niche breadth than *D. gudgeri* (2.2) (Table 3). ANOSIM demonstrated no significant difference between size classes (Fig. 5B).

Ordination using percentage mass contributions suggested prey overlap for the shelf species *Dipturus* sp. A, *D. cerva, D. lemprieri* and *D. whitleyi* (Fig. 6A). However ANOSIM ($P < 0.001$; global R statistic 0.34) suggested otherwise finding significant differences among all species. These differences were reflected by the MDVSIP values which showed *Dipturus* sp. A had the lowest dispersion (0.82) and *D. whitleyi* had the highest dispersion value (1.28). Pair-wise comparisons suggested that differences were greatest between *D. cerva* and *D. whitleyi* ($P < 0.001$; R statistic 0.52). SIMPER identified

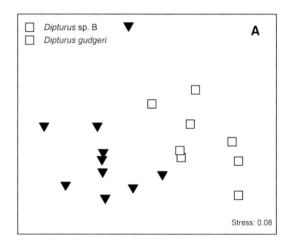

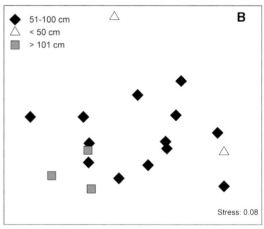

Fig. 5 (**A**) Non-parametric MDS ordination of mean percentage mass contributions for *Dipturus* sp. B and *D. gudgeri*. (**B**) Non-parametric MDS ordination of mean percentage mass contributions for *Dipturus* sp. B and *D. gudgeri* size classes)

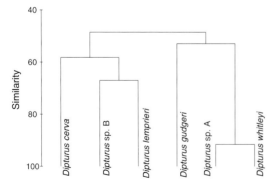

Fig. 4 Cluster analysis of the similarities in the dietary composition contributing towards trophic levels for six skate species

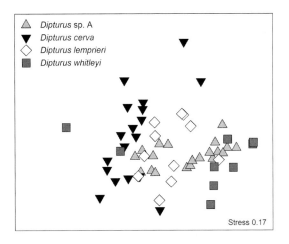

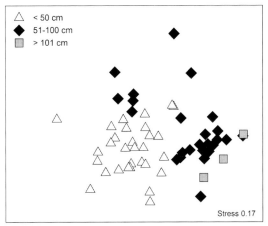

Fig. 6 (**A**) Non-parametric MDS ordination of mean percentage mass contributions for *Dipturus* sp. A, *D. cerva*, *D. lemprieri* and *D. whitleyi*. (**B**) Non-parametric MDS ordination of mean percentage mass contributions for *Dipturus* sp. A, *D. cerva*, *D. lemprieri* and *D. whitleyi* size classes

carideans as the prey item responsible for this separation. *Dipturus cerva* and *D. whitleyi* had the smallest and largest niche breadth, respectively (Table 3). MDS showed significant differences among the three size classes of all skate species (Fig. 6B).

Discussion

The six skate species studied here can be broadly divided into two groups based on different prey species and depth preferences. *Dipturus* sp. B and *D. gudgeri* occurred along the continental slope, whereas *Dipturus* sp. A, *D. cerva, D. lemprieri*

and *D. whitleyi* were common on the continental shelf, as was also observed by Last and Stevens (1994). Segregations of skate species has also been observed off the west coast of southern Africa where Ebert et al. (1991) found that 14 species were separated by depth zones. The fishing methods used to sample skates in this study were similarly divided between these depth zones with mostly droplines and some trawling occurring on the continental slope and Danish seine, trawling and gillnetting occurring on the shelf. This poses some difficulties in the interpretation of these data as past studies have shown that gear selectivity affects catch composition (Walker 2004) and can bias fish dietary studies (Morato et al. 2003). In the context of this study, biases were identified in stomach fullness data and our ability to identify prey species (from areas not sampled in this study). Both *Dipturus* sp. B and *D. gudgeri* were primarily collected by baited demersal dropline and both frequently had empty stomachs or contained only bait and/or hooks. Studies have shown that fishes feeding to satiation have a reduced response to bait (Lokkeborg et al. 1995), thus fishes with empty stomachs are more likely to be caught and form a disproportionate percentage of the catch. A high occurrence of empty stomachs has also been reported in other elasmobranch dietary studies using baited long-lines as a collection method (Morato et al. 2003; Lokkeborg et al. 1995).

A diverse range of prey species were found among the six skate species, with some species having greater prey diversity than others. Remarkable differences were detected between the two assemblages, with the shelf species having a more diverse diet than the slope species. In contrast, Ebert et al. (1991) found the South African slope species consumed a higher diversity of prey than the shallower shelf species. Niche breadth can give an indication of whether a species is a generalist or specialist feeder (Levins 1968). The continental slope species *Dipturus* sp. B and *D. gudgeri* both had low niche breadths (≤ 2.2), suggesting they may be specialist feeders but this study did not determine the range of potential prey available nor sample the populations widely (Gerking 1994), thus conclusions are difficult to draw.

Within the area studied *Dipturus* sp. B was found to be a crustacean specialist, preying mostly on *Dagnaudus petterdi* (antlered crab), and galatheids (squat lobsters). In contrast, the sympatric species *D. gudgeri*, a large skate attaining 185 cm TL predominantly targeted teleosts, at least one (*Lepidorhynchus denticulatus*) of which was benthopelagic. Different predator strategies are required to target these very different prey groups, the latter requiring more mobility and the former a capacity to crush robust exoskeletons. The large size of *D. gudgeri* is likely an advantage in preying on teleosts as it enhances the species capacity to move or potentially herd prey. Factors such as mobility, seasonal occurrence, abundance and distribution of potential prey, as well as ecomorphology, which are known to effect diet (Smale and Cowley 1992; Muto et al. 2001; Motta 2004) have not been assessed here.

Niche breadth (3.7–4.1) for the four continental shelf species indicated that they were more generalised feeders, but as is the case with the deeper water representatives studied here, it is unclear if this is a result of prey targeting or if it just represents the diversity of available prey species. The diet of *D. cerva* and *D. lemprieri* consisted almost exclusively of benthic crustaceans. *Dipturus cerva* largely targeted small, soft bodied crustaceans such as shrimp, *L. sydniensis*, while *D. lemprieri* was more opportunistic targeting a wider diversity of benthic crustaceans but predominantly brachyurans. Teleosts and cephalopods were sporadically preyed on by *D. lemprieri* suggesting a more opportunistic predatory behaviour. The diet of *D. lemprieri* was similar to *Dipturus* sp. A which also consumed benthic crustaceans, although other prey items including octopus and benthic teleosts were of greater importance in the latter species. The broader niche breadth of *Dipturus* sp. A suggests that this species is potentially more mobile or has a different foraging strategy providing greater opportunity to prey on more mobile organisms. *Dipturus whitleyi* had the most diverse diet of all shelf skates targeting a variety of crustaceans, octopus and teleosts. This emphasises the opportunistic nature of the feeding strategies employed by this species, and is likely a function of its large size in comparison to the other shelf skates examined here, being born at ~20 cm TL (Last and Stevens 1994) and attaining 200 cm TL. The four continental shelf skate species studied here targeted similar prey, but consumed different proportions of each, suggesting there is niche specialisation and separation amongst them.

Interestingly, the four shelf species preyed on small quantities of elasmobranch egg cases and juveniles which may be due to their similar spatial distributions. Studies have shown that embryos of egg-laying elasmobranchs are susceptible to predation (Cox and Koob 1993) because of the signal given off by the embryo circulating water around the egg case, this signal can be detected by the mechanoreceptive lateral line of certain predators (Luer and Gilbert 1985; Collin and Whitehead 2004). Skates have been documented to feed on juvenile elasmobranchs in other parts of the world, although they did not contribute significantly to the overall diet (Ebert et al. 1991; Smale and Cowley 1992; Orlov 1998; San Martin et al. 2007).

Intra-specific size-related changes in diet

Although ontogenetic changes occurred in all species examined in this study with the exception of *D. gudgeri*; cumulative prey curves for some size classes (*D. cerva* >56 cm TL, *D. gudgeri* >101 cm TL, *Dipturus* sp. B >83 cm TL, and *D. whitleyi* <75 cm and 76–100 cm TL) did not reach an asymptote. Consequently, more stomachs for these size classes may be needed to adequately describe the diet. Small benthic crustaceans were consumed in all species at smaller sizes but variations amongst them became more evident with increasing size. Some species made a transition to larger crustaceans with more robust exoskeletons (*Dipturus* sp. B, *D. cerva* and *D. lemprieri*), whereas others consumed larger proportions or larger teleosts (*Dipturus* sp. A, *D. whitleyi* and *D. gudgeri*). The small benthic shrimp *L. sydniensis* made up a large proportion of the diet of small individuals but only occurred in the shelf skate species as it does not occur in deeper waters (Poore 2004). The smaller continental slope species *Dipturus* sp. B preyed on small galatheids initially, an observation also noted in skate species in other parts of the world (Koen Alonso et al. 2001). Unquantified observations during this study indicated that juvenile skates have less developed teeth and smaller mouths than adults and it is unlikely that they have the ability to crush harder and large prey. The results of this study support those of Smale and Cowley (1992) and Fariasa et al. (2006) who found a relationship between predator body and mouth size and changes in diet to larger prey. The implications of these results are that prey species are potentially more a function of size rather than other life history characteristics. However, all factors (including tooth morphology between the species) were not thoroughly investigated in this study.

The diet of the continental slope skate *D. gudgeri*, unlike the other skates examined here, did not change from one prey group to another with increasing size, rather it remained feeding exclusively on teleosts with these increasing in proportion and size with increasing skate size. Similar patterns have been observed in other skate species where teleosts dominate the diet (Lucifora et al. 2000). This behaviour has been identified mostly in large skates where they are able to exploit a broader habitat range and capture larger more mobile prey items to meet their energetic requirements (Muto et al. 2001). This was further demonstrated in this study by the large shelf skate *D. whitleyi* which also preyed on teleosts throughout all size classes. Several bentho-pelagic prey species were found only in the diet of these two larger species (*D. gudgeri* and *D. whitleyi*). Studies have reported pelagic species such as anchovies, *Engraulis capensis*, and pilchards, *Sardinops ocellatus*, in the diet of skates (Ebert et al. 1991; Smale and Cowley 1992) but a lack of understanding of the spatial and temporal distribution of potential prey species makes drawing conclusions difficult. For example, pelagic prey may be targeted only when near the sea floor rather than skates feeding pelagically.

Trophic levels

Surprisingly few studies have focused on skate trophic levels and their placement within the food web (Morato et al. 2003; Braccini and Perez 2005). The diet information gained from this study has demonstrated that both the continental slope and shelf skate species examined are secondary and tertiary consumers. The tertiary consumers (>4.0) preyed more widely on eels, other teleosts, octopus and elasmobranchs which contributed high energy/trophic levels between 3.2 and 4.2. Similar results were found for the thornback ray, *Raja clavata*, in the northeastern Atlantic Ocean which had a TR of 4.1 and consumed a large quantity of teleosts (Morato et al. 2003), whereas the diet of the sand skate, *Psammobatis extenta*, in Argentinean waters consisted primarily of small invertebrates, and had a TR of 3.5 (Braccini and Perez 2005).

Trophic levels identified in this study were sensitive to how the prey items were classified. For example, in the present study eels (Anguillidae) were separated from other teleosts. Teleosts alone exhibited a TR of 3.2, whereas eels had a significantly higher TR of 4.2. If eels are grouped with other teleosts, the TR is lowered, as Cortés (1999) also noted a more precise trophic level is calculated if each prey item is identified to the lowest taxon.

There have been several conflicting issues on what factors contribute to higher trophic levels (Pauly et al. 1998b; Cortés 1999). Cortés (1999) found a positive correlation between body size and trophic levels for some shark species, although there are exceptions in large sharks such as the basking shark (*Cetorhinus maximus*), megamouth shark (*Megachasma pelagios*), and the whale shark (*Rhincodon typus*), that are filter feeders primarily feeding on zooplankton. From this study, it was determined that *Dipturus* sp. A (attaining ~70 cm TL) and *D. whitleyi* had the highest trophic levels (TR = 4.2) suggesting that size may not be the major determining factor but rather the prey ingested. Cluster analysis demonstrated similarities between species that fed on similar prey and that had similar trophic levels.

Overlap amongst species

The two sympatric continental slope skate species (*Dipturus* sp. B and *D. gudgeri*) demonstrate little overlap in prey. It has been hypothesised by McEachran et al. (1976) that a more abundant and less specialised skate may reduce the diversity of prey of a less abundant and more specialised species by out-competing the less abundant species for its preferred prey. This scenario may also be applicable to *Dipturus* sp. B and *D. gudgeri* on the continental slope as the latter is larger, more abundant and widespread, therefore probably more dominant in this environment. Our data indicates that, despite *D. gudgeri* being larger than *Dipturus* sp. B, it is unlikely that size can solely account for differences in prey species found between these two skates.

The four sympatric continental shelf skate species examined in this study (*Dipturus* sp. A, *D. cerva*, *D. lemprieri* and *D. whitleyi*) had limited overlap in prey items indicating some partitioning of prey resources reducing potential competition. Our results suggest that partitioning of prey resources among the shelf species was largely a function of skate size, and that there was little overlap of prey among similar sizes. Similar results have been reported by Smale and Cowley (1992).

Conclusion

Prior to this study, little information existed on the feeding habits of the skate species in Tasmanian waters, south-eastern Australia. This work has provided fundamental information on the diet of two sympatric skate species (*Dipturus* sp. B and *D. gudgeri*) from the continental slope and four sympatric species (*Dipturus* sp. A, *D. cerva*, *D. lemprieri* and *D. whitleyi*) from the continental shelf region. The species in these two assemblages showed variations in feeding strategies that exploited different food sources.

The slope assemblage consisted of specialised feeders, with *Dipturus* sp. B targeting anomurans and a particular species of brachyuran, whereas the diet of *D. gudgeri* consisted almost exclusively of teleosts. This feeding strategy could be influenced by less abundant and less diverse food resources in the deeper environment (Laptikhovsky 2005), or abundances of one species out competing the other. Although spatial and seasonal factors may greatly contribute to resource partitioning there are no indications of seasonal variability in fish diet on the continental slope (Bulman et al. 2002) and food has been recognised as the commonly partitioned resource in marine fish environments (Ross 1986).

Within the shelf species, the prey groups exploited were diverse comprising of crustaceans (carideans, achelates, anomurans, brachyurans and isopods), octopus, elasmobranchs and teleosts. Limited prey items were shared amongst the species suggesting partitioning of prey resources with skate body size being a contributing factor. Resource partitioning among these species may also be due to their morphological attributes that have developed independently as a result of prey use and environmental conditions, allowing multiple species to utilise the similar habitat (Connell 1980).

To date, our understanding of these benthic ecosystems is scant. To increase knowledge of higher predators within this region further study into benthic communities is essential.

Acknowledgments The authors would like to express their gratitude to W. White and M. Bracccini for providing assistance with statistical analyses. Thanks also to J. Bizzarro and S. Irvine for their helpful comments on the manuscript. This research was supported by Deakin University (School of Life and Environmental Sciences), Victoria, Australia and CSIRO Marine and Atmospheric Research, Hobart, Australia. Finally we thank the American Elasmobranch Society (AES) for travel funds to attend and present this paper at the 'Biology of Skates' symposium held in conjunction with the annual AES meeting.

References

Bonfil R (1994) Overview of world elasmobranch fisheries. FAO Fisheries Technical Paper 341:119 pp

Braccini JM, Gillanders BM, Walker TI (2005) Sources of variation in the feeding ecology of the piked spurdog (*Squalus megalops*): implications for inferring predator–prey interactions from overall dietary composition. ICES J Mar Sci 62:1076–1094

Braccini JM, Perez JE (2005) Feeding habits of the sandskate *Psammobatis extenta* (Garman, 1913): sources of variation in dietary composition. Mar Freshw Res 56:395–403

Bulman C, Althaus F, He X, Bax NJ, Williams A (2001) Diets and trophic guilds of demersal fishes of the south-eastern Australian shelf. Mar Freshw Res 52:537–548

Bulman CM, He X, Koslow JA (2002) Trophic ecology of the mid-slope demersal fish community off southern Tasmania, Australia. Mar Freshw Res 53:59–72

Clarke KR, Gorley RN (2001) Primer v5: user manual/tutorial. Primer-E, Plymouth Marine Laboratory

Collin SP, Whitehead D (2004) The functional roles of passive electroreception in non-electric fishes. Anim Biol 54:1–25

Connell JH (1980) Diversity and the coevolution of competitors, or the ghost of competition past. Oikos 35:131–138

Cortés E (1997) A critical review of methods of studying fish feeding based on analysis of stomach contents: application to elasmobranch fishes. Can J Fish Aquat Sci 54:726–738

Cortés E (1999) Standardized diet composition and trophic levels of sharks. ICES J Mar Sci 56:707–717

Cox DL, Koob TJ (1993) Predation on elasmobranch eggs. Environ Biol Fishes 38:117–125

Daley R, Webb H, Hobday A, Timmis T, Walker T, Smith T, Dowdney J, Williams A (2006) Ecological Risk Assessment for the Effects of Fishing. Report for the automatic longline sub-fishery of the Southern and Eastern Scalefish and Shark Fishery. Australian Fisheries Management Authority, Canberra

Dolgov AV (2005) Feeding and food consumption by the Barents sea skates. J Northwest Atl Fish Sci 35:495–503

Ebert DA, Compagno LJV (2007) Biodiversity and systematics of skates (Chondrichthyes: Rajiformes:Rajoidei). Environ Biol Fishes (this issue)

Ebert DA, Cowley PD, Compagno LJV (1991) A preliminary investigation of the feeding ecology of skates (Batoidea: Rajidae) off the west coast of Southern Africa. South Afr J Mar Sci 10:71–81

Edgar GJ (2000) Australian Marine life—The Plants and Animals of Temperate Waters. Reed New Holland, Australia, 544 pp

Ellis JR, Pawson MG, Shackley SE (1996) The comparative feeding ecology of six species of shark and four species of ray (Elasmobranchii) in the North-east Atlantic. J Mar Biol Ass UK 76:89–106

Fariasa I, Figueiredo I, Moura T, Serrano Gordo L, Neves ABS-P (2006) Diet comparison of four ray species (*Raja clavata*, *Raja brachyura*, *Raja montagui* and *Leucoraja naevus*) caught along the Portugese continental shelf. Aquat Living Resour 19:105–114

Ferry LA, Cailliet GM (1996) Sample size and data analysis: are we characterizing and comparing diet properly? In: MacKinlay D, Shearer K (eds) International Congress on the biology of fishes, San Francisco, California, pp 71–80

Furlani D, Gales R, Pemberton D (2007) Otoliths of common temperate Australian fish: a photographic guide. CSIRO Publishing, Australia

Gerking SD (1994) Feeding ecology of fish. Academic Press, San Diego, 416 pp

Gomon MF, Glover JCM, Kuiter RH (1994) The fishes of Australia's South Coast. State Print, Adelaide, Australia, 992 pp

Graham KJ, Andrew NL, Hodgson KE (2001) Changes in relative abundance of sharks and rays on Australian South East Fishery trawl grounds after twenty years of fishing. Mar Freshw Res 52:549–561

Hale HM (1927–1929) The crustaceans of South Australia. Parts 1 and 2. James Government Printer, South Australia, 201 pp

Hansson S (1998) Methods of studying fish feeding: a comment. Can J Fish Aquat Sci 55:2706–2707

Hyslop E (1980) Stomach contents analysis a review of methods and their application. J Fish Biol 17:411–429

Jones D, Morgan G (1994) A field guide to crustaceans of Australian waters. Reed books, New Holland, 224 pp

Koen Alonso M, Crespo EA, Garcia NA, Pedraza SN, Mariotti PA, Beron Vera B, Mora NJ (2001) Food habits of *Dipturus chilensis* (Pisces: Rajidae) off Patagonia, Argentina. ICES J Mar Sci 58:288–297

Laptikhovsky VV (2005) A trophic ecology of two grenadier species (Macrouridae, Pisces) in deep waters of the Southwest Atlantic. Deep-Sea Res Part 1 52:1502–1514

Last PR, Stevens JD (1994) Sharks and Rays of Australia. CSIRO, Hobart, 513 pp

Levins R (1968) Evolution in changing environments: some theoretical explorations. Princeton University Press, Princeton, New Jersey, 120 pp

Linke TE, Platell ME, Potter IC (2001) Factors influencing the partitioning of food resources among six fish species in a large embayment with juxtaposing bare sand and seagrass habitats. J Exp Mar Biol Ecol 266:193–217

Lokkeborg S, Olla BL, Pearson WH, Davis MW (1995) Behavioural responses of sablefish, *Anoplopoma fimbria*, to bait odour. J Fish Biol 46:142–155

Lucifora LO, Valero JL, Bremec CS, Lasta ML (2000) Feeding habits and prey selection by the skate *Dipturus chilensis* (Elasmobranchii: Rajidae) from the south-western Atlantic. J Mar Biol Ass UK 80:953–954

Luer CA, Gilbert PW (1985) Mating behaviour, egg deposition, incubation period, and hatching in the clearnose skate, *Raja eglanteria*. Environ Biol Fishes 13:161–171

MacDonald JS, Green RH (1983) Redundancy of variables used to describe importance of prey species in fish diets. Can J Fish Aquat Sci 40:635–637

McEachran JD, Boesch DF, Musick JA (1976) Food division within two sympatric species-pairs of skates (Pisces: Rajidae). Mar Biol 35:301–317

Morato T, Sola E, Gros MP, Menezes G (2003) Diets of thornback ray (*Raja clavata*) and tope shark (*Galeorhinus galeus*) in the bottome longline fishery of the Azores northeastern Atlantic. Fish Bull 101:590–602

Motta PJ (2004) Prey capture behaviour and feeding mechanisms of elasmobranchs. In: Carrier JC, Musick JA, Heithaus MR (eds) Biology of sharks and their relatives. CRC Press, New York, pp 165–202

Moura T, Figueiredo I, Bordalo-Machado P, Gordo LS (2005) Feeding habits of *Chimera monstrosa* L. (Chimaeridae) in relation to its ontogenetic development on the southern Portuguese continental slope. Mar Biol Res 1:118–126

Muto EY, Soares LSH, Gitein R (2001) Food resource utilization of the skates *Rioraja agassizii* (Muller & Henle, 1841) and *Psammobatis extenta* (Garman, 1913) on the continental shelf off Ubatuba, Southeastern Brazil. Rev Brazil Biol 61:217–238

Orlov AM (1998) On feeding of mass species of deep-sea skates (*Bathyraja* spp., Rajidae) from the pacific waters of the Northern Kurils and Southeastern Kamchatka. J Ich 38:635–644

Orlov AM (2003) Diets, feeding habits, and trophic relations of six deep-benthic skates (Rajidae) in the western Bering Sea J Ich Aquat Biol 7:45–60

Pauly D, Christensen V, Dalsgaard J, Froese R, Torres FJ (1998b) Fishing down marine food webs. Science 279:860–863

Pauly D, Trites AW, Capuli E, Christensen V (1998a) Diet composition and trophic levles of marine mammals. ICES J Mar Sci 55:467–481

Pielou EC (1966) The measurements of diversity in different types of biological collections. J Theor Biol 63:131–144

Pinkas LM, Oliphant S, Iverson ILK (1971) Food habits of albacore, bluefin tuna and bonito in Californian waters. California Fish Game 152:451–463

Poore GCB (2004) Marine Decapod Crustacea of Southern Australia—A Guide to Identification, 574 pp

Ross ST (1986) Resource partitioning in fish assemblages: a review of field studies. Copeia 1986:352–388

San Martin MJ, Braccini JM, Tamini LI, Chiaramonte GE, Perez JE (2007) Temporal and sexual effects in the feeding ecology of the marbled sand skate *Psammobatis bergi* Marini, 1932. Mar Biol *on line*:10.1007/s00227–006–0499–6

Smale MJ, Cowley PD (1992) The feeding ecology of skates (Batoidea: Rajidae) off the cape south coast, South Africa. South Afr J Mar Sci 12:823–834

University of Columbia (2005) Marine trophic index for LME: Patagonian Shelf, The Sea Around Us Project and the Convention on Biological Diversity, Columbia http://www.seaaroundus.org/TrophicLevel/LME-TrophicIndex.aspx?LME = 14&FAO = 0&country = Patagonian%20Shelf

Walker T, Dowdney J, Williams A, Fuller M, Webb H, Bulman C, Sporcic M, Wayte S (2006) Ecological Risk Assessment for the Effects of Fishing: report for the Shark gillnet component of the Gillnet Hook and Trap Sector of the Southern and Eastern Scalefish and Shark Fishery. Australian Fisheries Management Authority, Canberra

Walker TI (2004) Elasmobranch Fisheries Mangement Techniques. In: Musick J, Bonfil R (eds) Elasmobranch Fisheries Mangement Techniques. Asia Pacific Economic Cooperation. Singapore, pp 285–323

Wayte S, Dowdney J, Williams A, Bulman C, Sporcic M, Fuller M, Smith A (2006) Ecological Risk Assessment for the effects of fishing: report for the otter trawl subfishery of the Commonwealth trawl sector of the Southern and Eastern Scalefish and Shark Fishery. Australian Fisheries Management Authority, Canberra

White WT, Platell ME, Potter IC (2004) Comparisons between the diets of four abundant species of elasmobranchs in a subtropical embayment: implications for resource partitioning. Mar Biol 144:439–448

Comparative feeding ecology of four sympatric skate species off central California, USA

Joseph J. Bizzarro · Heather J. Robinson · Christopher S. Rinewalt · David A. Ebert

Originally published in the journal Environmental Biology of Fishes, Volume 80, Nos 2–3, 197–220.
DOI 10.1007/s10641-007-9241-6 © Springer Science+Business Media B.V. 2007

Abstract The big (*Raja binoculata*), California (*R. inornata*), longnose (*R. rhina*), and sandpaper (*Bathyraja kincaidii*), skates are commonly found on soft-bottom regions of the central California continental shelf and upper slope. The feeding ecology of this assemblage was compared to evaluate the degree of trophic separation among species, based on the results of previous species-specific diet studies. Specimens were collected from fishery independent trawl surveys conducted during September 2002–March 2003 at depths of 9–536 m. Using single and compound measures, diet composition of small (≤60 cm TL) and large (>60 cm TL) individuals were compared within continental shelf (≤200 m) and slope (>200 m) regions using traditional, multivariate, and novel techniques. Diet compositions within size classes were similar in both regions. Diet compositions between size classes generally differed, however, with fishes more important and crustaceans (especially shrimps and euphausiids) less important in the diets of large individuals. Crabs contributed substantially to skate diet compositions on the shelf, but were uncommon prey items at deeper depths, probably because of their relative scarcity in slope waters. Conversely, cephalopods were common prey items at slope depths, but were rarely ingested at shelf depths. The studied skate assemblage appears to consist primarily of generalist crustacean and fish predators that exhibit high dietary overlap at similar sizes. It is possible that resource competition among skates and groundfish species has been reduced because of considerable recent declines in the biomass of upper trophic level groundfishes. Skates may therefore play important roles in contemporary benthic food web dynamics off central California.

Keywords *Bathyraja kincaidii* · Diet composition · *Raja binoculata* · *Raja inornata* · *Raja rhina* · Trophic relationships

Introduction

A combination of biological and physical processes is responsible for patterns of distribution and abundance among sympatric marine species (e.g. Mills 1969; Bakun 1996; Menge 2004), with the relative influence of these factors varying both spatially and temporally (e.g. Livingston 1984; Foster et al. 1988; Munday 2002). Detailed oceanographic information is available for many marine regions. However, corresponding data on

J. J. Bizzarro (✉) · H. J. Robinson · C. S. Rinewalt · D. A. Ebert
Pacific Shark Research Center, Moss Landing Marine Laboratories, 8272 Moss Landing Rd., Moss Landing, CA 95039, USA
e-mail: jbizzarro@mlml.calstate.edu

trophic relationships are generally unknown or limited to a few species. This lack of quantitative dietary information precludes a comprehensive understanding of the mechanisms responsible for community regulation and, consequently, the successful development of multi-species management plans. Determining the feeding ecology of upper trophic level species is especially important because of their recent global declines and the potential for associated ecosystem-level effects on species composition and diversity (Pauly et al. 1998; Jackson et al. 2001).

Elasmobranch fishes have played important roles in marine ecosystems throughout their history, often occupying the highest trophic levels as apex predators (Compagno 1990; Cortés 1999). Overfishing and habitat destruction, however, have greatly reduced many elasmobranch populations and impacted their ecological value over broad regions (Musick et al. 2000; Stevens et al. 2000). In addition, shifts in relative abundance of elasmobranchs have occurred, often resulting in proportional and sometimes absolute increases in smaller, lower trophic level species (Walker and Hislop 1998; Rogers and Ellis 2000; Hoff 2006). Despite the fundamental importance of feeding relationships in understanding food web dynamics, community structure, and energy transfer in marine systems, little is known about the feeding ecology of most elasmobranch species (Heithaus 2004; Wetherbee and Cortés 2004). This is especially true of batoids, which have received considerably less scientific attention than sharks (Ishihara 1990; Motta 2004).

Skates typically occur on unconsolidated substrates in temperate and boreal regions (Compagno 1990), where they are thought to serve important trophic roles (Ebert et al. 1991; Orlov 2003). Although somewhat restricted in habitat, skates consume a variety of prey taxa, including polychaetes, cephalopods, amphipods, decapods, euphausiids, and fishes (Ebert and Bizzarro 2007). Stingray predation has been demonstrated to regulate abundance and possibly composition of prey resources in shallow water regions (VanBlaricom 1982; Thrush et al. 1994; Peterson et al. 2001). Skates may play a similar role in continental shelf and slope benthic communities, but their relative inaccessibility has precluded comparable studies.

Trophic separation is one of the most important factors determining distribution and abundance of broadly co-occurring marine fishes (Ross 1986). Ontogenetic dietary changes are almost universal in fishes (Gerking 1994), and have been demonstrated among skates (e.g., Holden and Tucker 1974; Ajayi 1982; Brickle et al. 2003). Depth differences in diet composition have commonly been reported for teleosts (e.g., Pearcy and Hancock 1978; Yang and Livingston 1988; Arkhipkin et al. 2003), but similar studies of skates are limited (Hacunda 1981; Templeman 1982; Morato et al. 2003). As policy shifts from managing single stocks towards the establishment of marine reserves for conservation and propagation of groundfish assemblages, information on size- and depth-related trophic differences becomes crucial for the development of effective management strategies.

Until recently, no published quantitative information on diet composition of any skate species existed from California waters. The completion of recent species-specific diet studies (Rinewalt et al. 2007; Robinson et al. 2007; Bizzarro and Ebert unpublished data), however, has enabled an assessment of the comparative feeding ecology of skate assemblages off central California. The overall goal of this study was to elucidate the ecological roles of the most common skate species (i.e., big skate, *Raja binoculata*; California skate, *R. inornata*; longnose skate, *R. rhina*; and sandpaper skate, *Bathyraja kincaidii*) in continental shelf and upper slope regions. Specific objectives were to compare intra- and interspecific ontogenetic diet compositions among skates from these physiographic regions to determine their trophic relationships and degree of dietary overlap.

Materials and methods

Data collection

Skates were collected from fishery-independent bottom trawl surveys of the National Marine Fisheries Service, Santa Cruz Laboratory (NMFS-SCL). Surveys occurred monthly during

September 2002 - March 2003 in Monterey Bay, California and adjacent regions to the north (Fig. 1). Fishing operations were conducted from 9 to 536 m over primarily soft substrate during daylight and early evening hours. All individuals of the aforementioned skate species that were caught during these trawls were identified, sexed, and weighed to the nearest 1.0 kg. Total length (TL), the linear distance from the tip of the snout to the tip of the tail, was measured to the nearest 1.0 mm. Stomachs were excised from each individual and stored frozen until processing.

Prey items were identified to the lowest possible taxonomic level, enumerated, and weighed. When the exact number of individuals of a prey taxon could not be positively determined, a conservative count was estimated. All specimens of each prey type were blotted dry and weighed collectively to the nearest 0.001 g. Incidentally ingested materials (e.g., plant matter, sediment, detritus), parasites, and unidentified organic matter were discarded.

Data analyses

All stomach samples of each skate species were pooled to investigate diet composition and food habits of the local population, and then apportioned for ontogenetic comparisons within two physiographic regions. Only stomachs containing prey items were utilized for calculations and analyses. To examine potential ontogenetic dietary differences, specimens of each species were separated into small (≤60 cm TL) and large size classes (>60 cm). Skates were additionally segregated between continental shelf (≤200 m) and continental slope (>200 m) habitats based on the average depth of each tow.

Sample size sufficiency of diet composition estimates was investigated using cumulative prey curves. A cumulative prey curve was generated by resampling species richness of 500 randomly selected stomach samples using Matlab (version 6.5.0.180913a, The MathWorks, Inc., Natick, MA) to calculate a mean and variability estimate for each sample (Adams 2004). The mean cumulative number of prey taxa present in each sample was then plotted against the randomly pooled number of stomach samples. Adequate sample size has been assumed if the resulting curve approached an asymptote and displayed a reduction in variability (Hurturbia 1973; Cailliet 1977; Ferry and Cailliet 1996). Although widely used (e.g., Gelseichter et al. 1999; Morato et al. 2003; Braccini and Perez 2005; Scenna et al. 2006), no quantitative criteria have been developed for this method. The use of qualitative and subjective criteria for the evaluation of sample size sufficiency limits the

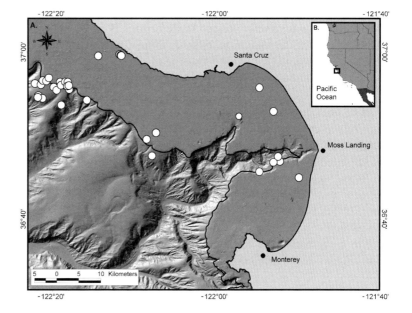

Fig. 1 Location of fishery-independent bottom trawls (*n* = 35; depicted by white circles) conducted in and adjacent to Monterey Bay, California from September 2002-March 2003 (**A**) and inset map depicting study region (**B**). Multibeam bathymetric imagery was collected and processed by the Monterey Bay Aquarium Research Institute. Black line depicts the 200 m isobath

utility of this technique and precludes definitive results that are comparable between studies.

A new method was developed using linear regression and the coefficient of variation to provide standard, quantitative measures of sample-size sufficiency. To determine if a cumulative prey curve reached an asymptote, the slope of the line generated from the curve endpoints was compared to a line of zero slope (horizontal asymptote). The endpoints consisted of the mean cumulative number of prey taxa generated for the final four stomach samples. A linear regression was conducted using these endpoints to determine if the slope of the corresponding best fit line was significantly different than a line of zero slope. Slopes were statistically compared using a Student's t-test, where $t = (b–0)/S_b$, b = slope of the best fit line (or regression coefficient), and S_b = standard error (SE) of the slope (Zar 1996). Slopes were not significantly different if $P > 0.05$, indicating that the curve reached an asymptote. It is possible to conduct this analysis using three instead of the recommended four curve endpoints. However, based on simulations involving actual and manufactured data sets that were unrelated to this study, the use of four endpoints reduced the influence of a single anomalous value and resulted in greater statistical power. The use of five or more points is also not recommended, as corresponding best fit lines do not adequately represent the curve terminus.

The slope of the line generated from the SE of the endpoints was similarly tested to determine if variation around the mean stabilized, indicating adequate precision ($P > 0.05$). The mean coefficient of variation (CV = (standard deviation (SD)/$\overline{X}$))*100) of the endpoints was additionally calculated to provide a standard measure of precision. The minimum number of samples necessary for sample size sufficiency was determined by applying these techniques sequentially backwards along the curve until tests indicated that mean species richness and SE estimates did not reach asymptotes. If these criteria were not met, diet and food habits were not quantitatively assessed.

The contribution of each prey taxon to diet composition was estimated with three relative measures of prey quantity (RMPQ) described by Hyslop (1980), and an additional, non-additive measure of occurrence (Mohan and Sankaran 1988). Prey items were grouped into the following seven categories to eliminate biases associated with comparisons based on variable levels of prey identification (Cortés 1997): polychaetes (POLY), gastropods and bivalves (MOLL), cephalopods (octopods and squids; CEPH), shrimp-like crustaceans (shrimps, euphausiids, and mysids; SHRIMP), brachyuran crabs (CRAB), other crustaceans (ostracods, anomurans, amphipods, isopods, and unidentified crustaceans; OCRUST), and fishes (teleosts and elasmobranchs; FISH). The numerical index ($\overline{\%N}$, (number of individuals of a prey category/total number of individuals among all prey categories)*100) and the gravimetric index based on wet weight ($\overline{\%W}$, (weight of individuals of a prey category/total weight of individuals among all prey categories)*100) were calculated for each sample to provide mean and variability estimates (Cailliet et al. 1986; Ferry and Cailliet 1996). Percentage frequency of occurrence (%FO) was considered as: (number of stomachs containing a prey taxon/total number of stomachs containing prey)*100. An additional RMPQ, the occurrence index (%O), was calculated and used as a proxy for the nonadditive %FO, in situations where an additive measure of prey occurrence was required (Assis 1996; Cortés 1997). The equation for %O is as follows (Mohan and Sankaran 1988): %$O_i = (n_i/\Sigma\ n_i)*100$, where O_i = percentage occurrence of prey category i in relation to the total occurrence of all prey categories, n_i = the total number of occurrences of prey category i, and Σn_i = number of occurrences of all prey categories.

Diet composition was further described using compound indices incorporating previously calculated RMPQ. The index of relative importance (IRI) was modified from Pinkas et al. (1971), with weight substituted for volume, and calculated for each prey category using overall species data as follows: $\overline{IRI} = (\overline{\%N} + \overline{\%W})*\%FO$. Results were expressed as a percentage to facilitate more apprehensible comparisons among food types and between skate species (Carrassón et al. 1992; Cortés 1997):

$$\overline{\%\text{IRI}}_i = (\text{IRI}_i * 100) / \sum_{i=1}^{n} \text{IRI}_i,$$

where %IRI$_i$ = index value for the ith prey category and n = the total number of prey categories.

The geometric index of importance (GII) was calculated for overall and apportioned species data and expressed as a percentage (Assis 1996):

$$\overline{\%\text{GII}}_j = \left(\sum_{i=1}^{n} V_i\right)_j / n,$$

where %GII$_j$ = index value for the jth prey category, V_i = the ith RMPQ (i.e., $\overline{\%N}$, $\overline{\%W}$, and %O) of the jth prey category, and n = number of RMPQ. Percentage IRI and %GII values were determined for each sample, generating a mean and SE estimate for all prey taxa.

Food habits and feeding behavior of the local skate populations were evaluated using all available data from each species. Because gastropods and bivalves were considered to be incidentally ingested for *B. kincaidii* (Rinewalt et al. 2007) and were rare or absent in the diet of other species, they were excluded from these and all subsequent analyses. Mean number and mean weight of prey items were compared using Kruskal–Wallis and Dunn's post-hoc tests for tied ranks (Dunn 1964). Dietary specialization was evaluated using Amundsen et al.'s (1996) method, by which %N (calculated as $\overline{\%N}$) among all stomachs containing a prey category ($\overline{\%Ni}$) was plotted against %FO for that prey category.

Diet composition was compared within and among species using qualitative similarity measures and null model simulations. Three common overlap measures, percentage overlap, simplified Morisita's, and Pianka's, were used for comparative purposes and calculated with Ecological Methodology software (version 5.1, Krebs 1999). Percentage overlap ranges from 0 to 100%, whereas the other indices range from zero (no dietary overlap) to 1.00 (identical prey composition). Because evaluation criteria for overlap indices are arbitrary (Zaret and Rand 1971; Langton 1982), additional statistical analyses were conducted to more rigorously quantify dietary comparisons. Observed (Pianka's) overlap values were compared to a distribution of expected overlap values based on null model simulations. Null model distributions were generated with EcoSim v. 7.72 (Goetelli and Entsminger 2005) from 1000 iterations of RA2, which randomizes the dietary contribution of utilized prey categories but retains the zero structure of unutilized categories (Winemiller and Pianka 1990). The observed index value was considered statistically significant from the null distribution if it was > or < 95% of the simulated indices ($P < 0.05$). Observed values >95% of simulated indices indicated significant dietary overlap ($P_{\text{sim}} < 0.05$), whereas observed values <95% of simulated indices signified significant trophic separation ($P_{\text{dif}} < 0.05$; Bethea et al. 2004). Intermediate values less than corresponding mean simulated index values indicated insignificant dietary differences, whereas observed values greater than simulated index values indicated insignificant dietary similarity.

Diet composition was additionally evaluated using multivariate techniques. Principal Components Analysis (PCA) was used to investigate sources of dietary variation among skate groups in each physiographic region and to substantiate results of dietary comparisons. Individual-based %GII values were calculated (i.e., (%N +%W)/2)) and analyzed using a covariance matrix with Multivariate Statistical Package software (MVSP; version 3.13, Kovach Computing Services, Wales, UK). Kaiser's rule was used to determine the number of interpretable axes (Legendre and Legendre 1983) and Euclidean bi-plots of corresponding principal component scores for each species group and prey group were generated. Cluster analysis was utilized to compare diet compositions among skate groups from both physiographic regions. Calculations were performed with MVSP using the unweighted pair-wise group mean average method with the simplified Morisita's index (C_H) as a measure of similarity. Similarity values among clusters that were ≥50% of the maximum overall similarity distance were considered to indicate major divisions and used to distinguish trophic guilds (Yoklavich et al. 2000).

Results

Overall comparisons

One thousand and fifty eight skates were collected among the four study species, of which the great majority (95.9%, n = 1015) contained prey items. The percentage of empty stomachs was comparable among species, R. binoculata (4.9%, n = 9), R. inornata (6.0%, n = 16), R. rhina (2.4%, n = 12) and B. kincaidii (5.4%, n = 6).

Depth composition differed among species (Table 1), with distinct shelf and slope assemblages present. Average tow depth ranged from 14–532 m (n_{shelf} = 22; n_{slope} = 13). All R. inornata and all but two R. binoculata were caught on the continental shelf. The distribution of R. rhina was widespread throughout both regions. Bathyraja kincaidii was taken primarily from slope waters.

Size composition also differed among species, with R. binoculata and R. rhina reaching greater lengths than B. kincaidii and R. inornata (Table 1). Total length of B. kincaidii individuals ranged from 32.7 to 57.0 cm. Raja inornata was slightly larger, with a maximum total length of 65.7 cm. Both small and large R. rhina were collected in substantial numbers, with no apparent depth differences in size composition. Although the largest specimens recorded were R. binoculata, most individuals of this species were < 50 cm TL (72.3%, n = 126).

Cumulative prey curves indicated sufficient sample size for overall diet composition of each species. All curves reached actual asymptotes and displayed no variability at endpoints (CV = 0.00%; Fig. 2). Relatively few stomachs were necessary to attain sufficient sample sizes for B. kincaidii (n_{min} = 39) and R. binoculata (n_{min} = 40), whereas considerably more were necessary for R. inornata (n_{min} = 139) and R. rhina (n_{min} = 241).

RMPQ indicated that shrimp-like crustaceans and fishes were the most important prey taxa in the overall diets of central California skates, with differences apparent in the relative importance of supplemental prey groups. Gastropods and bivalves were consumed by R. inornata and R. rhina, but were uncommon and contributed little to diet composition by all measures (Table 2). Polychaetes were an important prey group for B. kincaidii, occurring in 64.8% of stomachs and contributing 20.5 ± 1.7% ($\overline{X}$ ± SE) by weight, but were scarce or absent in the diets of other species. Shrimp-like crustaceans were the most important prey group numerically for all species but R. binoculata and were also dominant by $\overline{\%W}$ for B. kincaidii. Fishes were the primary prey taxon in the diets of R. binoculata and R. rhina and were also the most substantial prey group by $\overline{\%W}$ for R. inornata. Although commonly ingested (%FO = 58.1%), fishes contributed substantially less to the diet composition of B. kincaidii by other measures. Crabs were important supplemental prey items for R. binoculata and R. inornata, but were rarely taken by R. rhina or B. kincaidii (Table 2). Cephalopods and other crustaceans were minor prey groups for other skate species, but were relatively common in the diet of B. kincaidii (%FO_{CEPH} = 61.0, %FO_{OCRUST} = 34.3%). Most RMPQ yielded similar results, although shrimp-like crustaceans generally contributed more substantially to $\overline{\%N}$ than $\overline{\%W}$, with fishes exhibiting the opposite pattern.

Compound measures indicated the same general trends as RMPQ and were consistent in relative, but not absolute, importance of prey taxa

Table 1 Distribution of stomach samples of four co-occurring skate species collected off central California, including number of samples among depth (shelf = ≤200 m; slope = 200 m) and size (small = ≤60 cm TL; large = >60 cm TL) categories, total number of samples (n), depth range (m), and size range (cm) of individuals. Only stomach samples containing prey items are included

Species	Shelf small	Shelf large	Slope small	Slope large	Total n	Depth range (m)	Size range (cm)
Raja binoculata	132	39	1	1	173	14–532	17.6–135.7
Raja inornata	231	20	0	0	251	17–194	21.8–65.7
Raja rhina	171	118	118	79	486	29–532	23.8–109.0
Bathyraja kincaidii	9	0	96	0	105	132–532	32.7–57.0

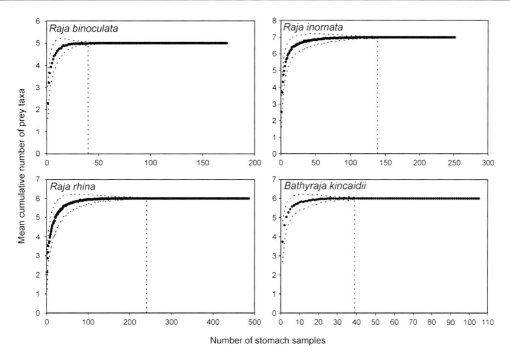

Fig. 2 Mean cumulative number of prey taxa per stomach sample for the four study species. Dashed lines represent mean standard deviation values. Drop-lines indicate the minimum number of samples necessary for adequate sample sizes

(Fig. 3). For *R. binoculata*, both indices indicated that fishes were the most important prey group ($\overline{\%\text{IRI}} \pm$ SE = 41.1 ± 44.8; $\overline{\%\text{GII}} \pm$ SE = 35.1 ± 3.9), followed by shrimp-like crustaceans ($\overline{\%\text{IRI}}$ = 34.6 ± 40.3; $\overline{\%\text{GII}}$ = 32.1 ± 4.0) and crabs ($\overline{\%\text{IRI}}$ = 21.9 ± 30.6; $\overline{\%\text{GII}}$ = 23.9 ± 3.6). These groups were also dominant in the diet of *R. inornata* by $\overline{\%\text{IRI}}$ (SHRIMP = 37.1 ± 41.6; FISH = 33.2 ± 40.1; CRAB = 27.0 ± 34.7) and $\overline{\%\text{GII}}$ (SHRIMP = 32.5 ± 3.9; FISH = 30.9 ± 4.1; CRAB = 27.5 ± 4.0). Fishes ($\overline{\%\text{IRI}}$ = 52.5 ± 52.3; $\overline{\%\text{GII}}$ = 47.4 ± 4.6) and shrimp-like crustaceans ($\overline{\%\text{IRI}}$ = 43.5 ± 45.8; $\overline{\%\text{GII}}$ = 42.0 ± 4.4) comprised over 89% of *R. rhina* diet composition by $\overline{\%\text{GII}}$ and 96% by $\overline{\%\text{IRI}}$. The diet of *B. kincaidii* was dominated by shrimp-like crustaceans ($\overline{\%\text{IRI}}$ = 65.7 ± 40.7; $\overline{\%\text{GII}}$ = 50.2 ± 2.5), with additional prey taxa, especially polychaetes ($\overline{\%\text{IRI}}$ = 13.4 ± 20.8; $\overline{\%\text{GII}}$ = 16.9 ± 2.2) and cephalopods ($\overline{\%\text{IRI}}$ = 9.9 ± 16.0; $\overline{\%\text{GII}}$ = 13.8 ± 1.8), also contributing substantially to dietary importance. Calculations of $\overline{\%\text{GII}}$ were much more precise than those of $\overline{\%\text{IRI}}$ (Fig. 3), with SE for the latter measure typically comparable or greater than mean index values.

Significant differences were detected in mean number (Kruskal–Wallis = 145.59, $P < 0.001$) and mean weight (Kruskal–Wallis = 48.04, $P < 0.001$) of prey items among the overall diets of skate species. *Bathyraja kincaidii* stomach samples contained significantly more prey items ($\overline{X} \pm$ SE: 22.5 ± 1.8) than those of *R. binoculata* (4.6 ± 0.3; Dunn's Q = 9.85; $P < 0.001$), *R. inornata* (4.5 ± 0.3; Dunn's Q = 11.30; $P < 0.001$), and *R. rhina* (7.2 ± 0.4; Dunn's Q = 11.13; $P < 0.001$). A maximum number of 97 prey items was recovered from *B. kincaidii* stomachs, as compared to 59 for *R. rhina*, 24 for *R. inornata*, and 19 for *R. binoculata*. Prey contents of *B. kincaidii* (2.1 ± 0.3 g) stomachs weighed significantly less than those of *R. binoculata* (25.1 ± 5.1 g; Dunn's Q = 3.70; $P = 0.001$) and *R. rhina* (13.4 ± 1.2; Dunn's Q = 5.98; $P < 0.001$). *Raja rhina* stomach contents were also considerably heavier than those of *R. inornata* (5.7 ± 0.7; Dunn's Q = 4.86; $P < 0.001$). Maximum weights of stomach contents were as follows: *R. binoculata*

Table 2 Mean and standard error (SE) of percentage number ($\overline{\%N}$) and percentage weight ($\overline{\%W}$); percentage frequency of occurrence (%FO) and percentage occurrence (%O) of prey taxa found in the stomachs of four species of co-occurring skates (see methods for prey group descriptions)

Prey Group	R. binoculata $\overline{\%N}$	$\overline{\%W}$	%FO	%O	R. inornata $\overline{\%N}$	$\overline{\%W}$	%FO	%O	R. rhina $\overline{\%N}$	$\overline{\%W}$	%FO	%O	B. kincaidii $\overline{\%N}$	$\overline{\%W}$	%FO	%O
POLY	–	–	–	–	0.4 (0.2)	<0.1 (<0.1)	2.0	1.0	–	–	–	–	11.9 (1.7)	20.5 (1.7)	64.8	19.9
MOLL	–	–	–	–	2.8 (0.6)	1.1 (0.4)	10.4	5.2	–	–	–	–	–	–	–	–
CEPH	5.3 (1.4)	5.7 (1.6)	11.6	5.8	4.3 (0.8)	6.0 (1.3)	15.9	8.0	0.6 (0.2)	0.2 (0.1)	4.1	2.4	5.4 (0.6)	20.2 (0.6)	61.0	18.7
CRAB	22.7 (2.3)	25.7 (2.8)	49.1	24.5	25.2 (2.2)	30.8 (2.6)	47.8	23.9	6.9 (0.7)	11.8 (1.2)	26.5	15.3	0.4 (0.2)	0.3 (0.2)	9.5	2.9
SHRIMP	31.4 (2.6)	32.5 (3.0)	59.0	29.4	37.2 (2.3)	25.0 (2.4)	59.4	29.6	0.8 (0.3)	0.9 (0.4)	2.1	1.2	67.8 (2.8)	37.3 (2.8)	98.1	30.1
OCRUST	6.4 (1.4)	3.6 (1.1)	13.9	6.9	3.4 (0.7)	2.2 (0.7)	11.2	5.6	49.3 (1.8)	34.0 (1.9)	66.0	38.0	9.1 (1.8)	6.8 (1.8)	34.3	10.5
FISH	34.1 (2.5)	32.5 (3.0)	67.1	33.4	26.7 (2.2)	34.7 (2.7)	53.8	26.8	41.5 (1.9)	52.0 (2.0)	71.0	40.8	5.4 (0.9)	14.9 (0.9)	58.1	17.8

(515.2 g), *R. rhina* (263.8 g), *R. inornata* (78.6 g), and *B. kincaidii* (14.7 g).

Overall diets of most skate species were somewhat specialized on shrimp-like crustaceans and, to a lesser extent, fishes. *Raja binoculata* and *R. inornata* exhibited similar feeding behavior (Fig. 4). Shrimp-like crustaceans and fishes were consumed by most individuals of these species and constituted at least half of all co-occurring prey items. Crabs were ingested by approximately half of all *R. binoculata* and *R. inornata* individuals, and were of similar prey-specific $\overline{\%N}$ to shrimp-like crustaceans and fishes. Other crustaceans and cephalopods were rare in the diet of both species, but were substantial in the diet of a limited number of *R. binoculata* individuals. *Raja rhina* was a specialized shrimp-like crustacean and fish predator, with other prey items rarely taken and relatively unimportant (Fig. 4). *Bathyraja kincaidii* was a shrimp-like crustacean specialist. Shrimp-like crustaceans were found in 98.1% of individuals of this species and comprised the great majority of prey items. Polychaetes, cephalopods, and fishes were commonly ingested by *B. kincaidii*, but were taken in relatively low abundance. Crabs were rarely consumed by *B. kincaidii* and did not contribute substantially to prey contents when present.

Ontogenetic comparisons: shelf assemblage

Cumulative prey curves indicated sufficient sample sizes for small and large specimens of most shelf species (Fig. 5). The curve generated for small *R. binoculata* reached a true asymptote and displayed no variability at the endpoints (CV = 0.00%, n_{min} = 103), as did curves for small *R. inornata* (CV = 0.00%, n_{min} = 133) and small *R. rhina* (CV = 0.00%, n_{min} = 132). Statistical tests conducted for large *R. binoculata* indicated that the prey curve reached an asymptote for both mean number (t = 3.16, P = 0.087, CV = 1.50 ± 0.42%) and SE (t = –3.06, P = 0.092). Sample size sufficiency was achieved at 32 samples. The cumulative prey curve for large *R. inornata* did not reach an asymptote for either species richness (t = 5.62, P = 0.030, CV = 9.95 ± 1.55%) or precision (t = –20.73, P = 0.002) estimates. A sufficient number of

Fig. 3 Diet composition of *Raja binoculata, R. inornata, R. rhina*, and *Bathyraja kincaidii* as determined with the index of relative importance (IRI) and geometric index of importance (GII). Mean values are expressed for each prey category as a percentage of total diet composition. Standard error of the measures is depicted by error bars. See methods for prey group descriptions

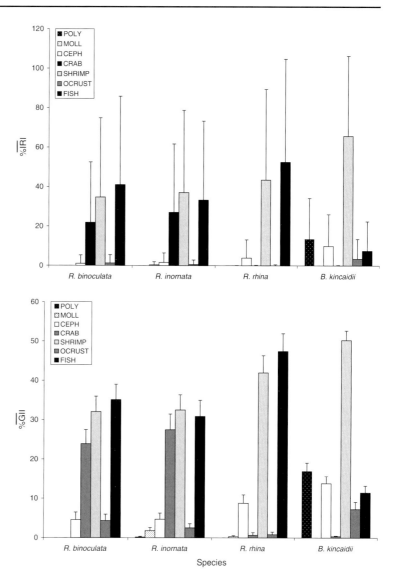

samples (n_{min} = 112) was obtained to adequately describe the diet of large *R. rhina* (mean number: t = 1.18, P = 0.359, CV = 2.93 ± 0.42%; SE: t = −1.17, P = 0.364).

Single indices and $\overline{\%\text{GII}}$ indicated that fishes were the most important prey group of large skates on the continental shelf, whereas shrimp-like crustaceans were the most important prey group of small skates. Fishes dominated diet composition of large *R. binoculata* and large *R. rhina* by all measures (Table 3), and comprised 66.1 ± 3.8 $\overline{\%\text{GII}}$ and 88.5 ± 3.6 $\overline{\%\text{GII}}$, respectively. Cephalopods (16.2 ± 3.3 %GII) and crabs (14.7 ± 2.6 %GII) were important supplemental prey items of large *R. binoculata*, whereas shrimp-like crustaceans (7.3 ± 2.5 %GII), the second most important prey group of large *R. rhina*, were relatively unimportant. Conversely, shrimp-like crustaceans dominated all dietary indices for small *R. binoculata* and small *R. rhina* and were also the most important prey group of *R. inornata* (Table 3). Diets of small skates were more diverse than those of large skates. Fishes were important in the diets of all small species groups and crabs were of comparable importance to small *R. binoculata* and small *R. inornata* (Table 3). Polychaetes, gastropods and bivalves, and other crustaceans were rare or absent in the diets of both small and large skates from shelf waters.

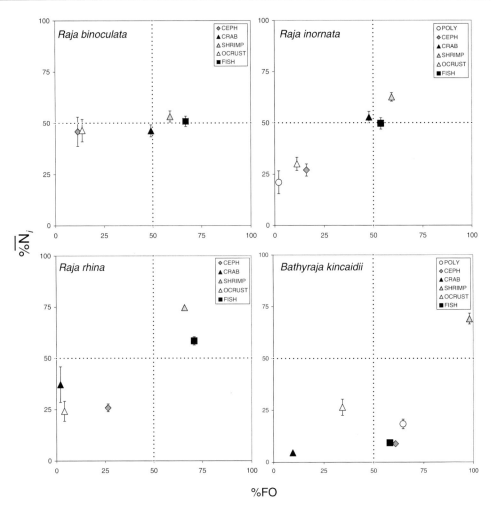

Fig. 4 Graphical representation of the feeding behavior of four species of sympatric skates; $\overline{\%N_i}$ = prey-specific percentage number, %FO = percentage frequency of occurrence

Differences were noted between the diets of small and large skates on the continental shelf, whereas diet compositions were similar within size classes. Dietary overlap values within size groups were considerably greater than those between size groups by all measures (Table 4). Accordingly, simulated Pianka's overlap values were greater than observed values for all pairwise comparisons involving like size groups and less than observed values for all large-small comparisons. Dietary overlap between large *R. binoculata* and small *R. inornata* was the only comparison for which qualitative overlap indices indicated similarity (>0.60) and null model simulations did not. Significant differences in diet composition were detected for the following species-pairs: small *R. binoculata*-small *R. inornata* (P_{sim} = 0.000), large *R. binoculata*-large *R. rhina* (P_{sim} = 0.047), small *R. rhina*-large *R. binoculata* (P_{dif} = 0.017), and small *R. rhina*-large *R. rhina* (P_{dif} = 0.032).

Inspection of PCA bi-plots and vector plots revealed patterns in diet composition of continental shelf skates that supported and further explained results of pairwise comparisons. Principal components (PCs) 1 and 2 were interpretable and accounted for 88.4% (PC 1 = 57.1%, PC 2 = 31.3%) of total variation in diet composition. Principal component 1 was strongly bipolar (i.e., influenced by the differential contribution of two

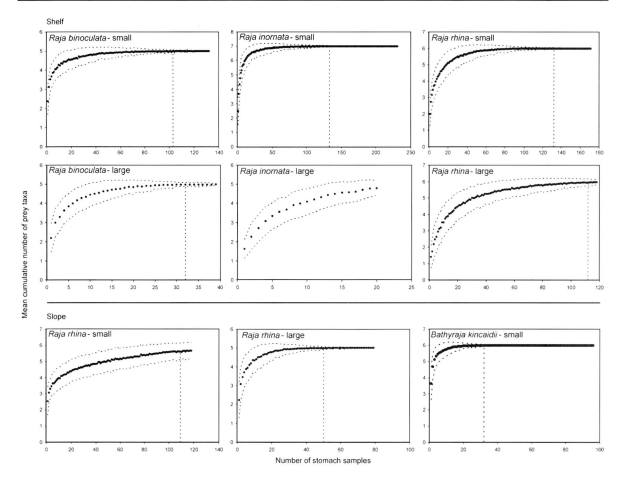

Fig. 5 Mean cumulative number of prey taxa per stomach sample for small (≤60 cm TL) and large (> 60 cm TL) individuals of the four study species in continental shelf (≤200 m) and slope (> 200 m) regions. Dashed lines represent mean standard deviation values. Drop-lines indicate the minimum number of samples necessary for adequate sample sizes

variables), with FISH (positive) and SHRIMP (negative) each contributing substantially, or loading heavily (Fig. 6A). CRAB (negative), and to a lesser extent, SHRIMP and FISH (positive) loaded heavily on PC 2 (Fig. 6A). Small *R. rhina* loaded primarily negatively with respect to PC 1 and positively with respect to PC 2, indicating the differential importance of shrimp-like crustaceans and, secondarily fishes, in the diet of these individuals. Conversely, large *R. rhina* loaded strongly to the positive on PC 1, with weaker positive loadings on PC 2 indicating the distant secondary importance of shrimp-like crustaceans (Fig. 6B). Large *R. binoculata* and large *R. rhina* loadings overlapped considerably, with negative dispersion on PC 2 suggesting the greater relative influence of crabs on the former species. Small *R. binoculata* and small *R. rhina* loadings were scattered throughout both PCs, indicating the more general feeding habits of these groups. Polychaetes, other crustaceans, and cephalopods contributed very little to the dietary variation of this assemblage.

Ontogenetic comparisons: slope assemblage

Cumulative prey curves indicated sufficient sample sizes for small and large specimens of all slope species (Fig. 5). The curves generated for small *B. kincaidii* (CV = 0.00%, n_{min} = 32) and large *R. rhina* (CV = 0.00%, n_{min} = 50) reached actual asymptotes and displayed no variability at the

Table 3 Mean and standard error (SE) of percentage number ($\overline{\%N}$) and percentage weight ($\overline{\%W}$); percentage occurrence (%O) and mean and SE of geometric index of importance ($\overline{\%GII}$) of prey taxa found in the stomachs of small (≤60.0 cm TL) and large (> 60.0 cm TL) individuals of three skate species from continental shelf (≤200 m) waters (see methods for prey group descriptions)

Prey Group	Size Class	R. binoculata %N	%W	%O	%GII	R. inornata %N	%W	%O	%GII	R. rhina %N	%W	%O	%GII
POLY	Small	–	–	–	–	0.5 (0.2)	< 0.1 (< 0.1)	1.1	0.2 (0.2)	–	–	–	–
	Large	–	–	–	–	–	–	–	–	–	–	–	–
MOLL	Small	–	–	–	–	3.0 (0.7)	1.2 (0.4)	5.5	2.6 (0.8)	0.3 (0.2)	< 0.1 (< 0.1)	1.7	0.6 (0.1)
	Large	–	–	–	–	–	–	–	–	0.2 (0.2)	< 0.1 (< 0.1)	0.7	0.1 (0.1)
CEPH	Small	1.9 (0.9)	1.9 (1.1)	1.8	1.5 (1.1)	3.9 (0.8)	5.7 (1.3)	7.6	4.4 (1.5)	4.5 (1.0)	8.8 (1.9)	12.4	6.1 (1.8)
	Large	16.6 (4.8)	18.5 (5.5)	19.7	16.2 (3.3)	–	–	–	–	2.4 (1.1)	2.2 (1.2)	4.9	1.8 (1.1)
CRAB	Small	24.7 (2.7)	29.9 (3.4)	25.0	26.1 (3.7)	27.3 (2.3)	33.5 (2.7)	25.1	29.7 (4.1)	0.9 (0.5)	1.1 (0.6)	1.7	0.8 (0.6)
	Large	16.5 (3.9)	12.6 (4.2)	22.5	14.7 (2.6)	–	–	–	–	1.6 (1.0)	1.6 (1.1)	2.1	1.2 (1.0)
SHRIMP	Small	39.6 (2.9)	41.9 (3.5)	36.0	41.8 (3.9)	39.1 (2.4)	26.7 (2.5)	30.4	34.1 (3.9)	71.1 (2.5)	55.5 (3.3)	50.5	64.8 (3.8)
	Large	2.6 (1.6)	< 0.1 (< 0.1)	4.2	1.1 (0.5)	–	–	–	–	11.2 (2.6)	6.9 (2.2)	12.7	7.3 (2.5)
OCRUST	Small	7.5 (1.8)	4.2 (1.4)	7.7	5.0 (1.8)	3.4 (0.7)	2.4 (0.8)	5.7	2.6 (1.1)	1.2 (0.4)	1.6 (0.7)	4.0	1.2 (0.6)
	Large	3.1 (1.8)	1.6 (1.3)	4.2	1.9 (0.9)	–	–	–	–	1.7 (1.0)	1.0 (0.9)	2.8	1.0 (0.9)
FISH	Small	26.3 (2.5)	22.1 (2.9)	29.4	25.6 (3.3)	22.8 (2.1)	30.4 (2.7)	24.7	26.5 (3.8)	21.9 (2.3)	32.9 (3.2)	29.8	26.6 (3.8)
	Large	61.2 (5.5)	67.3 (6.2)	49.3	66.1 (3.8)	–	–	–	–	82.9 (3.0)	88.3 (2.7)	76.8	88.5 (3.6)

endpoints. Statistical tests conducted for small *R. rhina* indicated that the prey curve reached an asymptote for mean number ($t = 0.22$, $P = 0.849$; CV = $8.98 \pm 0.14\%$) and SE ($t = 0.63$; $P = 0.592$). A minimum of 109 samples was necessary for sufficient sample size.

Mean percentage GII values and RMPQ indicated that shrimp-like crustaceans and fishes were the most dominant prey groups among continental slope skates. Cephalopods, polychaetes, and other crustaceans were also important prey taxa in this region. Shrimp-like crustaceans were the primary prey group of small *R. rhina* and small *B. kincaidii*, whereas fishes were the primary prey group of large *R. rhina*. Fishes ($\overline{X} \pm$ SE: 25.7 ± 3.3 %GII) and cephalopods (17.4 ± 2.7 %GII) were important supplemental prey taxa of small *R. rhina*. Diet composition of small *B. kincaidii* was more diverse, with polychaetes, cephalopods, fishes, and other crustaceans each contributing substantially (Table 5). Shrimp-like crustaceans (21.1 ± 2.8 %GII) and cephalopods (13.1 ± 2.2 %GII) were of considerable supplemental importance to large *R. rhina*. Crabs and shelled molluscs were either absent or rare in the diet of skates on the continental slope (Table 5).

Diet compositions of small *R. rhina* and small *B. kincaidii* were similar to each other, but of variable similarity to that of large *R. rhina*. Diet of small *R. rhina* and *B. kincaidii* exhibited 74.3% overlap, with index values ranging from 0.906 to 0.914. Diet composition of these groups was determined to be significantly similar based on null model simulations ($P = 0.048$). Diets of large *R. rhina* and small *B. kincaidii* were dissimilar (Table 4), but results of null model simulations were not significant ($P = 0.193$). Although null model simulations indicated that diet composition between small and large *R. rhina* were more different than observed ($P_{dif} = 0.353$), overlap indices indicated dietary similarity (Table 4).

Examination of PCA bi-plots and vector plots revealed patterns in diet composition of continental slope skates that supported and further elucidated results of pairwise comparisons. Principal components 1 and 2 were interpretable and accounted for 85.2% (PC 1 = 62.8%, PC 2 = 22.4%) of total variation in diet composition. Principal component 1 was strongly bipolar, with

Table 4 Pairwise comparisons of diet composition among small (lowercase species names) and large (uppercase species names) skates in two physiographic regions. Qualitative overlap indices and results of null model simulations are depicted. Null model simulations were conducted using Pianka's index, variance of mean simulated overlap values shown in parentheses; Obs-Sim = observed (calculated) overlap value compared to simulated value, Obs Sim = dietary similarity, Obs Sim = dietary dissimilarity. Significant dietary difference or similarity indicated P-value (0.05)

Pairwise comparisons	Overlap indices			Null model simulations		
	Percentage	Morisita's	Pianka's	Mean Overlap	Obs-Sim	P-Value
Shelf						
binoculata-BINOCULATA	44.8	0.541	0.555	0.770 (0.020)	Dissimilar	0.086
binoculata-inornata	91.2	0.987	0.988	0.705 (0.022)	Similar	**0.000**
binoculata-rhina	70.8	0.849	0.873	0.773 (0.020)	Similar	0.275
binoculata-RHINA	36.6	0.474	0.526	0.769 (0.020)	Dissimilar	0.067
BINOCULATA-inornata	49.1	0.604	0.623	0.699 (0.022)	Dissimilar	0.273
BINOCULATA-rhina	35.9	0.397	0.397	0.772 (0.200)	Dissimilar	**0.017**
BINOCULATA-RHINA	71.2	0.927	0.955	0.768 (0.019)	Similar	**0.047**
inornata-rhina	68.4	0.776	0.805	0.699 (0.022)	Similar	0.261
inornata-RHINA	38.3	0.498	0.561	0.702 (0.023)	Dissimilar	0.168
rhina-RHINA	37.7	0.443	0.456	0.772 (0.020)	Dissimilar	**0.032**
Slope						
Continental Slope						
rhina-RHINA	60.9	0.697	0.699	0.610 (0.035)	Similar	0.353
rhina-kincaidii	74.3	0.906	0.914	0.701 (0.022)	Similar	**0.048**
RHINA-kincaidii	45.7	0.481	0.491	0.634 (0.026)	Dissimilar	0.193

FISH (positive) and SHRIMP (negative) each loading heavily (Fig. 7A). CEPH (negative), and to a lesser extent, SHRIMP and FISH (positive) loaded heavily on PC 2 (Fig. 7A). Small *R. rhina* loaded primarily negatively with respect to PC 1 and PC 2, but considerable scatter indicated the differential importance of shrimp-like crustaceans, fishes, and cephalopods among individuals. Small *B. kincaidii* loadings displayed a similar pattern, with slightly heavier negative loadings on PCs 1 and 2 indicating the greater relative importance of cephalopods and, to a lesser extent, polychaetes and other crustaceans in the diet of these skates. Positive loadings of large *R. rhina* on PC 1 suggested the primary importance of fishes in the diet of most individuals, but negative loadings on PC 2 and considerable scatter along both axes signified the supplemental importance of cephalopods and shrimps (Fig. 7B). Crabs contributed very little to the dietary variation of this assemblage.

Ontogenetic and depth comparison

Cluster analysis based on diet compositions of all size and depth groupings revealed two distinct trophic guilds ($C_H > 0.752$). Size was the primary distinguishing factor between guilds, with little segregation by depth and species evident (Fig. 8). Large *R. binoculata* and *R. rhina* from the shelf and slope comprised one highly similar guild ($C_H > 0.927$), for which fishes were the dominant prey taxa. Small individuals of all species in shelf and slope habitats comprised a second trophic guild ($C_H > 0.787$). Shrimp-like crustaceans were dominant in the diets of all skate groups within this guild. Supplemental prey taxa were of greater relative importance to small skate groups as indicated by the distinction of some minor clades within this guild.

Discussion

Comparative diet composition and trophic ecology

Crustaceans and fishes were the dominant prey taxa in the diets of skates on soft bottom regions of the central California continental shelf and upper slope. Overall diet composition of the local population indicated that *B. kincaidii* specialized on

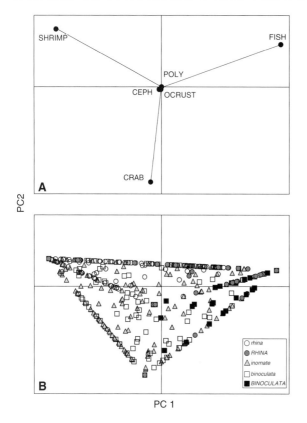 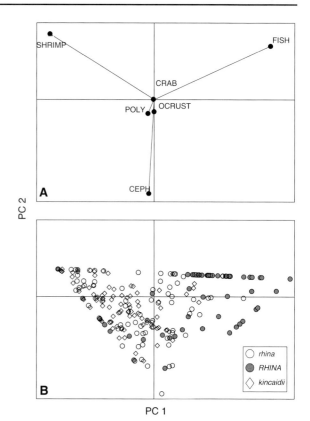

Fig. 6 Vector plot of first (PC 1) and second (PC 2) principal component scores calculated using individual-based geometric index of importance values for prey groups (**A**) and Euclidean bi-plot of PC 1 and PC 2 for small *Raja binoculata* (lowercase), large *R. binoculata* (uppercase), small *R. inornata* (lowercase), small *R. rhina* (lowercase), and large *R. rhina* (uppercase) from shelf waters (**B**). See methods for prey group descriptions

Fig. 7 Vector plot of first (PC 1) and second (PC 2) principal component scores calculated using individual-based geometric index of importance values for prey groups (**A**) and Euclidean bi-plot of PC 1 and PC 2 for small *Raja rhina* (lowercase), large *R. rhina* (uppercase), and small *Bathyraja kincaidii* (lowercase) from slope waters (**B**). See methods for prey group descriptions

shrimp-like crustaceans, mainly consuming euphausiids (Rinewalt et al. 2007). *Raja binoculata* and *R. inornata* ate a mixture of shrimps, fishes, and crabs, with fishes primary in the diet of the former species. Fishes were also of principal importance in the diet of *R. rhina*, and shrimp-like crustaceans were only slightly less important. The dominance of decapods and fishes in skate diets has also been demonstrated across broad taxa and geographic regions (e.g., Ebert et al. 1991; Orlov 2003; Dolgov 2005).

Additional prey taxa did not contribute substantially to the diet of most local skate species. Polychaetes were important supplemental prey items for *B. kincaidii*, but were rare or absent among other species. Cephalopods were of minor importance to all skates, contributing most to the diets of *R. rhina* and *B. kincaidii*. Molluscs were ingested whole, always occurred with other prey items, and were comprised of small, common species (Rinewalt et al. 2007; Robinson et al. 2007). Molluscs were also generally absent in the diets of other skate species and, when present, contributed little to diet composition (Ebert and Bizzarro 2007). Therefore it is likely that molluscs were incidentally ingested.

Interspecific differences in weight and number of prey contents were associated with variation in skate size compositions and depth distributions, respectively. Mean weight of prey contents increased with increasing maximum TL of skates, but results were greatly influenced by the presence of a few extremely large *R. binoculata*.

Table 5 Mean and standard error (SE) of percentage number (%N) and percentage weight (%W); percentage occurrence (%O) and mean and SE of geometric index of importance (%GII) of prey taxa found in the stomachs of small (≤60.0 cm TL) and large (>60.0 cm TL) individuals of two skate species from continental slope (>200 m) waters (see methods for prey group descriptions)

Prey Group	Size Class	R. rhina %N	%W	%O	%GII	B. kincaidii %N	%W	%O	%GII
POLY	Small	–	–	–	–	12.8 (1.9)	20.8 (2.8)	20.8	17.9 (2.2)
	Large	–	–	–	–				
MOLL	Small	0.5 (0.2)	0.2 (0.1)	2.8	0.3 (0.2)	–	–	–	–
	Large	2.0 (0.9)	0.6 (0.9)	4.6	1.3 (0.5)				
CEPH	Small	12.0 (1.8)	23.4 (2.9)	23.1	17.4 (2.7)	5.5 (0.7)	20.8 (2.9)	18.3	13.9 (1.9)
	Large	10.8 (2.0)	15.4 (3.1)	17.6	13.1 (2.2)				
CRAB	Small	0.3 (0.2)	0.6 (0.4)	0.8	0.3 (0.3)	0.5 (0.2)	0.3 (0.1)	3.2	0.4 (0.2)
	Large	–	–	–	–				
SHRIMP	Small	67.9 (2.8)	45.4 (3.8)	42.2	56.1 (3.5)	66.6 (3.0)	38.0 (3.3)	29.7	49.8 (2.5)
	Large	31.2 (3.6)	11.0 (2.7)	30.1	21.1 (2.8)				
OCRUST	Small	0.3 (0.3)	0.4 (0.4)	0.4	0.2 (0.3)	9.9 (2.0)	7.4 (2.0)	11.4	8.0 (1.9)
	Large	0.6 (0.4)	1.0 (0.4)	2.0	0.8 (0.5)				
FISH	Small	19.1 (2.6)	30.1 (3.3)	30.7	25.7 (3.3)	4.7 (0.8)	12.7 (2.3)	16.7	9.9 (1.5)
	Large	55.3 (4.1)	72.1 (4.2)	45.8	63.7 (3.8)				

When compared statistically, only prey weight of *R. rhina*, which contained a considerable proportion of large individuals, was significantly different from that of both *B. kincaidii* and *R. inornata*. Mean total prey weight of *R. microocellata* (12.4 g; Rousset 1987) was comparable to that of similar-sized *R. rhina*, whereas prey contents of larger *Dipturus chilensis* specimens were much heavier than prey contents of any study species ($\overline{X}$ = 269.7 g, Koen Alonso et al. 2001). Unlike prey weight, mean prey number was more strongly associated with depth distribution than size composition among central California skates. Euphausiids are common at shelf breaks and upper slopes, where swarms are often trapped during downward vertical migrations (Isaacs and Schwartzlose 1965; Ressler et al. 2005). *Raja rhina* and especially *B. kincaidii* are deeper dwelling species and consumed a greater proportion of these relatively small prey items than *R. inornata* and *R. binoculata*. In addition, *B. kincaidii* ingested a considerable number of amphipods, as previously reported among small skates (Capapé 1976; Braccini and Perez 2005).

Ontogenetic differences were the primary source of variation among skate diets; the diets of small skates were mainly comprised of shrimp-like crustaceans, whereas the diets of larger skates consisted primarily of fishes. Size-related differences in diet composition have been demonstrated for many skate species (e.g., Holden and Tucker 1974; Ajayi 1982; Brickle et al. 2003; Orlov 2003; Braccini and Perez 2005). The trend of crustaceans and fishes dominating diet composition of small and large individuals, respectively, has also been well documented for skates (e.g., Steven 1932; Abd El-Aziz 1986). In this study, 60 cm TL distinguished small from large specimens and related well to shifts in diet composition for *R. rhina*, as suggested by Ebert (2003), and *R. binoculata*. Insufficient sample size precluded a quantitative assessment of large *R. inornata* diet composition. However, fishes were dominant both numerically and gravimetrically and were present in stomach contents of 90% of large *R. inornata* individuals. Conversely, shrimp-like crustaceans were dominant in the diets of all small skate groups in both continental shelf and slope waters.

Depth-related dietary differences have also been documented for skates (Hacunda 1981; Templeman 1982; Morato et al. 2003) but were constrained in this study because sample size was too low to test size and depth variables concurrently. Intraspecific dietary differences between shelf and slope depths were detected at lower levels of taxonomic resolution for *R. rhina* (Robinson et al. 2007). Other species were largely confined to shelf (i.e., *R. binoculata*, *R. inornata*)

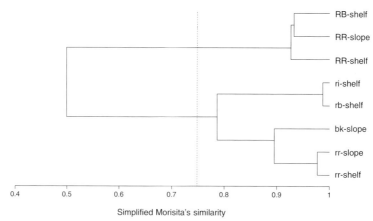

Fig. 8 Cluster analysis of diet compositions of skate species groups based on the geometric index of importance. Hierarchical clustering was performed using average linkage with the simplified Morisita's index as a measure of similarity. Uppercase notation refers to large (>60 cm TL) species groups; lowercase notation refers to small (≤60 cm TL) species groups. Dashed line indicates major trophic guilds

or slope (i.e., *B. kincaidii*) waters, precluding similar analyses. General prey type was consistent between physiographic regions, however, and large and small skates clustered distinctly regardless of depth.

For the shelf assemblage, most prey items consisted of common benthic crustaceans and fishes. Cancrid crabs (e.g., *Cancer gracilis*) and crangonid shrimps (e.g., *Neocrangon ressima*) were the dominant prey taxa within crab and shrimp-like crustacean groups (Robinson et al. 2007; Bizzarro and Ebert unpublished data). These prey taxa are abundant at shallow and mid-shelf depths, with crangonids ranging into deep shelf and upper slope regions (Schmitt 1972). Shrimps were important dietary components of all small skate groups, but crabs were conspicuously absent in the diet of *R. rhina*. This may have resulted from the deeper relative depth distribution of *R. rhina*, behavioral aspects, or morphological differences among species. Fishes (e.g., *Citharichthys* spp., *Chilara tayloria, Porichtys notatus, Sebastes* spp.), including several commercially exploited species, were important to all groups of small and especially large skates. With the exception of cephalopods (e.g., *Loligo opalescens*) in the diet of *R. binoculata*, other prey items were relatively unimportant to shelf skates.

The lack of significant dietary differences among comparably-sized shelf skates suggests that interspecific resource competition was not intense. Food and space are the primary limiting factors for fishes (Ross 1986). Slight depth differences were noted among *R. binoculata, R. inornata*, and *R. rhina*, but ranges overlapped considerably and all species co-occurred in trawls. Therefore, it does not appear that spatial segregation is evident within this region. It is more likely that prey resources are abundant (Schmitt 1972; Allen 2006) and that shelf skates are generalists, with large individuals mainly eating fishes and small skates mainly eating shrimps in relation to their availability. Differences in diet between size classes are likely attributable to metabolic requirements of larger individuals (Carlson et al. 2004) and not a function of resource partitioning. Groundfishes also generally consume crustaceans and fishes (Horn and Ferry-Graham 2006) and have greatly declined in shelf waters (Mason 2004), possibly resulting in the competitive release of skates. This situation has been described from the northwestern Atlantic (Link et al. 2002), but additional information is needed to properly evaluate the ecological interactions between skates and groundfishes off central California.

Diet composition of the upper slope skate assemblage was also comprised of commonly reported benthic species, including fishes, cephalopods, and polychaetes, but also contained midwater organisms. Cancrid crabs and true shrimps are relatively scarce in upper slope waters (Schmitt 1972), and were not typically observed among stomach contents (Rinewalt et al. 2007; Robinson et al. 2007). Although euphausiids are typically pelagic, swarms are often advected shoreward when in the water column. These swarms may become aggregated on the continental shelf and portions of the upper slope (Isaacs and Schwartzlose 1965; Ressler et al. 2005), a situation that has been reported locally at the headward parts of submarine canyons (Croll et al. 2005). Therefore, as suggested for teleosts (Pereyra et al. 1969; Chess et al. 1988), euphausiids are probably captured by skates when they are on or in close proximity to the benthos (Rinewalt et al. 2007). Skates can thereby serve as a source of energy transfer between pelagic and benthic environments. The great majority of identified cephalopods taken in this region consisted of the dominant local species, *Octopus rubescens* and *L. opalescens* (Hochberg 1998). Other crustaceans, comprised mainly of amphipods, and polychaetes were common in the diet of *B. kincaidii* and have been noted in the diets of other small skates (Muto et al. 2001; Braccini and Perez 2005). Myctophids were the main teleosts identified in the stomach contents of *B. kincaidii* and were likely taken at the seafloor in the same manner as euphausiids (Rinewalt et al. 2007). In addition to myctophids, larger fishes such as *Merluccius productus* and commercially important flatfishes (e.g., *Glyptocephalus zachirus*, *Microstomus pacificus*) were consumed in this region by *R. rhina* (Robinson et al. 2007).

The diet compositions of upper slope skates were not as strongly correlated with size differences as those of shelf species. Diets of small *B. kincaidii* and small *R. rhina* were significantly similar, largely because of the shared importance of shrimp-like crustaceans (e.g., euphausiids) and cephalopods (e.g., *O. rubescens*), but the relative contribution of supplemental prey items differed considerably. Whereas fishes were of greater importance to small *R. rhina*, polychaetes and other crustaceans (e.g., amphipods) were consumed almost exclusively by *B. kincaidii*. Ontogenetic shifts from small crustaceans to larger crustaceans have been reported for other small skate species and early life stages of larger species (Orlov 1998; Muto et al. 2001; Braccini and Perez 2005), and were demonstrated between immature and mature *B. kincaidii* (Rinewalt et al. 2007). The diet of large and small *R. rhina*, unlike most comparisons between size classes, was determined to be rather similar from this study. Although significant size-based differences were detected from the more expansive data set of Robinson et al. (2007), ontogenetic differences were not evaluated between depth ranges and were reported as less substantial at slope depths. Only slight overlap, attributable to the differential consumption of fishes, euphausiids, and cephalopods, was evident between large *R. rhina* and small *B. kincaidii*.

Skates on soft bottom habitats off California, like those of other regions (Ebert et al. 1991; Orlov 2003), are considered important predators (Allen 2006). Using all available samples, Ebert and Bizzarro (2007) determined that the four study species were upper-trophic level consumers (trophic levels = 3.54–3.86). If samples were segregated between small and large size classes, however, trophic levels would be more varied, with large specimens probably representing apex predators (trophic level ≥ 4.00). Similarly, based on overall data, *R. rhina* clustered within a fish-feeding clade, whereas the other species were included in a decapod-feeding clade (Ebert and Bizzarro 2007). When species were segregated by size in this study, however, small skates formed a distinct crustacean-feeding clade and large species formed a fish-feeding clade. Because upper trophic level groundfish populations have declined in the California Current ecosystem (Mason 2004; Essington et al. 2006), the importance of skates as predators in soft bottom regions and as potential predators and competitors with groundfishes has likely increased.

Prey composition of study species was somewhat restricted and probably reflected a limited diversity of feeding modes among skates. Durophagy has been reported for other chondrichthyan groups (Johnson and Horton 1972; Segura-Zarosa et al. 1997; Summers 2000; Motta 2004), but has not been documented for skates. Additionally, whereas the largest shark and ray species exploit planktonic food sources, these resources are unavailable to skates. Feeding behavior is also restricted among rajiforms. Some stingrays excavate feeding pits to capture infaunal organisms (Gregory et al. 1979; VanBlaricom 1982). This behavior has not been documented among skates, however, and likely limits their ability to feed within the benthos. Although polychaetes are reported as important prey items for some small skates, epibenthic species, not deep infaunal types, are generally reported (Mabragaña et al. 2005; Rinewalt et al. 2007).

Statistical techniques and caveats

General prey categories were considered more appropriate than lowest possible taxonomic designations for this study because a considerable portion of prey items were identified to high taxonomic levels and intra- and interspecific differences within established prey categories did not appear to be substantial (Rinewalt et al. 2007; Robinson et al. 2007; Bizzarro and Ebert unpublished data). Therefore, comparisons based on variable and dependent levels of identification (e.g., unidentified crab, *Cancer* sp., *Cancer gracilis*) would most likely have overestimated dietary variation more than the use of general prey categories overestimated dietary similarity. The inclusion of elasmobranchs and teleosts in a general fish category did not substantially reduce variation because elasmobranchs were insignificant dietary components (Ebert and Bizzarro 2007). Although shrimp-like crustaceans encompassed mysids, euphausiids, and true shrimps, variation in this category was limited because mysids were extremely rare in skate diets and euphausiids and shrimps were often indistinguishable and spatially segregated. Similarly, other crustaceans were composed mainly of unidentified crustaceans and the ostracod, *Euphilomedes carcharodonta*, at shelf depths and by gammarid amphipods, unidentified crustaceans, and the anomuran, *Mursia gaudichaudii*, at slope depths. Other crustaceans were also of minor importance to all skate species. The detection of both significantly similar and different results and the consistency of findings among species support the category distinctions used. The use of general prey categories, however, admittedly results in a simplified depiction of diet composition and increases the magnitude of dietary similarities among species. Therefore, the results of this study are valid for general prey categories, but more substantial differences in diet compositions may be evident at lower levels of taxonomic resolution.

Additional sources of variation, such as time of collection, sex, and prey size were mitigated when possible or not considered. Temporal dietary differences are known to occur among skates (McEachran et al. 1976; Pederson 1995). Samples were therefore limited to a seven-month period of generally consistent oceanographic conditions (Peterson and Schwing 2003). Because intergender variation in tooth morphology is not believed to modify diet (McEachran 1977; Motta 2004), sample size was limited, and no differences in diet are typically found for similar sized males and females (e.g., Morato et al. 2003; Scenna et al. 2006), including *R. rhina* (Robinson et al. 2007), sexual dietary differences were not evaluated. Differences in prey size have not been quantitatively evaluated for skates, but have been shown to vary among sharks (e.g., Bethea et al. 2004, 2005). Unfortunately, most prey items were present in advanced states of digestion, and reliable measurements were available for only a small portion of prey items, precluding analyses.

It is necessary to determine sample size sufficiency in diet composition studies to ensure that results obtained from the sampled population are representative of the population at large. However, precision of diet composition estimates was not evaluated in fish food habit studies prior to 1996 (Ferry and Cailliet 1996), and has been treated only qualitatively since. Original evaluation criteria for sample size sufficiency are also highly subjective (Hurturbia 1973; Cailliet 1977), leading many authors to analyze data that imprecisely represent diet composition. Because these criteria are solely visual, the dimensions of cumulative prey curve plots greatly influence results and can be manipulated to produce curves that appear to level off, although data may indicate otherwise. Recent attempts to improve assessment criteria are qualitative (Ferry and Cailliet 1996) or arbitrary (Koen Alonso et al. 2001) and therefore inappropriate. To properly evaluate sample size sufficiency, standard, quantitative techniques are necessary.

The techniques developed during this study were intended to quantitatively assess sample size sufficiency using the established criteria; cumulative prey curve stabilization at an asymptote, and a considerable reduction in the variability of randomized mean values (Ferry and Cailliet 1996). Statistical analyses were applied to cumulative species richness curves in this study, but also can be directly applicable to data generated from cumulative diversity curves. Standard error curves can also be constructed for visual inspection and should display the inverse pattern of cumulative prey curves, with SE decreasing with increasing sample size. Because prey taxa were grouped into a limited number of categories and sample size was considerable for most species, several curves reached true asymptotes. It should be noted, however, that this condition is not typical of most studies, especially those using lowest possible taxonomic designations and involving species with considerable dietary diversity. In situations such as these, the quantitative evaluation of dietary variability using SE may also indicate that significant variability exists although the prey curve has stabilized. Therefore, the use of SE as a determining factor in evaluating sample size sufficiency may be excessively restrictive for highly variable data sets and should be evaluated according to the needs and limitations of the researcher. Based on simulations using related data sets (Rinewalt et al. 2007; Robinson et al. 2007) and results of additional diet composition studies (Bizzarro 2005; Bizzarro and Ebert unpublished data), most estimates based on visual inspection slightly underestimated the asymptotic stabilization of the curve. It is suggested that at least 500 iterations are generated to increase precision of resampling estimates along the curve and best approximate the parameters to be tested.

The methods developed to evaluate cumulative prey curves are based on commonly employed and robust statistical analyses. This use of the *t*-statistic to compare regression coefficients was developed by Fisher (1922) and later described in detail by Zar (1996). The CV has commonly been used in the age and growth literature (Chang 1982; Campana 2001) to evaluate precision. Because precision of diet composition is highly influenced by the niche breadth of the predator population and accuracy and level of prey identification, no precision value can be considered adequate for all studies. The reporting of CV, however, will provide a standard method of comparison among diet studies, and values <10% are a general benchmark for adequate precision in age and growth studies (Campana 2001).

Compound indices have been criticized as redundant and unnecessary (MacDonald and Green 1983; Hansson 1998). However, their use is widespread in elasmobranch food habit studies. In addition, individual RMPQ have inherent biases and describe different aspects of the trophic ecology of a predator, such as feeding behavior (%N), nutrition (%W), and population-based food habits (%FO, %O) (Hyslop 1980; Bigg and Perez 1985). Therefore, to evaluate dietary importance and facilitate comparisons between studies, two compound indices, IRI (Pinkas et al. 1971) and GII (Assis 1996) were calculated and compared. Although the use of IRI has become almost universal in elasmobranch diet studies, it is far more affected by prey occurrence and less precise than GII. In IRI calculations, %FO is multiplied by aggregate %N and %W values (or $\overline{\%N}$ and $\overline{\%W}$ values), whereas %O, %N, and %W are summed in GII and therefore contribute equally. Additionally, %FO is not additive, and combined values from all prey categories customarily exceed 100%. Frequency of occurrence, therefore, is of greater influence in IRI calculations than %N and %W combined. As a result, IRI is biased against relatively uncommon prey taxa, resulting in inflated values of importance for common prey groups (Hansson 1998). Furthermore, because IRI values range from 0 to 20,000, whereas comparable GII values range from 0 to 173.2, precision is far greater for GII calculations. Unless the researcher wants to accentuate the contribution of population-wide food habits, GII incorporating %N, %W, and %O is recommended as the more appropriate measure of dietary importance.

PCA was used to investigate sources of variation in diet composition among species (Crow 1979; Paukert and Wittig 2002). Although assumptions of multivariate normality and equality of variance-covariance matrices were violated, even large departures from ideal data structure are tolerable when PCA is used for descriptive purposes (Gauch 1982). To minimize the effect of these assumptions, a sample size more than three times greater than the number of discriminating variables is recommended (McGarigal et al. 2000; Paukert and Wittig 2002). In this study, the ratio of sample size to prey variables was approximately 50:1 for the slope assemblage and more than 100:1 for the shelf assemblage. The most important assumption of PCA, linearity among variables, must be met for reliable results on the second and subsequent principal component axes, especially when PCA is used for inference or confirmation (Pielou 1984; McGarigal et al. 2000). Arch or compression effects were not evident in ordination plots for either skate assemblage, suggesting that any violations of this assumption were not pronounced (Gauch 1982; Pielou 1984). PCA is a more robust multivariate technique than its nonparametric equivalent, multidimensional scaling (MDS), and unlike MDS (unless post-hoc SIMPER tests are included), provides quantitative information on specific sources of variation, enabling interpretation of ordination plots (McGarigal et al. 2000).

Concluding remarks

The results of this study fill substantial information gaps concerning the feeding ecology and trophic relationships of skates in California waters. Prior to this and other skate diet studies included in this volume (Ebert and Bizzarro 2007; Rinewalt et al. 2007; Robinson et al. 2007), no published information was available on the food habits of skates in this region. Skates on soft bottom habitats of the central California continental shelf and upper slope show ontogenetic shifts in diet composition from crustaceans (primarily shrimps and euphausiids) to fishes, findings consistent with those of similar studies. Competition for prey resources did not appear to be intense among studied skates, at least at general levels of taxonomic identification, possibly because substantial recent declines in sympatric upper trophic level groundfish populations may have resulted in their competitive release.

Although considerable recent attention has been devoted to habitat or ecosystem-based management, baseline resource utilization studies are often generalized or lacking. It is anticipated that the results of this research will contribute to a greater understanding of food web dynamics in marine benthic communities off California, and the formulation of more effective multi-species management policies. Additional information on daily ration and local abundance is necessary to determine the amount of biomass removed by skates and their role in energy transfer among trophic levels. Quantitative information about habitat associations and long-term trends in diet composition (via stable isotope analysis) of skates and co-occurring groundfishes are also needed to fully understand the ecological interactions between these groups and the potential impacts of skates on the recovery of depleted groundfish populations.

Acknowledgements We thank Churchill Grimes, Don Pearson, John Field, E.J. Dick, and Alex MacCall of the National Marine Fisheries Service (NMFS), Santa Cruz Laboratory, for providing the skates used in this study and Brooke Flammang, Daniele Ardizzone, Aaron Carlisle, Chante Davis, Rob Leaf, Colleena Perez, and Tonatiuh Trejo of the Pacific Shark Research Center (PSRC), Moss Landing Marine Laboratories (MLML), for assisting with specimen processing. Thanks also to Jim Harvey, Lara Ferry-Graham, and Mike Graham of MLML for statistical advice, and to Stori Oates and two anonymous reviewers for their comments on earlier drafts of this manuscript. Funding for this research was provided by the National Oceanic and Atmospheric Administration/NMFS (to the National Shark Research Consortium and PSRC), and by the National Sea Grant College Program of the U.S. Department of Commerce's National Oceanic and Atmospheric Administration (NOAA) under NOAA Grant no. NA04OAR4170038, project number R/F-199, through the California Sea Grant College Program and in part by the California State Resources Agency.

References

Abd El-Aziz SH (1986) Food and feeding habits of *Raja* species (Batoidea) in the Mediterranean waters off Alexandria. Bull Inst Oceanog Fisheries 12:265–276

Adams J (2004) Foraging ecology and reproductive biology of Cassin's Auklet (*Ptychoramphus A aleuticus*) in the California Channel Islands. M.S. Thesis, Moss Landing Marine Laboratories, California State University, San Francisco, 119 pp

Ajayi TO (1982) Food and feeding habits of *Raja* species (Batoidei) in Carmarthen Bay, Bristol Channel. J Mar Biol UK 62:215–223

Allen MJ (2006) Soft substrata: continental shelf and upper slope. In: Allen LG, Pondella DJ, Horn MH (eds) Ecology of marine fishes: California and adjacent waters. University of California Press, Berkeley, pp 167–202

Amundsen PA, Gabler HM, Staldvik FJ (1996) A new approach to graphical analysis of feeding strategy from stomach contents data - modification of the Costello (1990) method. J Fish Biol 48: 607–614

Arkhipkin A, Brickle P, Laptikhovsky V (2003) Variation in the diet of the Patagonian toothfish with size, depth and season around the Falkland Islands. J Fish Biol 63:428–441

Assis CA (1996) A generalized index for stomach contents analysis in fish. Sci Mar 60:385–389

Bakun A (1996) Patterns in the ocean: ocean processes and marine population dynamics. California Sea Grant College System/National Oceanic and Atmospheric Administration in cooperation with Centro de Investigaciones Biológicas del Noroeste, La Jolla, CA, 323 pp

Bethea DM, Buckel JA, Carlson JK (2004) Foraging ecology of the early life stages of four sympatric shark species. Mar Ecol Prog Ser 268:245–264

Bethea DM, Carlson JK, Buckel JA, Satterwhite M (2005) Ontogenetic and site-related trends in the diet of the Atlantic sharpnose shark *Rhizoprionodon terraenovae* from the northeast Gulf of Mexico. Bull Mar Sci 78:287–307

Bigg MA, Perez MA (1985) Modified volume: a frequency-volume method to assess marine mammal food habits. In: Beddington JR, Beverton RJH, Lavigne DM (eds) Marine mammals and fisheries. George Allen and Unwin, London, pp 77–283

Bizzarro JJ (2005) Fishery biology and feeding ecology of rays in Bahía Almejas, Mexico. M.S. Thesis, Moss Landing Marine Laboratories, California State University, San Francisco, 468 pp

Braccini JM, Perez JE (2005) Feeding habits of the sandskate *Psammobatis extenta* (Garman, 1913): sources of variation in dietary composition. Mar Freshw Res 56:395–403

Brickle P, Laptikhovsky V, Pompert J, Bishop, A (2003) Ontogenetic changes in the feeding habits and dietary overlap between three abundant rajid species on the Falkland Islands' Shelf. J Mar Biol Ass UK 83:1119–1125

Cailliet GM (1977) Several approaches to the feeding ecology of fishes. In: Simenstad C, Lipovsky S (eds) Fish food habit studies: proceedings of the first Pacific Northwest technical workshop. Washington Sea Grant Program, University of Washington, Seattle, pp 1–13

Cailliet GM, Love MS, Ebeling AW (1986) Fishes: a field and laboratory manual on their structure, identification, and natural history. Wadsworth Publishing Co, Belmont, CA, 186 pp

Campana SE (2001) Accuracy, precision, and quality control in age determination, including a review of the use and abuse of age validation methods. J Fish Biol 59:197–242

Capapé C (1976) Contribution a la biologie des *Rajidae* de cotes tunisiennes. 8. *Raja melitensis* Clark, 1926, Regime alimentaire. Archives de l'Institut Pasteur de Tunis 53:39–45

Carlson JK, Goldman JK, Lowe CG (2004) Metabolism, energetic demand, and endothermy. In: Carrier JC, Musick JA, Heithaus MR (eds) Biology of sharks and their relatives. CRC Press, New York, pp 203–224

Carrassón M, Stefanescu C, Cartes JE (1992) Diets and bathymetric distributions of two Bathyal sharks of the Catalan deep sea (western Mediterranean). Mar Ecol Prog Ser 82:21–30

Chang WYB (1982) A statistical method for evaluating the reproducibility of age determination. Can J Fish Aquat Sci 39:1208–1210

Chess JR, Smith SE, Fischer PC (1988) Trophic relationships of the shortbelly rockfish, *Sebastes jordani*, off central California. CalCOFI Rep 29:129–136

Compagno LJV (1990) Alternative life history styles of cartilaginous fishes in time and space. Environ Biol Fishes 28:33–75

Cortés E (1997) A critical review of methods of studying fish feeding based on analysis of stomach contents: application to elasmobranch fishes. Can J Fish Aquat Sci 54:726–738

Cortés E (1999) Standardized diet compositions and trophic levels of sharks. ICES J Mar Sci 56:707–717

Croll DA, Marinovic B, Benson S, Chavez FP, Black N, Ternullo R, Tershy BR (2005) From wind to whales: trophic links in a coastal upwelling system. Mar Ecol Prog Ser 289:117–130

Crow ME (1979) Multivariate statistical analysis of stomach contents. In: Lipovsky SJ, Simenstad CA (eds) Fish food habit studies: proceedings of the second Pacific Northwest technical workshop. Washington Sea Grant Program, University of Washington, Seattle, pp 87–96

Dolgov A (2005) Feeding and food consumption by the Barents Sea skates. J Northw Atl Fish Sci 35:496–503

Dunn OJ (1964) Multiple contrasts using rank sums. Technometrics 6:241–252

Ebert DA (2003) Sharks, rays and chimaeras of California. University of California Press, Berkeley, 285 pp

Ebert DA, Bizzarro JJ (2007) Standardized diet compositions and trophic levels of skates. Environ Biol Fish (this volume)

Ebert DA, Cowley PD, Compagno LJV (1991) A preliminary investigation of the feeding ecology of skates (Batoidea: Rajidae) off the west coast of southern Africa. So Afr J Mar Sci 10:71–81

Essington TE, Beaudreau AH, Wiedenmann J (2006) Fishing through marine food webs. Proc Nat Acad Sci 103:3171–3175

Ferry LA, Cailliet GM (1996) Sample size sufficiency and data analysis: are we characterizing and comparing diet properly? In: MacKinlay D, Shearer K (eds) Feeding ecology and nutrition in fish: proceedings of the symposium on the feeding ecology and nutrition

in fish. International Congress on the Biology of Fishes, San Francisco, CA, pp 71–80
Fisher RA (1922) The goodness of fit of regression formulae and the distribution of regression coefficients. J Roy Statist Soc 85:597–612
Foster MS, DeVogelaere AP, Harrold C, Pearse JS, Thum AB (1988) Causes of spatial and temporal patterns in rocky intertidal communities of central and northern California. Mem Cal Acad Sci 9:1–45
Gauch HG (1982) Multivariate analysis in community ecology. Cambridge University Press, Cambridge, 298 pp
Gelsleichter J, Musick JA, Nichols S (1999) Food habits of the smooth dogfish, *Mustelus canis*, dusky shark, *Carcharhinus obscurus*, Atlantic sharpnose shark, *Rhizoprionodon terraenovae*, and the sand tiger, *Carcharias taurus*, from the northwest Atlantic Ocean. Environ Biol Fish 54:205–217
Gerking SD (1994) Feeding ecology of fish. Academic Press, San Diego, CA 416 pp
Goetelli NJ, Entsminger GL (2005) EcoSim: null models software for ecology. Version 7.72. Acquired Intelligence Inc and Kesey-Bear. Jericho, VT, http://garyentsminger.com/ecosim.htm
Gregory MR, Balance PF, Gibson GW, Ayling AM (1979) On how some rays (Elasmobranchia) excavate feeding depressions by jetting water. J Sed Petrol 49:1125–1130
Hacunda JS (1981) Trophic relationships among demersal fishes in a costal area of the Gulf of Maine. Fish Bull 79:775–788
Hansson S (1998) Methods of studying fish feeding: a comment. Can J Fish Aquat Sci 55:2706–2707
Heithaus MR (2004) Predator-prey interactions. In: Carrier JC, Musick JA, Heithaus MR (eds) Biology of sharks and their relatives. CRC Press, New York, pp 487–521
Hochberg FG (1998) Class Cephalopoda. In: Scott PV, Blake JA (eds) Taxonomic atlas of the benthic fauna of the Santa Marina Basin and the western Santa Barbara Channel. Volume 8: the Mollusca. Part 1: the Aplacophora, Polyplacophora, Scaphopoda, Bivalvia, and Cephalopoda. Santa Barbara Museum of Natural History, Santa Barbara, CA, pp 175–236
Hoff GR (2006) Biodiversity as an index of regime shift in the eastern Bering Sea. Fish Bull 104:226–237
Holden MJ, Tucker RN (1974) The food of *Raja clavata* Linnaeus 1758, *Raja montagui* Fowler 1910, *Raja naevus* Müller and Henle 1841 and *Raja brachyuran* Lafont 1873 in British waters. Journal du Conseil International pour l'Exploration de la Mer 35:189–193
Horn MH, Ferry-Graham LA (2006) Feeding mechanisms and trophic interactions. In: Allen LG, Pondella DJ, Horn MH (eds) Ecology of marine fishes:California and adjacent waters. University of California Press, Berkeley, pp 387–410
Hurturbia J (1973) Trophic diversity measurement in sympatric predatory species. Ecology 54:885–890
Hyslop EJ (1980) Stomach contents analysis: a review of methods and their application. J Fish Biol 17:411–429

Isaacs JD, Schwartzlose RA (1965) Migrant sound scatterers: interactions with the sea floor. Science 150:1810–1813
Ishihara H (1990) The skates and rays of the western North Pacific: an overview of their fisheries, utilization, and classification. In: Pratt HL Jr, Gruber SH, Taniuchi T (eds) Elasmobranchs as living resources: advances in the biology, ecology, systematics, and the status of fisheries. NOAA Tech Rep NMFS 90, pp 485–498
Jackson JBC, Kirby MX, Berger WH, Bjorndal KA, Botsford LW, Bourque BJ, Bradbury RH, Cooke R, Erlandson J, Estes JA, Hughes TP, Kidwell S, Lange CB, Lenihan HS, Pandolfi JM, Peterson CH, Steneck RS, Tegner MJ, Warner RR (2001) Historical overfishing and the recent collapse of coastal ecosystems. Science 293:629–638
Johnson AG, Horton H (1972) Length-weight relationships, food habits and sex and age determination of the ratfish, *Hydrolagus colliei* (Lay and Benett). Fish Bull 70:421–429
Koen Alonso M, Crespo EA, García NA, Pedraza SN, Mariotti PA, Berón Vera B, Mora NJ (2001) Food habits of *Dipturus chilensis* (Pisces: Rajidae) off Patagonia, Argentina. ICES J Mar Sci 58:288–297
Krebs CJ (1999) Ecological methodology. Benjamin Cummings, Menlo Park, CA, 620 pp
Langton RW (1982) Diet overlap between Atlantic cod, *Gahus morhua*, silver hake, *Merluccius bilinearis*, and fifteen other northwest Atlantic finfish. Fish Bull 80:745–759
Legendre L, Legendre P (1983) Numerical Ecology. Elsevier Scientific Publishing Company, New York, 419 pp
Link JS, Garrison LP, Almeida FP (2002) Ecological interactions between elasmobranchs and groundfish species on the Northeastern U.S. continental shelf. I. evaluating predation. North Am J Fish Management 22:550–562
Livingston RL (1984) Trophic response of fishes to habitat variability in coastal seagrass systems. Ecology 65:1258–1275
Mabragaña E, Giberto DA, Bremec CS (2005) Feeding ecology of *Bathyraja macloviana*(Rajiformes: Arhynchobatidae): a polychaete-feeding skate from the south-west Atlantic. Sci Mar 69:405–413
MacDonald JS, Green RH (1983) Redundancy of variables used to describe importance of prey species in fish diets. Can J Fish Aquat Sci 40:635–637
Mason JE (2004) Historical patterns from 74 years of commercial landings from California waters. CalCOFI Rep 45:180–190
McEachran JD (1977) Reply to "sexual dimorphism in skates (Rajidae). Evolution 31:218–220
McEachran JD, Boesch DF, Musick JA (1976) Food division within two sympatric species-pairs of skates (Pisces: Rajidae). Mar Biol 35:301–317
McGarigal K, Cushman S, Stafford S (2000) Multivariate statistics for wildlife and ecology research. Springer-Verlag, Inc, New York, 283 pp

Menge BA (2004) Bottom-up/top-down determination of rocky intertidal shorescape dynamics. In: Polis GA, Power ME, Huxel GR (eds) Food webs at the landscape level. The University of Chicago Press, Chicago, IL, pp 62–81

Mills EL (1969) The community concept in marine zoology, with comments on continua and instability in some marine communities: a review. J Fish Res Bd Canada 26:1415–1428

Mohan MV, Sankaran TM (1988) Two new indices for stomach content analysis of fishes. J Fish Biol 33:289–292

Morato T, Solà E, Grós MP, Menezes G (2003) Diets of thornback ray (*Raja clavata*) and tope shark (*Galeorhinus galeus*) in the bottom longline fishery of the Azores, northeastern Atlantic. Fish Bull 101:590–602

Motta PJ (2004) Prey capture behavior and feeding mechanisms of elasmobranchs. In: Carrier JC, Musick JA, Heithaus MR (eds) Biology of sharks and their relatives. CRC Press, New York, pp 165–202

Munday PL (2002) Does habitat availability determine geographical-scale abundances of coral-dwelling fishes? Coral Reefs 21:105–116

Musick JA, Burgess G, Cailliet G, Camhi M, Fordham S (2000) Management of sharks and their relatives (Elasmobranchii). Fisheries 25:9–13

Muto EY, Soares LSH, Goitein R (2001) Food resource utilization of the skates *Rioraja agassizii* (Müller and Henle, 1841) and *Psammobatis extenta* (Garman, 1913) on the continental shelf off Unatuba, south-eastern Brazil. Rev Brasil Biol 61:217–238

Orlov AM (1998) The diets and feeding habits of some deep-water benthic skates (Rajidae) in the pacific waters off the northern Kuril islands and southeastern Kamchatka. Alaska Fish Res Bull 5:1–17

Orlov AM (2003) Diets, feeding habits, and trophic relations of six deep-benthic skates (Rajidae) in the western Bering Sea. Aqua, J Ich Aquat Biol 7:45–60

Paukert CP, Wittig TA (2002) Applications of multivariate statistical methods in fisheries. Fisheries. 27:16–22

Pauly D, Christensen V, Dalsgaard J, Froese R, Torres Jr F (1998) Fishing down marine food webs. Science. 279:860–863

Pearcy WG, Hancock D (1978) Feeding habits of Dover sole, *Microstomus pacificus*; rex sole, *Glyptocephalus zachirus*; slender sole, *Lyopsetta exilis*; and Pacific sanddab; *Citharichthys sordidus*, in a region of diverse sediments and bathymetry off Oregon. Fish Bull 76:641–651

Pederson SA (1995) Feeding habits of the starry ray (*Raja radiata*) in West Greenland waters. ICES J Mar Sci 52:43–53

Pereyra WT, Pearcy WG, Carvey Jr FE (1969) *Sebastodes flavidus*, a shelf rockfish feeding on mesopelagic fauna, with consideration of the ecological implications. J Fish Res Board Can 26:2211–2215

Peterson CH, Fodrie FJ, Summerson HC, Powers SP (2001) Site-specific and density-dependent extinction of prey by schooling rays: generation of a population sink in top quality habitat for bay scallops. Oecologia 129:349–356

Peterson WT, Schwing FB (2003) A new climate regime in northeast pacific ecosystems. Geophys Res Let 30(17):1896,doi:10.1029/2003GL017528

Pielou EC (1984) The interpretation of ecological data: a primer on classification and ordination. John Wiley and Sons, Inc, New York, 263pp

Pinkas L, Oliphant MS, Iverson ILK (1971) Food habits of albacore, bluefin tuna and bonito in Californian waters. Calif Dept Fish and Game Fish Bull 152, 105 pp

Ressler PH, Brodeur RD, Peterson WT, Pierce SD, Vance PM, Røstad A, Barth JA (2005) The spatial distribution of euphausiid aggregations in the northern California Current during August 2000. Deep-Sea Res II 52:89–108

Rinewalt CS, Ebert DA, Cailliet GM (2007) The feeding habits of the sandpaper skate, *Bathyraja kincaidii* (Garman, 1908) in central California: seasonal variation in diet linked to oceanographic conditions. Environ Biol Fish (this volume)

Robinson CS, Cailliet GM, Ebert DA (2007) Food habits of the longnose skate, *Raja rhina* (Jordan and Gilbert, 1880), in central California waters. Environ Biol Fish (this volume)

Rogers SI, Ellis JR (2000) Changes in the demersal fish assemblages of British coastal waters during the 20th century. ICES J Mar Sci 57:866–881

Ross ST (1986) Resource partitioning in fish assemblages: a review of field studies. Copeia 2:352–388

Rousset J (1987) Feeding of the ray *Raja microocellata* (Montagu, 1818) in Bertheaume Bight (Brittany). Cah Bio Mar, Paris 28:199–206

Scenna LB, García de la Rosa SB, Díaz de Astarloa JM (2006) Trophic ecology of the Patagonian skate, *Bathyraja macloviana*, on the Argentine continental shelf. ICES J Mar Sci 63:867–874

Schmitt WL (1972) The marine decapod crustacea of California. University of California Press, Berkeley, 470 pp

Segura-Zarzosa JC, Abitia-Cárdenas LA, Galván-Magaña F (1997) Observations on the feeding habits of the shark *Heterodontus francisci* Girard 1854 (Chondrichthyes: Heterodontidae), in San Ignacio Lagoon, Baja California Sur, Mexico. Cien Mar 23:111–128

Steven GA (1932) Rays and skates of Devon and Cornwall. II. A study of the fishery; with notes on the occurrence, migrations and habits of the species. J Mar Biol Ass UK 18:1–34

Stevens JD, Bonfil R, Dulvy NK, Walker PA (2000) The effects of fishing on sharks, rays, and chimaeras (chondrichthyans), and the implications for marine ecosystems. ICES J Mar Sci 57:476–494

Summers AP (2000) Stiffening the stingray skeleton – an investigation of durophagy in myliobatid stingrays (Chondrichthyes, Batoidea, Myliobatidae). J Morph 243:113–126

Templeman W (1982) Stomach contents of the thorny skate, *Raja radiata*, from the northwest Atlantic. J Northw Atl Fish Sci 3:123–126

Thrush SF, Pridmore JE, Hewitt JE, Cummings VJ (1994) The importance of predators on a sandflat: interplay between seasonal changes in prey density and predator effects. Mar Ecol Prog Ser 107:211–222

VanBlaricom GR (1982) Experimental analyses of structural regulation in a marine sand community exposed to oceanic swell. Ecol Monogr 52:283–305

Walker PA, Hislop G (1998) Sensitive skates or resilient rays? Spatial and temporal shifts in ray species composition in the central and north-western North Sea between 1930 and the present day. ICES J Mar Sci 55:392–402

Wetherbee BM, Cortés E (2004) Food consumption and feeding habits. In: Carrier JC, Musick JA, Heithaus MR (eds) Biology of sharks and their relatives. CRC Press, New York, pp 225–246

Winemiller KO, Pianka ER (1990) Organization and natural assemblages of desert lizards and tropical fishes. Ecol Monogr 60:27–55

Yang MS, Livingston PA (1988) Food habits and daily ration of Greenland halibut, *Reinhardtius hippoglossoides*, in the eastern Bering Sea. Fish Bull 86:675–690

Yoklavich MM, Greene HG, Cailliet GM, Sullivan DE, Lea RN, Love MS (2000) Habitat associations of deep water rockfishes in a submarine canyon: example of a natural refuge. Fish Bull 98:625–641

Zar JH (1996) Biostatistical analysis. Prentice Hall, Upper Saddle River, NJ, 662 pp

Zaret TM, Rand AS (1971) Competition in tropical stream fishes: support for the competitive exclusion principle. Ecology 52:336–342

Standardized diet compositions and trophic levels of skates (Chondrichthyes: Rajiformes: Rajoidei)

David A. Ebert · Joseph J. Bizzarro

Originally published in the journal Environmental Biology of Fishes, Volume 80, Nos 2–3, 221–237.
DOI 10.1007/s10641-007-9227-4 © Springer Science+Business Media B.V. 2007

Abstract Skates by virtue of their abundance and widespread occurrence appear to play an influential role in the food webs of demersal marine communities. However, few quantitative dietary studies have been conducted on this elasmobranch group. Therefore, to better understand the ecological role of skates, standardized diet compositions and trophic level (TL) values were calculated from quantitative studies, and compared within and among skate and shark taxa. Prey items were grouped into 11 general categories to facilitate standardized diet composition and TL calculations. Trophic level values were calculated for 60 skate species with TL estimates ranging from 3.48 to 4.22 (mean TL = 3.80 ± 0.02 SE). Standardized diet composition results revealed that decapods and fishes were the main prey taxa of most skate species followed by amphipods and polychaetes. Correspondingly, cluster analysis of diet composition data revealed four major trophic guilds, each dominated by one of these prey groups. Fish and decapod guilds were dominant comprising 39 of 48 species analyzed. Analysis of skate families revealed that the Arhynchobatidae and Rajidae had similar TL values of 3.86 and 3.79 (t-test, P = 0.27), respectively. The Anacanthobatidae were represented by a single species, *Cruriraja parcomaculata*, with a TL of 3.53. Statistical comparison of TL values calculated for five genera (*Bathyraja*, *Leucoraja*, *Raja*, *Rajella*, *Rhinoraja*) revealed a significant difference between *Bathyraja* and *Rajella* (t-test, P = 0.03). A positive correlation was observed between TL and total length (L_T) with larger skates (e.g. >100 cm L_T) tending to have a higher calculated TL value (>3.9). Skates were found to occupy TLs similar to those of several co-occurring demersal shark families including the Scyliorhinidae, Squatinidae, and Triakidae. Results from this study support recent assertions that skates utilize similar resources to those of other upper trophic-level marine predators, e.g. seabirds, marine mammals, and sharks. These preliminary findings will hopefully encourage future research into the trophic relationships and ecological impact of these interesting and important demersal predators.

Keywords Skates · Elasmobranchs · Feeding ecology · Prey composition · Trophic guilds

D. A. Ebert (✉) · J. J. Bizzarro
Pacific Shark Research Center, Moss Landing Marine Laboratories, 8272 Moss Landing Road, Moss Landing, CA 95039, USA
e-mail: debert@mlml.calstate.edu

Introduction

Skates (Chondrichthyes: Rajiformes: Rajoidei) are the most diverse group of living cartilaginous fishes, with approximately 245 species recognized worldwide (Ebert and Compagno 2007). This total represents nearly 25% of the described species of chondrichthyan fishes (Compagno 2005). In addition, skates appear to exhibit a fairly high degree of endemism, with many species having somewhat localized regional distributions (McEachran and Miyake 1990; Compagno et al. 1991; Menni and Stehmann 2000; Last and Yearsley 2002). A morphologically conservative group, skates are most commonly found along outer continental shelves and upper slopes. They are the only cartilaginous fish taxon to exhibit a great diversity of species at higher latitudes.

Skates, by virtue of their abundance and species diversity, may play influential roles in the food webs of demersal marine communities. It has been hypothesized that skates may negatively impact commercially valuable groundfishes via ecological interactions such as competition and predation (Murawski 1991; Mayo et al. 1992; Fogarty and Murawski 1998; Link et al. 2002; Orlov 2004). Quantifiable studies, however, on the diet composition and trophic relationships of skates are few (Garrison and Link 2000; Bulman et al. 2001; Davenport and Bax 2002). Additionally, although some skate species are considered top predators (Macpherson and Roel 1987; Ebert et al. 1991; Orlov 1998, 2003; Link et al. 2002) surprisingly few quantitative estimates of trophic level values exist to substantiate this perception (Morato et al. 2003; Braccini and Perez 2005). Furthermore, the impact that skates may have on associated demersal marine communities is still relatively unknown.

If skates are upper trophic level predators, as demonstrated for sharks (Cortés 1999; Estrada et al. 2003), marine mammals (Bowen 1997; Pauly et al. 1998), and seabirds (Sanger 1987; Hobson 1993), they may influence the relative abundance and diversity of co-occurring demersal species (Beddington 1984; Rogers et al. 1999). Therefore, to better understand the ecological role of skates, we present standardized diet compositions and trophic level values calculated from quantitative studies. We then compare results within skate taxa and among skate and shark taxa to investigate similarities and differences in these parameters between elasmobranch groups.

Materials and methods

Diet composition data were summarized from peer-reviewed journal articles, graduate theses, and gray literature (see Appendix I). Only sources that included quantitative information in the form of indices or provided sufficient information to facilitate index calculations were utilized. The following search engines were initially used to locate citations: Aquatic Sciences and Fisheries Abstracts, Biosis, Web of Science, and Zoological Record. After relevant articles were obtained, their literature cited sections were perused for additional references. Colleagues were also contacted to obtain supplemental literature and especially unpublished graduate theses that may have been overlooked during previous searches. All pertinent literature was accumulated and examined, regardless of publication language.

Prey items were grouped into 11 general categories to facilitate standardized diet composition and trophic level calculations (Table 1). Trophic levels for generalized prey categories were directly incorporated when available or estimated by calculating a mean value from information contained in the following sources: Fauchauld and Jumars (1979); Hobson and Welch (1992); Hobson (1993); Pauly and Christensen (1995); Pauly et al. (1998); Cortés (1999); and Nyssen et al. (2002). Because multiple diet studies existed for some skates, an index of standardized diet composition, weighted by relative sample size, was calculated after Cortés (1999) to determine the proportion of each prey category in the diet of a species (P_j):

$$P_j = \sum_{i=1}^{n} (P_{ij} * N_i) / \sum_{j=1}^{11} * \sum_{i=1}^{n} (P_{ij} * N_i),$$

where P_{ij} = the proportion of prey category j from source i, N_i = the number of stomach samples containing food that were used to calculate P_{ij} for source i, n = number of sources, j = total number of prey categories, and $\sum P_j = 1$.

The proportion of each prey category (P_{ij}) was determined for each source using the following hierarchical criteria. Compound indices (e.g., index of relative importance, geometric index of importance, index of absolute importance) were

Table 1 Prey categories used to calculate standardized diet compositions and trophic levels of skates

Group code	Description	Trophic level
INVERT	Other invertebrates and unidentified invertebrates	2.5
POLY	Polychaetes and other marine worms	2.60
MOLL	Molluscs (excluding cephalopods) and unidentified mollusks	2.1
SQUID	Squids	3.2
CEPH	Octopi, cuttlefishes, and unidentified cephalopods	3.2
AMPH	Amphipods and isopods	3.18
EUPH	Euphausiids and mysids	2.25
DECA	Decapod crustaceans	2.52
OCRUST	Other crustaceans and unidentified crustaceans	2.4
CHOND	Chondrichthyan fishes	3.65
FISH	Teleost and agnathan fishes	3.24

used if available. If no compound index was available but more than one single index was used, a geometric index of importance was calculated on a percentage basis by averaging all available indices (e.g., (%N + %W)/2), as described by Assis (1996). Single indices were used when multiple indices were not available. Incidentally ingested materials (e.g., plant matter, sediment, detritus) and undeterminable items (e.g., offal, unidentified organic matter) were not incorporated in analyses. In addition, all non-additive indices, such as percent frequency of occurrence (%FO) or the original index of relative importance (Pinkas et al. 1971), were standardized so that the contribution of each index to diet composition calculations was equal. When only the total number of occurrences of each prey item was available, the occurrence index (O) was calculated after Mohan and Sankaran (1988). When only proportions (%FO) were provided, they were summed and standardized in the same manner.

Trophic levels were then calculated for each species using the following equation, after Cortés (1999):

$$TL_k = 1 + \left(\sum_{j=1}^{n} P_j * TL_j \right),$$

where TL_k = trophic level of species k, P_j = proportion of prey category j in the diet of species k, n = total number of prey categories, and TL_j = trophic level of prey category j. Mean trophic level (TL) values were also calculated for the order, families, and genera using TL estimates from individual species. When data were normally distributed and of equal variances, TL values were compared between and among groupings with t-tests and ANOVAs (SYSTAT, version 10, SPSS, Inc., Chicago, IL). If either of these assumptions were violated, data were log-transformed and re-evaluated prior to analysis. Because it was not possible to generate precision estimates from the available data (Ferry and Cailliet 1996), a minimum sample size of 20 individuals was arbitrarily chosen as the limit for inclusion of a species in all analyses.

Cluster analysis was utilized to compare standardized diet compositions among skate species. Calculations were preformed in SYSTAT (version 10, SPSS, Inc., Chicago, IL) using the unweighted pairwise group mean average method (UPGMA) with Euclidean distance as a measure of dissimilarity. Dissimilarity values among clusters that were ≥50% of the maximum overall dissimilarity distance were considered to indicate major divisions and used to distinguish trophic guilds of skates (Root 1967; Yoklavich et al. 2000).

Results

Standardized diet compositions and trophic levels were calculated for 60 skate species (Table 2). Overall, trophic level estimates ranged from 3.48 (*Rajella caudaspinosa*) to 4.22 (*Dipturus chilensis*). Eleven skate species were estimated to have TL values ≥4, with the genera *Bathyraja* and *Dipturus* each having four species represented. Two species, *Amblyraja radiata* and *Raja clavata*, had the most quantitative diet studies with nine each, followed by *Raja montagui* with six studies, and *Bathyraja aleutica* and *Leucoraja naevus* with five studies each. Three species had a total of four different diet studies, whereas six species and 14 species had three and two studies each, respectively. Thirty-two of the species included in the present study had only a single diet study. The species with the most stomachs examined was *Leucoraja erinacea* (*n* = 19,738), whereas *Rajella dissimilis* (*n* = 2) had the fewest. Six species had >1,000 stomachs examined from various combined studies, 31 species had between 134 and 952 stomachs examined, and 23 species had <100 stomachs examined.

Standardized diet composition results revealed that decapod crustaceans and secondarily fishes were the main prey taxa of most skate species with sufficient sample sizes (Table 2). Overall, decapods comprised 36.35 ± 3.06 SE of mean percent diet composition, were present in the diets of all species examined, and ranged from 0.26% of dietary composition in *Rhinoraja maclovania* to 77.16% for *Rioraja agassizii*. Fishes contributed 28.61 ± 3.31 to the mean diet composition of the skate species studied and comprised >50% of the diet composition of nine species (Table 2). Fishes were of greatest dietary importance to *Dipturus batis* (70.76%) and *D. chilensis* (81.76%), but in contrast were not present in the diets of *Leucoraja melitensis* or *R. maclovania*. Polychaetes and amphipods were of supplemental importance in skate diets, only contributing >50% of the diet composition of *Bathyraja griseocauda* (52.54%), *Rhinoraja albomaculata* (51.53%), *R. maclovania* (89.24%), and *R. taranetzi* (52.71%). Squids (3.36 ± 0.87), euphausids and mysids (6.07 ± 1.55), and other crustaceans (3.68 ± 0.79) were minor prey items. Shelled molluscs, octopi and cuttlefishes, and chondrichthyans were not important prey items for skates, with maximum reported diet composition values of 10.92% (*Leucoraja erinacea*), 14.01% (*Rostroraja alba*), and 5.00% (*Leucoraja fullonica*), respectively. Other invertebrates and unidentified invertebrates were only reported from 35.4% of the species studied and contributed no more than 0.82% to diet composition (*A. radiata*).

Cluster analysis of calculated diet composition estimates for sufficiently sampled species revealed four major trophic guilds with Euclidean distances (ED) > 13.7, or 50% of maximum dissimilarity (Fig. 1). These guilds were dominated by the following prey taxa: polychaetes, amphipods, fishes, and decapods. The polychaete guild (ED = 18.62) included only two species, both in the genus *Rhinoraja*; *R. albomaculata* and *R. macloviana*. The amphipod guild (ED = 16.86) was comprised of seven relatively small species; four of which are found within two genera, *Bathyraja* and *Rhinoraja*. Amphipods were the dominant prey taxon of this guild (41.90 ± 1.66), but decapods were also important (28.74 ± 2.68) and were the main taxon of *Bathyraja violacea* and *L. erinacea*. The fish guild (ED = 15.53), which contained many of the largest skate species within this study, had the greatest number of species at 20; half of which were represented by two genera, *Bathyraja* (*n* = 6) and *Raja* (*n* = 4). All representative species within the genera *Amblyraja* (*n* = 2) and *Dipturus* (*n* = 3) were found within this guild. Decapods were of substantial supplemental importance within the fish guild and were the primary prey taxon of one species (*Raja straeleni*). The decapod guild (ED = 13.9) contained 19 species, with the genus *Raja* represented by eight species. Decapods were dominant in the diets of all species within this guild, with fishes typically of secondary importance. Both the fish and decapod guilds had seven skate genera represented. Although no genera were represented among all four guilds, three genera, *Bathyraja*, *Leucoraja*, and *Rajella*, had representatives among three of the four trophic guilds.

The mean TL for all skate species combined was 3.80 ± 0.02 SE (*n* = 60). A breakdown of the three skate families revealed that the Arhynchobatidae and Rajidae had similar trophic levels of 3.86 and 3.79, respectively (Table 3). The Anacanthobatidae were represented by a single species, *Cruriraja parcomaculata*, with a TL of 3.53. Estimated TL values for the two families (Arhynchobatidae, *n* = 18 and Rajidae, *n* = 29) for which multiple samples were available showed normal distributions (Kolmogorov-Smirnov, $P > 0.05$) and equal variances (*F*-test, $P > 0.05$). Comparison of these two families using a two sample *t*-test revealed no significant difference ($t = 1.11$; $P = 0.27$).

The family Rajidae had eight genera represented, followed by the Arhynchobatidae with four and a single genus for the Anacanthobatidae.

Table 2 Standardized diet composition and trophic levels of skates

Family	Genus and species	TL	n	N	INVERT	POLY	MOLL	SQUID	CEPH	AMPH	EUPH	DECA	OCRUST	CHOND	FISH
Anacanthobatidae	*Cruriraja parcomaculata*	3.53	2	47	0.00	0.88	0.00	0.00	0.82	17.08	53.97	18.11	3.43	0.00	5.70
Arhynchobatidae	*Bathyraja aleutica*	3.96	5	416	0.34	3.44	0.00	11.46	3.04	8.76	1.47	31.39	0.00	0.16	39.94
	Bathyraja brachyurops	4.08	1	198	0.61	4.66	2.31	15.39	0.98	20.27	2.63	6.98	1.37	0.74	44.05
	Bathyraja griseocauda	4.09	1	154	0.37	4.33	2.43	5.23	0.74	52.54	5.18	1.85	1.01	3.00	23.32
	Bathyraja isotrachys	3.71	1	22	0.00	0.00	0.00	22.26	0.00	0.00	0.00	72.40	0.00	0.00	5.33
	Bathyraja kincaidii	3.54	2	134	0.00	18.22	0.00	2.27	0.30	4.83	31.46	37.35	0.10	0.00	5.48
	Bathyraja maculata	4.02	3	577	0.01	4.81	0.15	24.39	3.18	23.59	0.60	21.10	0.01	0.00	22.14
	Bathyraja matsubarai	3.87	2	257	0.00	2.41	0.31	19.69	2.73	0.63	0.12	46.71	0.00	0.00	27.41
	Bathyraja minispinosa	3.75	2	72	0.00	0.00	0.00	4.17	0.52	6.87	0.00	67.90	0.00	0.00	20.55
	Bathyraja parmifera	3.97	4	285	0.13	1.22	0.17	6.12	1.33	4.00	0.16	35.22	0.00	0.12	51.53
	Bathyraja smithii	3.91	1	32	0.00	0.00	0.00	0.00	5.66	0.00	5.60	37.93	0.00	0.00	50.81
	Bathyraja spinicauda	4.02	2	25	0.00	4.63	0.00	0.00	0.00	8.44	8.01	11.64	3.56	1.41	62.31
	Bathyraja trachura	3.78	1	4	0.00	37.02	0.00	6.57	1.61	16.07	10.82	11.82	2.19	0.00	13.90
	Bathyraja violacea	3.81	2	139	0.00	6.75	0.00	3.73	1.01	31.36	0.00	51.04	0.00	0.00	6.11
	Psammobatis extenta	3.84	2	693	0.00	4.24	0.00	0.00	0.00	48.42	0.00	46.61	0.68	0.00	0.06
	Rhinoraja albomaculata	3.85	1	201	0.82	51.53	0.00	5.70	0.00	36.17	1.25	2.38	0.61	0.00	1.54
	Rhinoraja interrupta	3.90	3	433	0.32	14.17	0.00	4.46	0.00	49.02	0.36	29.46	0.22	0.00	1.99
	Rhinoraja macloviana	3.66	2	186	0.01	89.24	0.00	0.00	0.00	10.47	0.00	0.26	0.01	0.00	0.00
	Rhinoraja taranetzi	3.91	1	63	0.20	17.36	0.00	3.67	0.17	52.71	10.06	9.05	2.59	0.00	4.19
	Rioraja agassizi	3.63	2	702	0.00	5.15	0.00	0.00	0.00	2.30	0.00	77.16	1.98	0.00	13.41
Rajidae	*Amblyraja hyperborea*	3.98	2	61	0.00	2.13	0.00	0.00	0.00	0.91	11.34	16.90	1.86	0.00	66.85
	Amblyraja radiata	3.82	9	8,381	0.82	18.78	0.20	0.63	3.21	9.65	4.29	22.55	9.87	0.11	29.89
	Dipturus batis	4.06	1	35	0.00	1.71	0.00	0.00	1.22	3.27	3.29	18.76	0.00	1.00	70.76
	Dipturus chilensis	4.22	1	353	0.02	0.04	0.02	11.55	0.26	5.08	0.00	0.87	0.33	0.07	81.76
	Dipturus laevis	3.52	1	3	0.00	0.00	0.00	0.00	0.00	0.00	0.00	100.00	0.00	0.00	0.00
	Dipturus lineatus	3.85	1	4	6.33	0.00	0.00	0.00	0.00	12.86	0.00	46.34	0.00	0.00	34.47
	Dipturus nidarosiensis	4.01	1	5	0.00	0.00	0.00	0.00	0.00	0.00	18.33	0.00	18.33	26.67	36.67
	Dipturus pullopunctatus	3.84	2	172	0.00	0.00	0.00	0.00	0.85	0.00	1.96	41.29	9.86	0.00	46.04
	Dipturus springeri	4.07	1	5	0.00	0.00	0.00	0.00	0.00	0.00	0.00	0.00	20.00	0.00	80.00
	Leucoraja erinacea	3.70	4	19,738	0.00	13.35	10.92	0.00	0.00	24.16	0.01	39.29	3.40	0.00	8.86
	Leucoraja fullonica	4.06	1	636	0.00	1.20	0.00	0.00	4.37	3.70	0.00	22.15	3.12	5.00	60.46
	Leucoraja garmani	3.54	1	15	0.00	0.72	0.00	0.00	2.69	1.03	3.52	84.59	5.69	0.00	1.76
	Leucoraja melitensis	3.65	1	473	0.00	0.00	0.00	0.00	0.00	23.40	0.00	57.71	18.88	0.00	0.00
	Leucoraja naevus	3.91	5	926	0.13	12.19	0.59	0.00	2.05	8.08	0.67	20.31	10.42	0.12	45.45
	Leucoraja ocellata	4.04	1	595	0.00	8.64	8.94	6.40	0.71	2.24	0.00	4.88	0.41	0.51	67.28
	Leucoraja wallacei	3.68	1	40	0.00	0.00	0.62	0.00	0.62	0.00	1.24	75.75	0.00	0.62	21.16
	Malacoraja senta	3.57	2	307	0.00	2.05	0.00	0.00	0.00	2.15	30.27	35.53	11.56	0.00	18.43
	Malacoraja spinacidermis	3.70	1	3	0.00	0.00	0.00	0.00	0.00	0.00	0.00	75.01	0.00	0.00	24.99

Table 2 continued

Family	Genus and species	TL	n	N	INVERT	POLY	MOLL	SQUID	CEPH	AMPH	EUPH	DECA	OCRUST	CHOND	FISH
	Neoraja stehmanni	3.52	1	3	0.00	0.00	0.00	0.00	0.00	0.00	0.00	100.00	0.00	0.00	0.00
	Raja binoculata	3.84	1	214	0.00	0.00	0.00	0.08	1.12	0.40	0.04	55.83	0.03	0.01	42.49
	Raja brachyura	3.82	4	137	0.00	3.31	1.29	0.00	0.37	16.03	18.67	25.58	0.43	0.00	34.32
	Raja clavata	3.69	9	3,424	0.02	3.37	2.75	0.44	3.01	2.73	1.51	55.37	9.78	0.21	20.82
	Raja eglantaria	3.68	3	651	0.00	0.36	7.81	0.28	0.04	0.93	4.13	57.39	2.09	0.04	26.93
	Raja inornata	3.76	1	264	0.00	0.01	0.40	0.69	0.22	0.02	2.71	62.06	0.20	0.00	33.69
	Raja microocellata	3.88	2	952	0.00	1.73	0.21	0.76	0.00	0.21	1.23	46.39	0.25	0.21	49.00
	Raja miraletus	3.67	3	1,331	0.00	1.09	1.45	0.39	1.41	12.07	3.10	58.93	9.97	0.00	11.59
	Raja montagui	3.59	6	1,533	0.14	10.03	0.45	0.03	0.25	6.05	4.71	65.11	7.07	0.00	6.17
	Raja pulchra	3.77	1	564	0.06	0.00	0.12	4.57	0.76	0.00	0.00	65.07	0.06	0.00	29.36
	Raja radula	3.79	1	801	0.19	2.25	0.66	1.87	6.65	7.40	0.00	54.21	3.84	0.00	22.94
	Raja rhina	3.86	2	567	0.00	0.00	0.05	2.04	4.34	0.00	12.59	34.94	0.05	0.00	46.00
	Raja straeleni	3.74	1	421	0.00	0.55	0.55	0.18	0.37	0.00	1.11	43.29	20.40	0.00	33.53
	Rajella barnardi	3.55	1	80	0.00	3.74	0.00	0.00	1.71	8.09	18.33	56.63	8.09	0.00	3.41
	Rajella bathyphila	3.80	1	4	0.00	16.75	0.00	0.00	0.00	47.27	6.05	4.59	25.34	0.00	0.00
	Rajella bigelowi	3.79	1	11	0.00	32.37	0.00	0.00	0.00	42.11	4.70	6.94	13.88	0.00	0.00
	Rajella caudaspinosa	3.48	1	48	0.00	13.44	0.00	0.00	1.75	0.00	22.91	38.84	19.42	0.00	3.64
	Rajella dissimilis	3.69	1	2	0.00	0.00	0.00	0.00	20.00	0.00	40.00	20.00	0.00	0.00	20.00
	Rajella fyllae	3.78	3	210	0.20	29.67	0.00	0.00	1.49	35.09	5.68	23.87	1.75	0.00	2.24
	Rajella leopardus	3.77	1	137	0.00	1.54	0.00	0.26	0.51	2.03	20.61	31.08	4.05	0.77	39.15
	Rajella ravidula	3.55	1	8	0.08	33.20	0.00	0.00	0.25	0.00	0.00	66.39	0.00	0.00	0.08
	Rostroraja alba	4.15	3	1,168	0.00	0.00	0.00	2.57	14.01	0.00	0.00	9.57	3.82	4.85	65.18

See Table 1 for definitions of prey categories; n = number of studies incorporated for each species, N = total number of stomachs analyzed among studies. Species are listed in alphabetical order within families. A complete list of the original studies is given in Appendix I

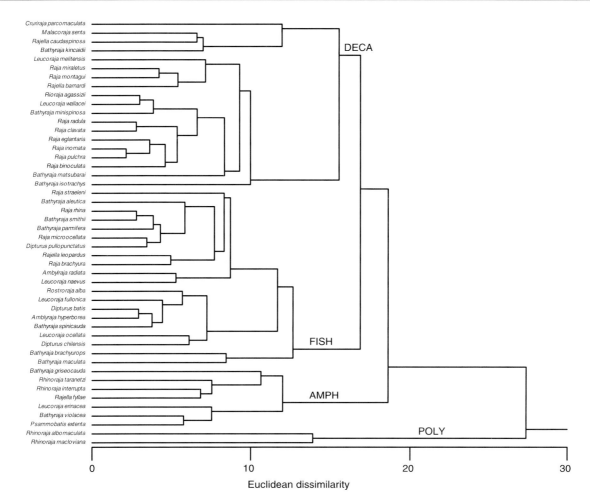

Fig. 1 Cluster analysis of standardized diet compositions of skate species ($n = 48$). Hierarchical clustering was performed using average linkage with Euclidean distance as a measure of dissimilarity. Major trophic guilds are indicated (see Table 2 for description of prey categories)

Of the 13 genera included in this study, five were represented by a single species. The remaining eight genera had at least two or more species studied within the genus. *Bathyraja* ($n = 13$) had the most species studied followed by the *Raja* ($n = 12$). All genera, except for the monotypic *Rostroraja* (TL = 4.15), had a mean TL between 3.5 and 3.9. The genera with the lowest TLs were *Cruriraja* and *Neoraja* at 3.53 and 3.52, respectively.

Statistical comparison of log-transformed TL values calculated for five genera (*Bathyraja, Leucoraja, Raja, Rajella, Rhinoraja*) for which sample size was reasonably high (Table 3) revealed a significant difference among samples (ANOVA, $F = 2.85$; $P = 0.04$). *Bathyraja* ($n = 12$) was found to feed at a higher TL than *Rajella* ($n = 4$) (Tukey's, $P = 0.04$). There were no significant differences found in comparisons between any of the other genera.

A positive relationship was observed between TL and total length (L_T) of skates (Fig. 2). Skates with TL values >3.9 ($n = 15$) were usually the larger species, e.g. those exceeding 100 cm L_T, however, *Bathyraja brachyurops* (73 cm L_{Tmax}) was an exception (TL = 4.08). Of those skates with TLs between 3.8 and 3.9 ($n = 17$) most were <100 cm L_T; three species (*Dipturus pullopunctata, Raja brachyura, R. rhina*) grow to between 120 and 137 cm L_T and one (*Raja binoculata*)

Table 3 Trophic levels of skates by family (in **bold**) and genera

Taxonomic group	n	Mean	UCL	LCL	Max	Min
Anacanthobatidae	**1**	**3.53**				
Cruriraja	1	3.53				
Arhynchobatidae	**19**	**3.86**	**3.93**	**3.78**	**4.09**	**3.54**
Bathyraja	13	3.89	3.98	3.79	4.09	3.54
Psammobatis	1	3.84				
Rhinoraja	4	3.83	4.02	3.65	3.91	3.66
Rioraja	1	3.63				
Rajidae	**40**	**3.79**	**3.84**	**3.73**	**4.22**	**3.48**
Amblyraja	2	3.90	4.92	2.88	3.98	3.82
Dipturus	7	3.94	4.15	3.73	4.22	3.52
Leucoraja	7	3.80	3.99	3.61	4.06	3.54
Malacoraja	2	3.64	4.46	2.81	3.70	3.57
Neoraja	1	3.52				
Raja	12	3.76	3.81	3.70	3.88	3.59
Rajella	8	3.68	3.79	3.57	3.80	3.48
Rostroraja	1	4.15				

n = number of species. Mean, 95% upper (UCL) and lower (LCL) limits, maximum, and minimum values are given

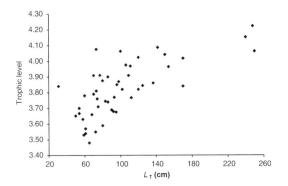

Fig. 2 Relationship between total length (L_T) and trophic level of skates for 48 species representing three families and 12 genera

exceeds 200 cm L_T. At the other extreme, *Psammobatis extenta*, at 31 cm L_{Tmax}, appears to have a much higher TL (3.84) than would be expected for a skate of this size. All skates with TLs <3.8, except for two species (*Amblyraja radiata* and *Raja pulchra*), have a maximum L_T of <100 cm.

Comparison of TL values at the ordinal level (Table 4) between skates and sharks showed that skates (Rajiformes) had lower trophic levels than all but two shark orders; the Heterodontiformes and Orectolobiformes. Log-transformed trophic level values for skate families (Table 4) relative to values for sympatric demersal shark families (Scyliorhinidae, Squalidae, Squatinidae, and Triakidae) revealed significant differences (ANOVA, $F = 10.04$; $P < 0.001$). The combined data for Arhynchobatidae and Rajidae were not significantly different from those of Scyliorhinidae, Squatinidae, and Triakidae (Tukey's, $P < 0.05$), but were significantly lower than the TL value calculated for Squalidae (Tukey's, $P < 0.001$). The Pristiophoridae, a small group of demersal sharks, were represented by a single species (*Pliotrema warreni*), that appeared to feed at a higher TL (4.2) than skates.

Discussion

Of the 245 described skate species (Ebert and Compagno 2007) <24% have had any quantitative dietary information reported, with a single study conducted for most of these species (53.3%). Furthermore, only 37 (14.6%) of 253 skate species have had >100 stomachs examined. In contrast to the relatively small percentage of individual skate species that have been studied, coverage at the generic level has been relatively broad, with quantitative diet information reported for 13 of 27 recognized skate genera (Ebert and Compagno 2007). Although significant differences were found between some skate genera, it appears that phylogeny is a less important predictor of TL than the prey categories used to calculate TL values, size, and possibly regional differences.

Table 4 Trophic levels of skates and sharks by order (in **bold**) and family

Order	Family	n	Mean	UCL	LCL	Max	Min
Rajiformes		**60**	**3.8**	**3.8**	**3.7**	**4.2**	**3.5**
	Anacanthobatidae	1	3.5				
	Arhynchobatidae	19	3.9	3.9	3.8	4.1	3.5
	Rajidae	40	3.8	3.9	3.7	4.2	3.5
Carcharhiniformes		**90**	**4.0**	**4.1**	**3.9**	**4.3**	**3.2**
	Carcharhinidae	39	4.1	4.2	4.1	4.3	3.8
	Hemigaleidae	2	4.2	4.3	4.1	4.3	4.3
	Proscyllidae	2	4.1	4.1	4.0	4.1	4.0
	Pseudotriakidae	1	4.3				
	Scyliorhinidae	21	3.9	4.0	3.8	4.2	3.5
	Sphyrnidae	6	3.9	4.2	3.6	4.3	3.2
	Triakidae	19	3.8	3.9	3.7	4.2	3.5
Lamniformes		**8**	**4.0**	**4.4**	**3.7**	**4.5**	**3.2**
	Alopiidae	2	4.2	4.2	4.2	4.2	4.2
	Cetorhinidae	1	3.2				
	Lamnidae	3	4.3	4.5	4.2	4.5	4.2
	Megachasmidae	1	3.4				
	Odontaspididae	1	4.4				
Orectolobiformes		**6**	**3.6**	**3.9**	**3.4**	**4.1**	**3.1**
	Ginglymostomidae	2	4.0	4.2	3.8	4.1	3.8
	Hemiscyllidae	2	3.6	3.8	3.5	3.7	3.5
	Rhincodontidae	1	3.6				
	Stegostomatidae	1	3.1				
Hexanchiformes		**5**	**4.3**	**4.5**	**4.5**	**4.7**	**4.2**
	Chlamydoselachidae	1	4.2				
	Hexanchidae	4	4.3	4.5	4.5	4.7	4.2
Pristiophoriformes		**1**	**4.2**				
	Pristiophoridae	1	4.2				
Squatiniformes		**6**	**4.1**	**4.2**	**4.0**	**4.2**	**4.0**
	Squatinidae	6	4.1	4.2	4.0	4.2	4.0
Squaliformes		**32**	**4.1**	**4.2**	**4.0**	**4.4**	**3.5**
	Echinorhinidae	1	4.4				
	Squalidae	31	4.1	4.2	4.0	4.3	3.5
Heterodontiformes		**1**	**3.2**				
	Heterodontidae	1	3.2				

Values for sharks taken from Cortés (1999). Mean, 95% upper (UCL) and lower (LCL) limits, maximum, and minimum values are given

Determination of prey categories used to calculate standardize diet compositions may influence the TL of individual species. *Psammobatis extenta*, for example, was found to have a lower TL (3.53) (Braccini and Perez 2005) based on the more general categories established for sharks by Cortés (1999) as compared to our higher estimated TL (3.84) using slightly different prey categories than we determined from a review of published skate diet studies. We attributed this difference mainly to our use of amphipods, a primary prey taxon of this and other skate species, as a distinct prey category. Based on published stable isotope studies (Hobson and Welch 1992; Nyssen et al. 2002), amphipods had a much higher calculated TL value (Table 1) than was estimated using the broad category of marine invertebrates (TL = 2.5; Cortés 1999). Therefore, the actual TL of this species was likely underestimated by Braccini and Perez (2005). Similarly, Morato et al. (2003) collapsed their data for *Raja clavata* into the same categories used by Cortés (1999) and then compared trophic levels among size classes and regions; probably affecting the accuracy of their results. These examples highlight the importance of calculating and perhaps refining prey categories that are more realistic to the predator in question.

Of the four guilds that characterized the diet of skates, fish and decapod guilds were dominant comprising 39 of 48 species. This finding is consistent with other studies that reveal skates to be primarily benthopelagic piscivores or epibenthic predators specializing on invertebrates and small crustaceans (Garrison and Link 2000; Bulman et al. 2001; Davenport and Bax 2002). It is noteworthy that 13 of 20 species in the fish guild had TL values >3.90, and with no species having a TL < 3.77. Fifteen of 20 species in this guild exceed 100 cm L_{Tmax}, with three species >2 m L_{Tmax}. This is not unexpected as larger fish species, including skates, generally have a higher proportion of fish in their diet (Garrison and Link 2000; Davenport and Bax 2002). By comparison, the decapod guild had a TL range of 3.48–3.87 and was composed of smaller species, with 17 of 19 species <100 cm L_{Tmax}.

Total length and TL were positively correlated, with larger skates tending to have a higher calculated TL value. This positive trend between body length and TL was similar to Cortés' (1999) findings for sharks, but differed from Pauly et al.'s (1998) who found that body length and TL were inversely related for marine mammals. Many marine mammals, e.g. baleen whales, are plankton feeders and like the two largest shark species, i.e. *Cetorhinus maximus* and *Rhincodon typus*, tend to feed at a lower TL; usually within a TL range of 3.2–3.4 (Cortés 1999; Pauly et al. 1998). Cortés (1999) suggested that body mass may be a better predictor of TL, but this variable was not available for many species.

Trophic level comparisons among species are best made at identifiable and comparable life history stages or at a similar maximum body size (Jennings 2005). Although ontogenetic shifts in diet are well documented among individual skate species (e.g., Ajayi 1982; Ellis et al. 1996; Muto et al. 2001; Brickle et al. 2003; Robinson et al. 2007; Treloar et al. 2007), only one study has attempted to calculate TL values at different size classes (Morato et al. 2003). Several studies have shown a shift from a diet primarily consisting of amphipods and crustaceans in smaller individuals to a diet primarily consisting of teleosts in larger skates (Smale and Cowley 1992; Yeon et al. 1999; Robinson et al. 2007). Using a fixed TL for an individual species that does not take into account an increase in body size, coupled with a likely shift in diet, will not appropriately describe the structure of aquatic food webs (Jennings 2005). Different size classes within a species may be considered functionally different species in terms of trophic dynamics (Ross 1986; Garrison and Link 2000). Therefore, calculations and comparisons of diet composition for different size classes or life history stages are more appropriate than those using all available data and result in more accurate estimates of TL and determinations of trophic roles within food webs.

The importance of skates to regional ecosystems in terms of abundance and biomass, and their subsequent impact via predation on commercially important groundfish has been the subject of much debate. On Georges Bank off the northeast coast of the U.S., for example, it has been suggested that predation by skates and other elasmobranchs may negatively influence recruitment of potentially important groundfish species (Murawski 1991; Mayo et al. 1992; Fogarty and Murawski 1998). However, there is little evidence to support this hypothesis, and in fact the predatory impact by skates on commercially valuable groundfish on the Georges Bank is considered by some to be insignificant (Link et al. 2002). Conversely, in the Aleutian Islands and Bering Sea skates are considered to be major predators of commercially important groundfish species (Orlov 2003; Gaichas et al. 2005).

Trophic level analysis of allopatric skate species provides little insight into the true role of these fish as predators or competitors within an ecosystem. Conversely, examination of co-occurring skate species will provide a more accurate understanding of these high TL predators. Although skates are secondary or tertiary consumers, their TL appears to vary between, and within, different ecosystems. *Raja clavata*, perhaps one of the best studied skate species (Ajayi 1982; Ellis et al. 1996; Holden and Tucker 1974; Quiniou and Andriamirado 1979), had a significantly higher TL in the Azores than in other regions of the northeastern Atlantic. This difference was attributable to a higher proportion of teleosts in the diet of individuals from the Azores (Morato et al. 2003).

Individual skate species within an ecosystem often occupy different TLs. However, when treated as a species complex, skates may occupy different TLs among different ecosystems. For example, in the eastern Bering Sea, skates appear to occupy somewhat higher mean TLs (3.88, $n = 9$ species) than those in the Benguela Current (3.73, $n = 9$) and California Current (3.75, $n = 4$) ecosystems. Reasons for these differences are unclear, but there are very few other demersal elasmobranch competitors, e.g. sharks, in the eastern Bering Sea (Mecklenberg et al. 2002). Since diets of demersal sharks typically include a greater proportion of fishes than skate diets (Cortés 1999) the higher TL observed for skates in the eastern Bering Sea may be reflective of a lack of other demersal sharks and the associated niche expansion of skates in this region. By comparison, the Benguela Current Ecosystem has one of the most diverse elasmobranch faunas with >60 different species (Compagno et al. 1991). Many of the shark families found within the Benguela Current Ecosystem, e.g. the Squalidae and Scyliorhinidae, have higher TL values, as calculated by Cortés (1999), relative to sympatric skates. Likewise, the California Current Ecosystem has several demersal scyliorhinid and squalid species (Ebert 2003) that, at least at the family level, have higher TL values (Cortés 1999) than the four skate species reported here. Conversely, the eastern Bering Sea has no scyliorhinids and only a single squalid species that is considered rare within that ecosystem (Mecklenberg et al. 2002).

Comparison of the skates to the eight major shark taxonomic groups at the ordinal level is somewhat misleading given the variety of external body morphologies, habitats, life-styles, foraging strategies, and prey items consumed by sharks (Compagno 1990). As an example, skates were found to occupy a higher TL than two of the shark orders (Cortés 1999; Table 4). Both of these orders have representatives that differ morphologically and generally occupy distinctly different habitats than most skate species. The Heterodontiformes are small, stout-bodied sharks with one living family and genus. Unlike skates, heterodontids are primarily durophagous, feeding on such hard-bodied prey items as sea urchins and gastropods (Compagno et al. 2005). The Orectolobiformes contain seven families and 14 genera of mostly small benthic, warm temperate to tropical sharks. This order includes the whale shark *Rhincodon typus*, the largest living fish and a plankton feeder (Compagno et al. 2005). Although the heterodontids somewhat overlap the skates in portions of their distribution, the orectolobids generally do not spatially overlap with skates in terms of habitat (Compagno 1990). It would have been of interest to compare the skates to other batoid groups, but unfortunately no comparable study, to the best of our knowledge, exists.

At the family level, skates were found to occupy TLs similar to those of several co-occurring shark families including the Scyliorhinidae, Squatinidae, and Triakidae (Cortés 1999; Table 4). Of the 23 shark families that Cortés (1999) examined, 14 had TL values ≥4 whereas the remaining 9 had TLs ≤3.9. Six of these nine families occupy distinctly different habitats from that of skates. Three of them (Cetorhinidae, Megachasmidae, Rhincodontidae) are plankton feeders and three (Hemiscyllidae, Sphyrnidae, Stegostomidae) are considered shallow warm temperate to tropical groups (Compagno et al. 2005). Five of these families (Cetorhinidae, Hemiscyllidae, Megachasmidae, Rhincodontidae, Stegostomidae) have TLs <3.6. The similarity of TL between the squatinids and skates is not unexpected as both occupy a similar habitat, co-occur in many areas, and are of similar size. Dietary studies on squatinids have shown them to be consumers of benthic and epibenthic teleosts, cephalopods, and crustaceans (Capape 1975; Ellis et al. 1996; Ebert 2003; Volger et al. 2003). Squatinids, with a mean TL of 4.1, are similar to many of the larger skates (e.g. *Dipturus batis*, *D. chilensis*, *Rostroraja alba*), mainly those with a $L_T > 100$ cm, whose TL is ≥4. Although some larger skate species may occupy similar TLs to that of squatinids, it appears based on limited evidence that these co-occurring species do not overlap in their dietary preferences. In a comparative study of the feeding ecology of six sharks (including members of the families Scyliorhinidae, Squatinidae, and Triakidae) and four skate species, Ellis et al. (1996) found a wide variety of food preferences with little overlap. In fact, only two of the skate species (*Raja clavata* and

R. montagui) in this study exhibited slight dietary overlap with a single shark species, the triakid shark *Mustelus asterias* (Ellis et al. 1996). In diet studies conducted along the west coast of southern Africa it was determined that most skates preferred crustaceans and small benthic teleosts whereas scyliorhinids preferred myctophids (Ebert et al. 1991, 1996). Therefore, though skates may occupy similar habitats, and have similar TLs, to scyliorhinids, squatinids, and triakids, they appear to exhibit very little dietary overlap with these shark families. It is interesting to note that the Squalidae and Pristiophoridae both had significantly higher TLs at 4.1 and 4.2, respectively, than skates of the families Arhynchobatidae and Rajidae; Anacanthobatidae was represented by one species and not statistically tested. Members of both these aforementioned shark families occupy similar habitats to those of many skate species. Numerous diet studies on squalids have shown that members of this family tend to feed mainly on teleosts and cephalopods (Ebert et al. 1992; Ellis et al. 1996; Cortés 1999; Link et al. 2002). On a species level the most frequently studied squalid, *Squalus acanthias*, had a calculated TL value of 3.9 (Cortés 1999), similar to many of the studied skates.

Skates are versatile colonizers of bottom habitats, with an abundant and diverse fauna on continental shelves and insular slopes in cool temperate to boreal and deepsea environs (Compagno 1990). They are among the top predators in demersal marine habitats and appear to play important trophic roles (Ebert et al. 1991; Orlov 2003). Results from this study support recent assertions that skates utilize similar resources to those of other upper trophic-level predators, e.g. seabirds, marine mammals, large teleosts, and some sharks (Sanger 1987; Bowen 1997; Cortés 1999; Garrison 2000; Davenport and Bax 2002). However, skates, unlike these other high TL marine vertebrates, have historically often been overlooked as top predators. This is most evident by the lack of quantitative diet studies on skates. All quantitative diet studies used in our analysis, with one exception, have been published since 1972 (see Appendix I). Reasons for the lack of quantitative studies are several and include a lack of adequate systematic knowledge of the group, a lack of resources to study non-target species or those of little economic value, and because skates often live in habitats (e.g. deepsea benthic habitats) that are difficult to study.

Skates as top predators and potential competitors with groundfishes may play an influential role in structuring demersal marine communities in which they occur. It has been well documented that some batoid groups, e.g. myliobatids and rhinobatids, do in fact play an influential role in shaping infaunal communities on soft-bottom substrates (Lasiak 1982; VanBlaircom 1982; Rossuow 1983; Smith and Merriner 1985; Gray et al. 1997; Ebert and Cowley 2003). However, similar studies on skates are still wanting. It is hoped that these preliminary findings will encourage future research, perhaps incorporating the use of stable isotopes, into the trophic relationships and ecological impact of these interesting and important demersal predators.

Acknowledgments We thank Simon Brown, Pacific Shark Research Center (PSRC), Moss Landing Marine Laboratories (MLML), for his help in summarizing the articles used in this paper and inputting data, and Joan Parker and the library staff at MLML for their invaluable help in obtaining literature. Funding for this research was provided by NOAA/NMFS to the National Shark Research Consortium and PSRC, and in part by the National Sea Grant College Program of the U.S. Department of Commerce's National Oceanic and Atmospheric Administration under NOAA Grant no. NA04OAR4170038, project number R/F-199, through the California Sea Grant College Program and in part by the California State Resources Agency.

Appendix I

List of all skate diet references used for this study.

Ajayi TO (1982) Food and feeding habits of *Raja* species (Batoidei) in Carmarthen Bay, Bristol Channel. J Mar Biol Assoc UK 62: 215–223

Berestovsky EG (1989) Feeding in the Skates, *Raja radiata* and *Raja fyllae*, in the Barents and Norwegian Seas. Voprosy Ikhtiologii 6:994–1002

Bjelland O, Bergstad OA, Skjaeraasen JE, Meland K (2000) Trophic ecology of deepwater fishes associated with the continental slope of the eastern Norwegian Sea. Sarsia 85:101–117

Bowman RE, Stillwell CE, Michaels WL, Grosslein MD (2000) Food of northwest Atlantic fishes and two common species of squid. NOAA Tech Memo NMFS-NE-155

Braccini JM, Perez JE (2005) Feeding habits of the sandskate *Psammobatis extenta* (Garman, 1913): sources of variation in dietary composition. Marine Freshw Res 56:395–403

Brickle P, Laptihovsky V, Pompert J, Bishop A (2003) Ontogenetic changes in the feeding habits and dietary overlap between three abundant rajid species on the Falkland Island's shelf. J Mar Biol Assoc UK 83:1119–1125

Capapé C (1975) Contribution a la biologie des Rajidae des cotes Tunisiennes. 4. *Raja clavata* (Linne 1758): regime alimentaire. Ann Inst Michel Pacha 8:16–32

Capapé C (1976) Contribution a la biologie des Rajidae de cotes tunisiennes. 8. *Raja melitensis* Clark, 1926, Regime alimentaire. Arch Inst Pasteur Tunis 53:39–45

Capapé C (1977) Contribution a la biologie des Rajidae des cotes tunisiennes. 12. *Raja alba* Lacepede, 1803, Regime alimentaire. Arch Inst Pasteur Tunis 54:85–95

Capapé C, Azouz A (1975) Etude du regime alimentaire de deux raies communes dans le golfe de Tunis: *Raja miraletus* Linne, 1758 et *R. radula*, Delaroche, 1809. Arch Inst Pasteur Tunis 52:233–250

Chuchukalo VI, Napazakov VV (2002) Feeding and trophic status of abundant skate species (Rajidae) of the western Bering Sea. Izvestiya TINRO 130:422–428

Dolgov A (2005) Feeding and food consumption by the Barents Sea skates. J Northw Atl Fish Sci 35:496–503

DuBuit MJ (1972) The role of geographical and seasonal factors in the diet of *R. naevus* and *R. fullonica*. Trav Lab Biol Halieutique, Rennes 6:35–50

Ebert DA, Cowley PD, Compagno LJV (1991) A preliminary investigation of the feeding ecology of skates (Batoidea: Rajidae) off the west coast of southern Africa. S Afr J Mar Sci 10:71–81

Ellis JR, Pawson MG, Shackley SE (1996) The comparative feeding ecology of six species of shark and four species of ray (Elasmobranchii) in the North-East Atlantic. J Mar Biol Assoc UK 76:89–106

Ezzat A, Abd El-Aziz SH, El-Charabawy MM, Hussein MD (1987) The food of *Raja miraletus* Linnaeus 1758 in Mediterranean waters off Alexandria. Bull Inst Oceanogr Fish. Cairo 13:59–74

Fitz ES, Daiber FC (1963) An introduction to the biology of *Raja eglanteria* Bosc 1802 and *Raja erinacea* Mitchill 1825 as they occur in Delaware Bay. Bull Bingham Oceanogr Collection 18:69–96

Glubokov AI, Orlov AM (2000) Some morphophysiological indices and feeding peculiarities of the Aleutian skate *Bathyraja aleutica* from the western Bering Sea. Russian Federal Research Institute of Fisheries and Oceanography 1:126–149

Gordon JDM, Duncan JAR (1989) A note on the distribution and diet of deep-water rays (Rajidae) in an area of the Rockall Trough. J Mar Biol Assoc UK 69:655–658

Hacunda JS (1981) Trophic relationships among demersal fishes in a coastal area of the Gulf of Maine. Fish Bull 79:775–788

Holden MJ, Tucker RN (1974) The food of *Raja clavata* Linnaeus 1758, *Raja montagui* Fowler 1910, *Raja naevus* Müller and Henle 1841, and *Raja brachyura* LaFont 1873 in British waters. J Cons Inst Explor Mer 35:189–193

Koen Alonso M, Crespo EA, Garcia NA, Pedraza SN, Mariotti PA, Beron Vera B, Mora NJ (2001) Food habits of *Dipturus chilensis* (Pisces: Rajidae) off Patagonia, Argentia. ICES J Mar Sci 58:288–297

Link JS, Almeida FP (2000) An overview and history of the food web dynamics program of the Northeast Fisheries Science Center, Woods Hole, Massachusetts. NOAA Tech Memo NMFS-NE-159

Lucifora LO, Valero JL, Bremec CS, Lasta ML (2000) Feeding habits and prey selection by the skate *Dipturus chilensis* (Elasmobranchii: Rajidae) from the south-western Atlantic. J Mar Biol Assoc UK 80:953–954

Mabragana E, Giberto DA, Bremec CS (2005) Feeding ecology of *Bathyraja macloviana* (Rajiformes: Arhynchobatidae): a polychaete-feeding skate from the south-west Atlantic. Sci Mar 69:405–413

Monteiro-Marques V, Ré P (1978) Regime alimentaire de quelques Rajidae des cotes Portugaises. Arq Mus Bocage 6:1–8

Morato T, Solà E, Grós MP, Menezes G (2003) Diets of thornback ray (*Raja clavata*) and tope shark (*Galeorhinus galeus*) in the bottom longline fishery of the Azores, northeastern Atlantic. Fish Bull 101:590–602

Muto EY, Soares LSH, Goitein R (2001) Food resource utilization of the skates *Rioraja agassizii* (Mueller and Henle, 1841) and *Psammobatis extenta* (Garman, 1913) on the continental shelf off Ubatuba, south-eastern Brazil. Brazilian J Biol 61:217–238

Olaso I, Sanchez F, Rodriguez-Cabello C, Velasco F (2002) The feeding behaviour of some demersal fish species in response to artificial discarding. Sci Mar 66:301–311

Orlov AM (1998) The diets and feeding habits of some deep-water benthic skates (Rajidae) in the Pacific waters off the northern Kuril Islands and southeastern Kamchatka. Alaska Fish Res Bull 5:1–17

Orlov AM (2003) Diets, feeding habits, and trophic relations of six deep-benthic skates (Rajidae) in the western Bering Sea. J Ichthyol Aquat Biol 7:45–59

Packer DB, Zetlin CA, Vitaliano JJ (2003) Essential Fish Habitat Source Document: Thorny skate, *Amblyraja radiata*, life history and habitat characteristics. NOAA Tech Mem NMFS-ME-178:39

Pederson SA (1995) Feeding habits of the starry ray (*Raja radiata*) in West Greenland waters. ICES J Mar Sci 52:43–53

Quiniou L, Andriamirado GR (1979) Variations of the diet of three species of rays from Douarnenez Bay (*Raja montagui* Fowler, 1910; *Raja brachyura* Lafont, 1873; *Raja clavata* L., 1758). Cybium 3:27–39

Rinewalt CS, Ebert DA, Cailliet GM (2007) The feeding habits of the sandpaper skate, *Bathyraja kincaidii* (Garman, 1908) in central California: seasonal variation in diet linked to oceanographic conditions. Environ Biol Fish (this volume)

Robinson HJ, Cailliet GM, Ebert DA (2007) Food habits of the longnose skate, *Raja rhina* (Jordan and Gilbert, 1880), in central California waters. Environ Biol Fish (this volume)

Rousset J (1987) Feeding of the ray *Raja microocellata* (Montagu, 1818) in Bertheaume Bight (Brittany). Cah Bio Mar, Paris 28:199–206

Serrano A, Velasco F, Olaso I (2003) Macrobenthic crustaceans in the diet of demersal fish in the Bay of Biscay in relation to abundance in the environment. Sarsia 88:36–48

Scenna LB, García de la Rosa SB, Díaz de Astarloa JM (2006) Trophic ecology of the Patagonian skate, *Bathyraja macloviana*, on the Argentine continental shelf. ICES J Mar Sci 63:867–874

Skjaeraasen JE, Bergstad OA (2000) Distribution and feeding ecology of *Raja radiata* in the northeastern North Sea and Skagerrak (Norwegian Deep). ICES J Mar Sci 57:1249–1260

Smale MJ, Cowley PD (1992) The feeding ecology of skates (Batoidea: Rajidae) off the Cape South Coast, South Africa. S Afr J Mar Sci 12:823–834

Soares, LSH, de Moraes Vazzoler AEA, Roberto Correa A (1999) Diel feeding chronology of the skate *Raja agassizii* (Müller and Henle) (Pisces, Elasmobranchii) on the continental shelf off Ubatuba, southeastern Brazil. Rev Bras Zool 16:201–212

Templeman W (1982) Stomach contents of the thorny skate, *Raja radiata*, from the Northwest Atlantic. J Northw Atl Fish Sci 3:123–126

Wakefield WW (1984) Feeding relationships within assemblages of nearshore and mid-continental shelf benthic fishes off Oregon. Unpubl. Masters Thesis, Oregon State University

Yang M-S (2003) Food habits of the important groundfishes in the Aleutian Islands in 1994 and 1997. Alaska Fish Sci Cent, Natl Mar Fish Serv AFSC Progress Rept 2003-07:231

Yeon IJ, Hong SH, Cha HK, Kim ST (1999) Feeding habits of *Raja pulchra* in the Yellow Sea. Bull Nat'l Fish Res Dev Inst Korea 57:1–11

References

Ajayi TO (1982) Food and feeding habits of *Raja* Species (Batoidei) in Carmarthen Bay, Bristol Channel. J Mar Biol Assoc UK 62:215–223

Assis CA (1996) A generalized index for stomach contents analysis in fish. Sci Mar 60:385–389

Beddington JR (1984) The response of multispecies systems to perturbations. In: May RM (ed) Exploitation of marine communities. Springer, Berlin, pp 209–225

Bowen WD (1997) Role of marine mammals in aquatic ecosystems. Mar Ecol Prog Ser 158:267–274

Braccini JM, Perez JE (2005) Feeding habits of the sandskate *Psammobatis extenta* (Garman, 1913): sources of variation in dietary composition. Mar Freshw Res 56:395–403

Brickle P, Laptihovsky V, Pompert J, Bishop A (2003) Ontogenetic changes in the feeding habits and dietary overlap between three abundant rajid species on the Falkland Island's shelf. J Mar Biol Assoc UK 83:1119–1125

Bulman C, Althaus F, He X, Bax NJ, Williams A (2001) Diets and trophic guilds of demersal fishes of the south-eastern Australian shelf. Mar Freshw Res 52:537–548

Capapé C (1975) Contribution a la biologie des Rajidae des cotes tunisiennes. 4. *Raja clavata* (Linne 1758): regime alimentaire. Ann Inst Michel Pacha 8:16–32

Compagno LJV (1990) Alternate life history styles of cartilaginous fishes in time and space. Environ Biol Fish 28:33–75

Compagno LJV (2005) Checklist of living chondrichthyes. In: Hamlett WC (ed) Reproductive biology and phylogeny of Chondrichthyes: sharks, batoids, and chimaeras. Science Publishers, Inc., pp 501–548

Compagno LJV, Ebert DA, Cowley PD (1991) Distribution of offshore demersal cartilaginous fishes (Class Chondrichthyes) off the west coast of southern Africa, with notes on their systematics. S Afr J Mar Sci 10:71–81

Compagno LJV, Dando M, Fowler S (2005) A field guide to the sharks of the world. Harper-Collins, London, 416 pp

Cortés E (1999) Standardized diet compositions and trophic levels in sharks. ICES J Mar Sci 56:707–717

Davenport SR, Bax NJ (2002) A trophic study of a marine ecosystem off southeastern Australia using stable isotopes of carbon and nitrogen. Can J Fish Aquat Sci 59:514–530

Ebert DA (2003) The sharks, rays and chimaeras of California. University California Press, Berkeley, California, USA, 285 pp

Ebert DA, Cowley PD (2003) Diet, feeding behaviour and habitat utilization of the blue stingray *Dasyatis chrysonota* (Smith, 1828) in South African waters. Mar Freshw Res 54:957–965

Ebert DA, Compagno LJV (2007) Biodiversity and systematics of skates (Chondrichthyes: Rajiformes: Rajoidei). Environ Biol Fish (this volume)

Ebert DA, Cowley PD, Compagno LJV (1991) A preliminary investigation of the feeding ecology of skates (Batoidea: Rajidae) off the west coast of southern Africa. S Afr J Mar Sci 10:71–81

Ebert DA, Compagno LJV, Cowley PD (1992) A preliminary investigation of the feeding ecology of squaloid sharks off the west coast of southern Africa. S Afr J Mar Sci 12:601–609

Ebert DA, Cowley PD, Compagno LJV (1996) A preliminary investigation of the feeding ecology of catsharks (Scyliorhinidae) off the west coast of southern Africa. S Afr J Mar Sci 17:233–240

Ellis JR, Pawson MG, Shackley SE (1996) The comparative feeding ecology of six species of shark and four species of ray (Elasmobranchii) in the North-East Atlantic. J Mar Biol Assoc UK 76:89–106

Estrada JA, Rice AN, Lutcavage ME, Skomal GB (2003) Predicting trophic position in sharks of the north-west Atlantic Ocean using stable isotope analysis. J Mar Biol Assoc UK 83:1347–1350

Fauchauld K, Jumars PA (1979) The diet of worms: a study of polychaetes feeding guilds. Oceanogr Mar Biol Annu Rev 17:193–284

Ferry LA, Cailliet GM (1996) Sample size sufficiency and data analysis: are we characterizing and comparing diet properly? In: MacKinlay D, Shearer K (eds) Feeding ecology and nutrition in fish: proceedings of the symposium on the feeding ecology and nutrition in fish. International Congress on the Biology of Fishes, San Francisco, CA, 14–18 July 1996, pp 71–80

Fogarty MJ, Murawski SA (1998) Large-scale disturbance and the structure of marine systems: fishery impacts on Georges Bank. Ecol Appl 8:S6–S22

Gaichas S, Matta B, Stevenson D, Hoff J (2005) Bering Sea and Aleutian Islands skates. In: Stock assessment and fishery evaluation report for the groundfish resources of the Bering Sea/Aleutian Islands regions. North Pacific Fisheries Management Council, Anchorage, AK, Section 16.3:825–857

Garrison LP (2000) Spatial and dietary overlap in the Georges Bank groundfish community. Can J Fish Aquat Sci 57:1679–1691

Garrison LP, Link JS (2000) Fishing effects on spatial distribution and trophic guild structure of the fish

community in the Georges Bank region. ICES J Mar Sci 57:723–730
Gray AE, Mulligan TJ, Hannah RW (1997) Food habits, occurrence, and population structure of the bat ray, *Myliobatis californica*, in Humboldt Bay, California. Environ Biol Fish 49:227–238
Hobson KA (1993) Trophic relationships among high Arctic seabirds: insights from tissue-dependent stable-isotope models. Mar Ecol Prog Ser 95:7–18
Holden MJ, Tucker RN (1974) The food of *Raja clavata* Linnaeus 1758, *Raja montagui* Fowler 1910, *Raja naevus* Müller and Henle 1841, and *Raja brachyura* LaFont 1873 in British waters. J Cons Inst Explor Mer 35:189–193
Hobson KA, Welch HE (1992) Determination of trophic relationships within a high Arctic marine food web using $\delta^{13}C$ and $\delta^{15}N$ analysis. Mar Ecol Prog Ser 84:9–18
Jennings S (2005) Size-based analyses of aquatic food webs. In: Belgrano A, Scharler UM, Dunne J, Ulanowicz RE (eds) Aquatic food webs. Oxford University Press, Oxford, pp 86–97
Lasiak TA (1982) Structural and functional aspects of the surf zone fish community in the Eastern Cape. PhD dissertation, University of Port Elizabeth, Port Elizabeth, South Africa
Last PR, Yearsley GK (2002) Zoogeography and relationships of Australasian skates (Chondrichthyes: Rajidae). J Biogeogr 29:1627–1641
Link JS, Garrison LP, Almeida FP (2002) Ecological interactions between elasmobranchs and groundfish species on the Northeastern U.S. Continental Shelf. I. Evaluating predation. N Am J Fish Manage 22:550–562
Macpherson E, Roel BA (1987) Trophic relationships in the demersal fish community off Namibia. In: Payne AIL, Gulland JA, Brink KH (eds) The Benguela and comparable ecosystems. S Afr J Mar Sci 5:585–596
Mayo RK, Fogarty MJ, Serchuk FM (1992) Aggregate fish biomass and yield on Georges Bank, 1960–87. J Northwest Atl Fish Sci 14:59–78
McEachran JD, Miyake T (1990) Zoogeography and bathymetry of skates (Chondrichthyes, Rajoidei). In: Pratt HL, Gruber SH, Taniuchi T (eds) Elasmobranchs as living resources: advances in the biology, ecology, systematics, and the status of the fisheries. NOAA Tech Rept 90, pp 305–326
Mecklenberg CW, Mecklenberg TA, Thorsteinson LK (2002) Fishes of Alaska. American Fisheries Society, Bethesda, MD, USA, 1037 pp
Menni RC, Stehmann MFW (2000) Distribution, environment, and biology of batoid fishes off Argentina, Uruguay, and Brazil. A review. Rev Mus Argentino Cienc Nat 2:69–109
Mohan MV, Sankaran TM (1988) Two new indices for stomach content analysis of fishes. J Fish Biol 33:289–292
Morato T, Solà E, Grós MP (2003) Diets of thornback ray (*Raja clavata*) and tope shark (*Galeorhinus galeus*) in the bottom longline fishery of the Azores, northeastern Atlantic. Fish Bull 101:590–602

Murawski SA (1991) Can we manage our multispecies fisheries? Fisheries 16:5–13
Muto EY, Soares LSH, Goitein R (2001) Food resource utilization of the skates *Rioraja agassizii* (Müller and Henle, 1841) and *Psammobatis extenta* (Garman, 1913) on the continental shelf off Ubatuba, southeastern Brazil. Brazil J Biol 61:217–238
Nyssen F, Brey T, Lepoint G, Bouquegneau JM, De Broyer C, Dauby P (2002) A stable isotope approach to the eastern Weddell Sea trophic web: focus on benthic amphipods. Polar Biol 25:280–287
Orlov AM (1998) The diets and feeding habits of some deep-water benthic skates (Rajidae) in the Pacific waters off the northern Kuril Islands and southeastern Kamchatka. Alaska Fish Res Bull 5:1–17
Orlov AM (2003) Diets, feeding habits, and trophic relations of six deep-benthic skates (Rajidae) in the western Bering Sea. J Ichthyol Aquat Biol 7:45–59
Orlov AM (2004) Trophic interrelationships in predatory fishes of Pacific waters circumambient the northern Kuril Islands and southeastern Kamchatka. Hydrobiol J 40:106–123
Pauly D, Christensen V (1995) Primary production required to sustain global fisheries. Nature 374:255–257
Pauly D, Trites AW, Capuli E, Christensen V (1998) Diet composition and trophic levels of marine mammals. ICES J Mar Sci 55:467–488
Pinkas LM, Oliphant S, Iverson ILK (1971) Food habits of albacore, bluefin tuna and Bonito in Californian waters. Calif Fish Game 152:1–105
Quiniou L, Andriamirado GR (1979) Variations of the diet of three species of rays from Douarnenez Bay (*Raja montagui* Fowler, 1910; *Raja brachyura* Lafont, 1873; *Raja clavata* L., 1758). Cybium 3:27–39
Robinson HJ, Cailliet GM, Ebert DA (2007) Food habits of the longnose skate, *Raja rhina* (Jordan and Gilbert, 1880), in central California waters. Environ Biol Fish (this volume)
Rogers SI, Clarke KR, Reynolds JD (1999) The taxonomic distinctness of coastal bottom-dwelling fish communities of the North-east Atlantic. J Anim Ecol 68:769–782
Root RB (1967) The niche exploitation pattern of the blue-gray gnatcatcher. Ecol Monogr 37:317–350
Ross ST (1986) Resource partitioning in fish assemblages: a review of field studies. Copeia 2:352–368
Rossuow GJ (1983) The biology of the lesser sandshark. *Rhinobatos annulatus* in Algoa Bay with notes on other elasmobranchs. PhD dissertation, University of Port Elizabeth, Port Elizabeth, South Africa
Sanger GA (1987) Trophic levels and trophic relationships of seabirds in the Gulf of Alaska. In: Croxall JP (ed) Seabirds: feeding ecology and role in marine ecosystems. Cambridge University Press, Cambridge, pp 229–257
Smale MJ, Cowley PD (1992) The feeding ecology of skates (Batoidea: Rajidae) off the Cape South Coast, South Africa. S Afr J Mar Sci 12:823–834
Smith JW, Merriner JV (1985) Food habits and feeding behavior of the cownose ray, *Rhinoptera bonasus*, in Lower Chesapeake Bay. Estuaries 8:305–310

Treloar MA, Stevens JD, Laurenson LJB (2007) Dietary comparisons of seven species of skates (Rajidae) in southeastern Australian waters. Environ Biol Fish (this volume)

VanBlaircom GR (1982) Experimental analyses of structural regulation in a marine sand community exposed to oceanic swell. Ecol Monogr 52:283–305

Volger R, Milessi AC, Quinones RA (2003) Trophic ecology of *Squatina guggenheim* on the continental shelf off Uruguay and Northern Argentina. J Fish Biol 62:1254–1267

Yeon IJ, Hong SH, Cha HK, Kim ST (1999) Feeding habits of *Raja pulchra* in the Yellow Sea. Bull Nat'l Fish Res Dev Inst Korea 57:1–11

Yoklavich MM, Greene HG, Cailliet GM, Sullivan DE, Lea RN, Love MS et al (2000) Habitat associations of deep-water rockfishes in a submarine canyon: an example of a natural refuge. Fish Bull 98:625–641

Normal embryonic development in the clearnose skate, *Raja eglanteria*, with experimental observations on artificial insemination

Carl A. Luer · Cathy J. Walsh · Ashby B. Bodine · Jennifer T. Wyffels

Originally published in the journal Environmental Biology of Fishes, Volume 80, Nos 2–3, 239–255.
DOI 10.1007/s10641-007-9219-4 © Springer Science+Business Media B.V. 2007

Abstract Clearnose skates, *Raja eglanteria*, are a common species of skate found seasonally along the Atlantic coast of North America from Cape Cod to mid-Florida and in the Gulf of Mexico from mid-Florida to eastern Texas. Clearnose skates inhabit the west central coast of Florida during winter months, when Gulf temperatures are conducive to mating and egg-laying (approximately 16°C–22°C). Specimens collected during this time will breed in captivity and, if maintained at 20°C, mated females will store sperm and continue to lay fertile eggs for up to six months. With fertile eggs available from day of laying until hatching, the first complete description of batoid elasmobranch embryonic development with a timetable has been established and is presented here. If eggs are maintained at 20°C, appearance and progression of embryonic events proceed over a reproducible timeline. The timeline has been divided into discrete time periods into which developmental events have been grouped. These include cleavage and expansion of the blastodisc (day 1–day 4), embryonic axis and neural tube formation (day 4–day 7), pharyngeal pouches and gill filament development (day 10–day 28), expansion of fins and transition from external filaments to internal gills (week 4–week 7), and epidermal pigmentation and yolk absorption (week 8–week 12). Fully formed offspring hatch after an incubation period of approximately 12 weeks (85 ± 6 days). To examine possible mechanisms of transport and storage of sperm in the reproductive tract of females following copulation, experiments resulting in the first artificial insemination for any elasmobranch fish are described. These experiments provide evidence for functional roles of the alkaline gland, whose secretions stimulate sperm motility and may enhance migration of sperm to the oviducal glands, and the clasper gland, whose secretions may provide nutrition during storage of viable sperm in the oviducal gland.

Keywords Class Chondrichthyes · Elasmobranch · Oviparity · Sperm viability · Sperm motility · Sperm storage · Alkaline gland · Clasper gland

C. A. Luer (✉) · C. J. Walsh
Mote Marine Laboratory, 1600 Ken Thompson Parkway, Sarasota, FL 34236, USA
e-mail: caluer@mote.org

A. B. Bodine · J. T. Wyffels
Department of Animal and Veterinary Sciences, Clemson University, Clemson, SC 29634, USA

Introduction

The evolutionary success of elasmobranch fishes (subclass Elasmobranchii: sharks, skates, rays, sawfish, and guitarfish) can be attributed in part to their complex reproductive adaptations that are in many ways as advanced as those of birds and mammals. Although copulation and internal fertilization are common to all elasmobranchs, numerous variations in strategies of embryonic nutrition and development exist, ranging from oviparity to viviparity. Oviparity, considered the primitive or ancestral condition in elasmobranchs, is found in only three families of sharks (Heterodontidae, Scyliorhinidae, and Hemiscylliidae). Among the batoids, oviparity is restricted to the skates (Order Rajiformes) where it occurs throughout all extant families.

Classically, life history information related to reproduction (i.e., mating behavior, oviparity versus viviparity, embryonic/fetal development, duration of incubation or gestation) for a particular species of elasmobranch has been compiled through a variety of means, including anecdotal observations in the wild, dissection of randomly collected fresh specimens or opportunistically available museum specimens, inference from closely related species, or more recently, through observing reproductive activity of animals in captivity. The relative ease with which the clearnose skate, *Raja eglanteria*, can be maintained in captivity has resulted in the opportunity to document virtually every aspect of the annual reproductive cycle for this oviparous species.

Raja eglanteria are not reproductively active year-round, with the annual cycle divided into a reproductive period (mating and egg-laying) and a period of reproductive quiescence. Timing of the cycle appears to be related to water temperature rather than time of year, since breeding activity for *R. eglanteria* in the eastern Gulf of Mexico is a winter event (Luer and Gilbert 1985), while breeding activity in the Delaware Bay occurs in the spring (Fitz and Daiber 1963). Mature *R. eglanteria* collected from the eastern Gulf of Mexico during the winter will breed in captivity and have provided the opportunity for copulatory behavior to be described in great detail (Luer and Gilbert 1985). Prior to copulation, a male will approach a female from behind, bite onto the trailing margin of one of her pectoral fins, and hold her firmly in his jaws. No preference for left or right pectoral fins was ever observed. Two sets of dermal spines on the dorsal surface of males, malar spines located lateral to the eyes and spiracles and alar spines distributed in rows over the distal portions of each pectoral fin, assist in holding the female in place. The male then rotates his pelvic region beneath the tail and pelvic fin of the female, flexes one of his claspers medially and inserts it into the female's cloaca where the clasper's distal end expands, anchoring it firmly inside the lower portion of the female's reproductive tract. Unlike sharks, skates do not possess siphon sacs and must rely on a more passive transfer of seminal fluid from male to female. As a result, mating episodes have been observed to last up to four hours.

Following copulation, sperm move up the female's paired uterine horns and, as in their shark relatives, are stored in each of two prominent glands located between the uterine horns and the paired oviducts. These specialized glands have three accepted names, oviducal glands (from their proximity to the oviducts), nidamental glands (from the analogy of egg capsule formation to providing material for a nest), and shell glands (from their role in synthesizing the shell-like egg cases that encapsulate the fertilized ova) (Carrier et al. 2004).

Ova are released from the ovaries in pairs (presumably one from each ovary), enter a common ostium, separate, and each ovum travels down one of the two oviducts to the shell glands, where fertilization and encapsulation of ova take place. Formation of the egg case around the fertilized ova occurs in stages as described in detail by Fitz and Daiber (1963). The freshly encapsulated fertile eggs remain in the paired uteri until oviposition, at which time the female arches her back slightly, contracts the posterior lobes of the pelvic fins ventrally, shakes the pelvic region from side to side, and expels a single egg, often partially buried in the holding tank substrate. The second egg of the pair can be laid from several minutes to several hours later (Luer and Gilbert 1985).

In addition to providing detailed observations of copulatory behavior and oviposition, captive breeding of *R. eglanteria* and subsequent laying of fertile eggs has contributed new information for this species regarding numbers of eggs produced per individual, egg laying intervals, incubation periods, hatching, and body dimensions of freshly hatched offspring (Luer and Gilbert 1985; Luer 1989). The captive setting has also created the unique opportunity to measure chronic fluctuations in serum steroid hormones from individual animals throughout the annual reproductive cycle (Rasmussen et al. 1999). Missing from the literature, however, is a complete, well-defined, timed series of developmental stages for any batoid.

Although embryonic development of cartilaginous fishes has been an object of interest since the time of Aristotle (Aristotle 343 BC) most studies have been fragmentary due to the difficulty in obtaining embryonic material. The first modern attempt to establish a series of developmental stages was the monograph of Balfour (1878), based primarily on the small spotted cat shark, *Scyliorhinus canicula* (formerly *Scyllium*), with attempts to fill gaps using specimens of *Torpedo* and *Galeus* (formerly *Pristiurus*). Scammon (1911) provided a partial set of normal stages for the spiny dogfish, *Squalus acanthias*, with an in-depth analysis of morphogenesis and organ formation based on three-dimensional reconstructions of serial sections. Incomplete descriptions of batoid development have been published for *Torpedo* (Ziegler and Ziegler 1892), *Raja* (Clark 1927) and *Rhinobatos* (Melouk 1949). Development of chimaeroid fishes has been described by Dean (1906) and nearly a century later by Didier and co-workers (1998), neither of which included timetables. The most complete recent account of elasmobranch development is described for the oviparous shark, *S. canicula*, in a set of 34 well-defined, time-based stages (Ballard et al. 1993). With fertile eggs available for *R. eglanteria* from day of laying until hatching, the first complete description of batoid elasmobranch embryonic development with a timetable is presented here.

While embryonic development occurs externally in oviparous species such as *R. eglanteria*, fertilization occurs internally in all elasmobranchs. That elasmobranch females can store sperm following copulation and fertilize ova at some later time was first suggested nearly a century ago when aquarium-maintained female skates continued to lay fertile eggs after being segregated from males (Clark 1922). First described in *S. canicula* (Metten 1939), the shell-secreting oviducal glands are the presumed sites of sperm storage in skates and in those sharks in which sperm storage has been documented (Pratt 1993). Even though oviducal gland morphology and its role in egg case formation are characterized for several elasmobranch species (Hamlett et al. 1998), mechanisms of sperm storage within the female reproductive tract are not well understood (Hamlett et al. 2002). To examine possible mechanisms of transport and storage of sperm, captive populations of *R. eglanteria* provide the opportunity for the controlled introduction of seminal fluid into the female reproductive tract without involving copulation. Such artificial insemination has been applied with tremendous success in farm animals, for which protocols are well established (Perry 1968). Based on artificial insemination techniques routinely applied to the poultry industry (Sexton 1979), experiments resulting in the first artificial insemination for any elasmobranch fish are described here.

Methods

Collection and maintenance of animals

Adult specimens of *R. eglanteria* inhabit near-shore waters off Sarasota, FL during winter months when water temperatures range between approximately 16°C–22°C. As described previously (Luer and Gilbert 1985), animals are collected using gill nets, trot lines, or rod and reel, and transported to Mote Marine Laboratory in live wells. For captive breeding, approximately 6–8 males and 8–12 females have been collected each winter since 1981. Animals are maintained in recirculating natural seawater systems equipped with both biological and particle filtration. Holding tanks range from 425 l for holding individual animals or breeding pairs to 6400 l for holding multiple specimens. Seawater system salinity is maintained between 30‰ and 35‰ by addition of water that has been de-ionized by reverse osmosis. Periodic exchanges with natural seawater are performed to supply trace elements and to keep nitrate accumulation below 20 mg l^{-1}. Water temperature is maintained at 20°C using chillers equipped with titanium heat exchangers. Skates are fed three times per week with a varied diet of thread herring, shrimp, and squid.

Captive skates are observed daily to document copulatory activity and to confirm dates of laying for freshly oviposited eggs. Egg pairs are incubated in partitioned racks in tanks subjected to the same conditions described for maintaining adults. Partitioned racks allow for egg pairs to be separated by their dates of laying so that the relative ages of embryos can be determined at any time during the developmental process (Luer and Gilbert 1985; Luer 1989).

Embryo photography

Photographic images of embryos were collected over a period of 10–15 years from more than 150 specimens using macrophotography, light microscopy, and scanning electron microscopy. Blastodisc and early stage embryos were photographed through a Unitron Model ZSB stereo zoom-dissecting microscope after cutting an opening in one side of the egg case. Macrophotography of embryos from day 17 through hatching was achieved following removal of the embryo and intact yolk from the egg case. These embryos were anesthetized with tricaine methanesulfonate, MS-222 (Crescent Research Chemicals, Phoenix, AZ) dissolved in seawater at 100 mg l^{-1}, or approximately half the dosage used for adult elasmobranchs (Gilbert and Wood 1957), placed in a glass box filled with filtered seawater to eliminate an air/water interface, photographed, and revived.

For scanning electron microscopy, embryos were removed from the yolk and immersed in either 10% neutral buffered formalin in elasmobranch-modified phosphate buffered saline (E-PBS) (Walsh and Luer 2004) or cold elasmobranch-modified Sorenson's buffer (2.5% glutaraldehyde, 3.2% paraformaldehyde in 0.1 M sodium phosphate, pH 7.3, 0.35 sucrose with trace amounts of CaCl$_2$) and fixed overnight at 4°C. Tissues were post-fixed in 2% osmium tetroxide for 2 h at room temperature in elasmobranch-modified Sorenson's buffer, washed twice in distilled water (5 min each) and dehydrated gradually through ascending concentrations of ethanol until completely dehydrated. Embryos were critical point dried in a SPI-DRY™ CPD JUMBO (Structure Probe, Inc, Westchester, PA), sputter coated with gold in a Hummer X (Anatech, Springfield, VA), and examined in a JEOL 848 or Hitachi 3500 electron microscope.

Accession of male-derived fluids/secretions

During the winters of 1994 and 1995, seminal fluid, alkaline gland fluid, and clasper gland secretion for biochemical analyses, artificial insemination and wet mounts for sperm motility studies were obtained from 10 mature male skates. Animals were euthanized with MS-222 dissolved in seawater at ≥250 mg l^{-1} (Beaver et al. 2001). After exposing the abdominal cavity from the ventral surface, seminal fluid and alkaline gland fluid were aspirated directly from the seminal vesicles and alkaline gland, respectively, using a 1 ml syringe and 26 gauge needle. Both seminal fluid and alkaline gland fluid were stored at 4°C until used. Clasper glands were exposed by dissecting claspers along the upper portion of the ventral surface. The clasper gland secretion, discharged from the midventral groove following mild electrical stimulation of the gland using a Model 340 induction stimulator (Harvard Apparatus, Dover, MA), was aspirated using a 20 gauge needle. The clasper gland secretion was diluted with an equal volume of E-PBS prior to storage at 4°C. For histology, clasper glands were fixed in 10% formalin in E-PBS, embedded in paraffin, sectioned at 6 μm, and stained with hematoxylin and eosin.

Wet mounts of fresh seminal fluid as well as seminal fluid combined with equal volumes of seawater, semen extender, clasper gland secretion, and/or alkaline gland fluid were made on alcohol cleaned slides and cover-slipped. Wet mounts were photographed both unstained and stained with 1% methylene blue in E-PBS. Some of the seminal fluid wet mounts were recorded through an Olympus BH2 compound microscope using a Sony 8 mm camcorder and converted to VHS tape. Wet mount video recordings provided the basis for a relative sperm motility scale: minimal movement of tails (+); gentle undulating motion of tails (++); rapid beating of tails (+++).

To visualize mitochondria, sperm preparations were stained with a 5 mg ml^{-1} solution of MTT (1-(4,5-dimethylthiazol-2-yl)-3,5-diphenylformazan) (Mosmann 1983) prepared in E-PBS.

Biochemical analyses

Preliminary biochemical analyses of seminal plasma, clasper gland secretion and alkaline gland fluid included determinations of total protein, amino acids, urea, and fatty acids. Seminal plasma was obtained as the supernatant following centrifugation of seminal fluid at $600 \times g$ for 20 min at 4°C. Total protein was determined in duplicate using the dye-binding method of Bradford (Bio-Rad, Hercules, CA) with bovine IgG as standard. To perform protein assays, seminal plasma and clasper gland secretion were diluted 500-fold with E-PBS, while alkaline gland fluid was undiluted. All samples were filtered through low protein binding 0.45 micron filters to remove any debris.

Amino acid analyses were performed after the particular biological fluid was de-proteinated using 0.01 N perchloric acid. Following removal of the protein precipitate, fluids were neutralized with 0.01 N KOH and the resulting KClO$_4$ removed by filtration. Free amino acids were analyzed by cation exchange using citrate buffer elution and ninhydrin for detection using a modification of the method of Spackman et al. (1958). Norleucine was used as internal standard. Because urea is present in elasmobranchs in such high concentrations, the peak attributed to urea in the samples was verified by running duplicate samples, one treated with urease to show disappearance of the peak, and one spiked with 100 mM urea. The peak position was confirmed by additivity. Fatty acids were analyzed by gas–liquid chromatography using the methanolic HCl derivatization method (Suhija and Palmquist 1988). Fatty acids were separated on a fused silica capillary column and detected by flame ionization. Heptadecanoic acid was used as an internal standard.

Sperm viability assay

Seminal fluid was diluted with an equal volume of Beltsville poultry semen extender (Sexton 1982) modified for elasmobranch osmotic conditions with the addition of 0.54 g NaCl, 0.78 g trimethylamine oxide, and 2.16 g urea per 100 ml. Sperm counts were estimated using a hemacytometer. Sperm viability was assessed using a fluorescent assay modified from a procedure routinely utilized with avian sperm (Bilgili and Renden 1984). Briefly, 50 µl of the diluted semen was added to 100 µl of the assay reagent (12.5 µg ml^{-1} ethidium bromide, EtBr; 7.5 µg ml^{-1} carboxyfluorescein diacetate, CFDA, Molecular Probes, Inc., Eugene, OR) in E-PBS and allowed to incubate in the dark for 30 min at 20°C. Sperm were pelleted via centrifugation at $200 \times g$ for 10 min, and washed once in E-PBS. Following the E-PBS wash, sperm were resuspended in 50 µl fresh assay reagent and viewed with fluorescence microscopy (470 nm excitation, 520 nm emission). Viable sperm possess an esterase that converts CFDA to a fluorescent green analog, while EtBr penetrates dead sperm and intercalates between DNA bases, resulting in a red fluorescence. A small drop of stained sperm was placed onto a clean microscope slide and a coverslip was carefully placed over the sample. Using a fluorescence microscope, five fields totaling at least 300 sperm were counted. The percentages of live sperm (green fluorescence) and dead sperm (red fluorescence) for each field were calculated and the average viability of the three fields reported.

Artificial insemination

During the winter of 1994, three mature females collected early in the winter breeding season and bearing no visible breeding wounds were maintained individually. Once eggs laid by these females were verified to be infertile, these females were used in artificial insemination studies. An approximate volume of 300 µl of seminal fluid was obtained from each of three mature males and diluted with an equal volume of elasmobranch-modified semen extender as described previously. Sperm concentration was determined using a hemacytomer and a total number of approximately $125–140 \times 10^6$ sperm with greater than 95% viability were introduced into females anesthetized with MS-222 dissolved in seawater at a concentration of 200 mg l^{-1} (Gilbert and Wood 1957). The sperm mixture was introduced into each female using a syringe and 28 gauge needle extended with PE 10 polyethylene tubing. The mixture was directed into the right uterine horn of the first female, into the left uterine horn of the second female, and into the common cloacal receptacle of the third female. Inseminated females were revived, and maintained individually. Eggs laid subsequent to this procedure were monitored for fertility.

Results

Embryonic development

A freshly deposited egg case is shown in Fig. 1a. Paired anterior and posterior horns can be seen to extend from corners of the case (Fitz and Daiber 1963). The end of the case with the longer horns (anterior end) is the first portion of the case to be formed and, consequently, is the end that protrudes initially from the cloaca during oviposition. The posterior end possesses string-like tendrils branching from the base of the posterior horns. Attached to these tendrils is a sticky, mucous membrane that assists in anchoring the egg case to its surrounding substrate. Along the lateral edge of each horn are pre-formed grooves, termed respiratory canals, which are not open to the environment at the time of oviposition. Within each egg pair, the posterior horn produced from the lateral side of the oviducal gland has a greater degree of curvature than the posterior horn from the medial side. By placing a freshly laid pair of eggs side-by-side, one can readily determine from which side of the reproductive tract each egg originated.

Cleavage and expansion of the blastodisc: Day 1–Day 4

By carefully removing a portion of the case from a freshly laid egg, the yolk can be examined. On the day the egg is laid, a distinct blastodisc is already clearly visible (Fig. 1b), but since the time between fertilization of the ova in the oviducal glands and deposition of the encapsulated egg can take a few days, the blastodisc is already undergoing early cleavage. With the exact time of fertilization as an unknown, day of laying will be used as the initial reference point for developmental ages in this discussion. Cleavage and thickening of the blastodisc continue during the first 24–48 h after oviposition (Figs. 1c and d), and by 72 h, confluence of cells at the posterior margin of the blastodisc forms an overhang with actively migrating cells creating a characteristic v-shaped notochordal triangle (Fig. 1e). By 96 h (4 days after oviposition), the embryonic axis is apparent with the anterior third of the axis beginning to expand laterally as a prelude to formation of the head enlargement (Fig. 1f).

Embryonic axis and neural tube formation: Day 4–Day 7

During day 4, the head enlargement thickens (Fig. 2a) then flattens into a medullary plate (Fig. 2b), and paired somites are visible for the first time. Prominent neural folds are elevated on either side of a deepening neural groove, which extends from the plate to the posterior lobes. Fusion of the folds to form the neural tube occurs during day 5. The initial closure occurs just posterior to the plate (Fig. 2b), with continued closure of the neural tube progressing along the dorsal midline both anteriorly and posteriorly. The anterior neuropore is visible briefly (Fig. 2c) prior to complete closure of the neural tube at the cranial end (Fig. 2d). Closure of the neural tube is complete by day 6, with the tail bud being the last portion of the tube to close (Fig. 2e). By day 7, cranial curvature, bulging of the optic vesicles, dorsal curvature of the trunk, and rhythmic side-to-side movement of the head are characteristic features at this stage of development (Fig. 2f).

Pharyngeal pouches and gill filament development: Day 10–Day 28

In the day 10 embryo (Fig. 3a), the otic placode has begun to invaginate to form the endolymphatic sac, the lens placode is visible, and the simple, uniform, tubular heart is undergoing

Fig. 1 Day of laying–Day 4. Freshly deposited egg case showing characteristic horns at the four corners, and the sticky, mucous membrane attached to tendrils on the posterior horns (**a**). A window cut into the side of a freshly laid egg case reveals the yolk with a clearly visible blastodisc, B (**b**). During the subsequent 24–48 h, cells on the blastodisc can be seen in various stages of cleavage (**c, d**), by 72 h, confluence of cells at the posterior margin of the blastodisc forms the characteristic v-shaped notochordal triangle (**e**), and by 96 h (4 days after oviposition), the embryonic axis and early indication of the head enlargement are apparent (**f**)

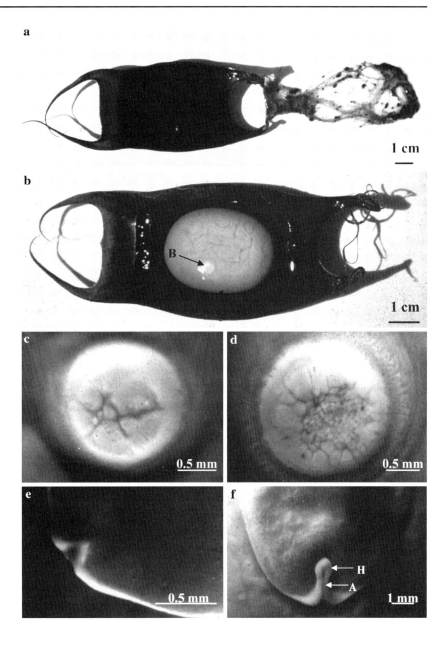

regional expansion. Pharyngeal arches are present, with the second pharyngeal pouch opening first, followed by the first pharyngeal pouch, or spiracle. The remaining pouches (3–6) open in successive order. By 14 days (Fig. 3b), buds of external gill filaments appear in successive order on those arches that have opened beginning with gill arch 2 (Fig. 3c). By day 17, (Figs. 3d and 4a), all gill slits are open and buds of gill filaments appear on all arches except the spiracle. Olfactory placodes have begun to form nasal pits and repositioning of the mandibular arches is reshaping the mouth from a diamond to an oval opening. A gut tube has formed and a definitive yolk stalk is present with blood circulating between the embryo and the network of blood vessels on the yolk surface. Posterior to the gill arches, pectoral fins are starting to develop, initially as a small flap on the trunk just dorsal to the yolk stalk (Fig. 3d).

Until the third to fourth week following oviposition, the yolk is not structurally self-supportive, but is stabilized by a clear albumen-

Fig. 2 Day 4–Day 7. During day 4, the head enlargement thickens (**a**) then flattens into a medullary plate, MP (**b**) and external appearance of somites, S, is evident for the first time. During day 5, closure of the neural tube along the dorsal midline continues both anteriorly and posteriorly, with the anterior neuropore, AN, visible briefly (**c**) prior to complete closure of the neural tube at the cranial end (**d**). Closure of the neural tube is complete by day 6, with the tail bud being the last portion of the tube to close (**e**). By day 7, cranial curvature, bulging of the optic vesicles, OV, and dorsal curvature of the trunk are visible (**f**)

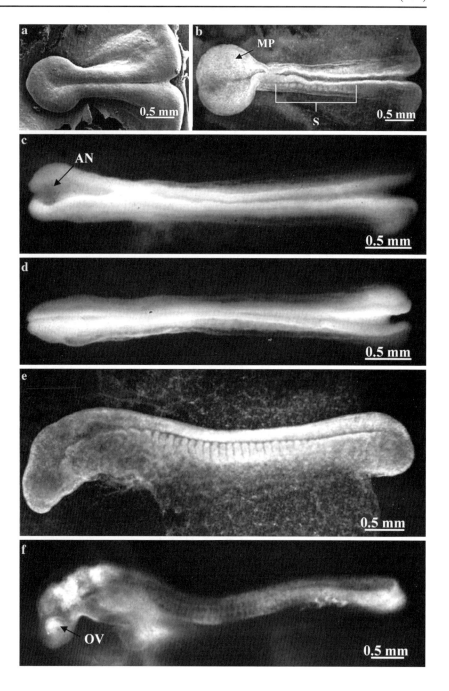

like gel that fills the space in the egg case not occupied by the yolk. This gel also functions to plug the egg case respiratory canals. During the latter stages of the completion of the yolk sac membrane, the gel becomes noticeably thinner, either by gradual dissolving or by enzymatic degradation. Once the gel disappears, the respiratory canals are no longer plugged, allowing seawater to enter freely through either or both ends of the egg case. When eggs are maintained at 20°C, respiratory pores typically are unplugged between 25 days and 28 days after oviposition (Luer and Gilbert 1985). From this time until hatching, the developing embryo facilitates movement of water through the egg case with the constant beating of its tail.

Fig. 3 Day 10–Day 17. In the 10 day embryo, the otic placode has begun to invaginate to form the endolymphatic sac, ES, the lens placode, LP, is visible, chambers of the linear heart, H, are forming, and pharyngeal arches, PA, are present (**a**). By 14 days, buds of external gill filaments appear on the open arches beginning with gill arch 2 (**b, c**), and by day 17, the mouth is being shaped by the mandibular arches and the pectoral fins are beginning to develop on the lateral edges of the trunk just dorsal to the yolk stalk (**d**).

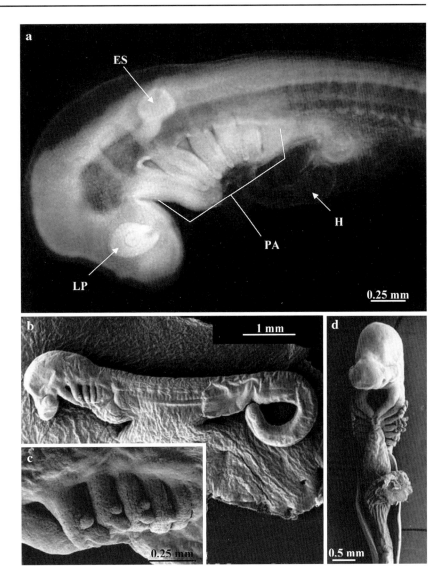

Expansion of fins and transition from external filaments to internal gills: week 4–week 7

During the fourth through the seventh weeks (Figs. 4b–f), pectoral fins will gradually expand in both lateral and dorsoventral directions to form the characteristic batoid pectoral disc, pelvic fins of male embryos show their first signs of differentiation into claspers, and first and second dorsal fins begin to arise midway down the tail. By week 4 (Fig. 4b), the eye is surrounded by a circumferential ring of pigment, the upper and lower jaws of the mouth flatten to form the characteristic straight line morphology, the rostrum appears ventral to the brain and anterior to the mouth, but is shorter than the forebrain. By week six (Fig. 4d), the rostrum extends forward past the forebrain, the pectoral disc is complete and the external gill filaments reach their maximum length. Internally, organs of the alimentary tract are forming and lobes of developing liver can be seen through the translucent abdominal wall. By seven weeks, internal gills are more prominent, external gill filaments begin their process of resorption, mouth and nares are easily distinguishable, size of the yolk mass is greater or equal to the disc width of the embryo, and the earliest indications of dermal pigmentation are visible in isolated areas on the dorsal surface (Figs. 4e, f).

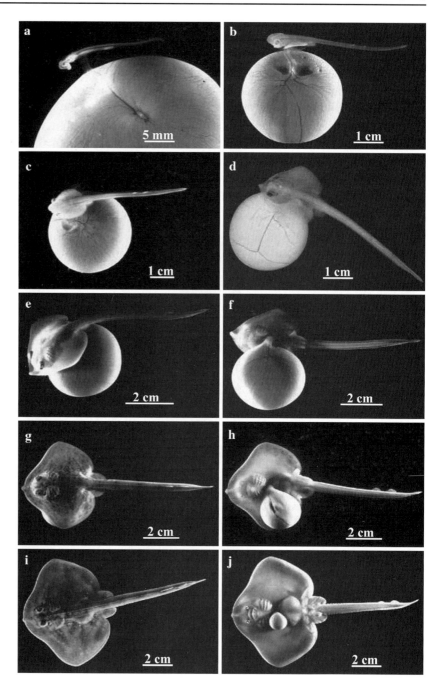

Fig. 4 Day 17–Week 10. Embryos at 17 d (**a**), 4 weeks (**b**), 5 weeks (**c**), 6 weeks (**d**), 7 weeks (**e**, **f**), 9 weeks (**g**, **h**), 10 weeks (**i**, **j**) of development are depicted. The batoid pectoral disc is complete by week 6, claspers begin to differentiate during the fourth to fifth weeks (visible in **f** and **j**), dermal pigmentation is extensive by week 9, and external yolk is nearly gone by week 10, although internal yolk will remain until a few days after hatching

Epidermal pigmentation and yolk absorption: week 8–week 12

During the eighth and ninth weeks, embryos develop their full dorsal pigmentation and external gill filaments have been completely resorbed (Figs. 4g, h). During the last 3 weeks prior to hatching, changes in the embryo's appearance primarily involve growth resulting from nutritional utilization of the external yolk mass. Also during this time, the embryo will have oriented itself so that its rostrum faces the anterior end of the egg case. By 10 weeks, space within the egg case has become so limiting that the pectoral fins fold dorsally over the body and the pelvic girdle bends laterally allowing the tail to reflect nearly

180°. In this configuration, the beating tail more efficiently circulates seawater through the respiratory canals and past the gill region of the embryo. The abdominal cavity bulges with internal yolk that will help to nourish the young skates for a short time after hatching (Figs. 4i, j). When the external yolk is gone, the fully formed offspring will bring the pelvic fin and tail back to their normal positions, providing enough force to poke through the seam between the anterior horns of the egg case, and, after unfolding its pectoral fins, the offspring will swim free of its egg case. If eggs are incubated at a constant temperature of 20°C, offspring will hatch after an average (± SE) of 12 weeks (85 ± 6 days; range, 74–94 days, $n = 67$). Total length at hatching is 13.7 ± 0.5 cm (range, 12.6–14.9 cm, $n = 112$) and disc width is 9.4 ± 0.5 cm (range, 7.5–10.6 cm, $n = 112$) with no statistical difference between frequency of male and female hatchlings (Luer 1989).

Artificial insemination

In an effort to understand the process of sperm movement from male to female during copulation, movement of sperm in the reproductive tract of the female following copulation, and storage of sperm in the oviducal glands, attempts to inseminate female skates artificially were performed. Seminal fluid containing sperm with greater than 95% viability was suspended in an elasmobranch-modified semen extender and introduced into three adult females. These females had been isolated from males since being collected from the wild and had been laying infertile eggs for several weeks. The seminal fluid and semen extender mixture was directed into the common cloacal receptacle of the first female (female A), into the right uterine horn of the second female (female B), and into the left uterine horn of the third female (female C). Introduction of seminal fluid into the common cloacal receptacle simulated the presumed location of seminal fluid discharged during copulation, while introduction into either the right or left uterine horn was performed to determine whether only eggs from the inseminated side of the reproductive tract would produce fertile eggs.

Following the insemination procedure, the first pair of eggs laid by each of the three skates was infertile. During the subsequent 6 weeks, female A laid eight pairs of eggs, all of which were fertile. In the same span, female B laid 13 pairs of eggs. In seven of the 13 pairs, both eggs were fertile, while in the other six pairs, only one egg was fertile. Of these six pairs, the curvature of the posterior horns confirmed that the fertile egg was produced in the right side of the female's reproductive tract. Female C continued to lay only infertile eggs.

Between 6 and 10 weeks after insemination, female A laid five additional pairs of eggs, with only a single egg being fertile. Female B continued to lay six additional egg pairs. Of the six pairs, two pairs were infertile, while four pairs had a single fertile egg. In all four of these pairs, the fertile egg was produced in the right uterine horn. Beyond 10 weeks, females A and B each laid two pairs of eggs, none of which were fertile. Although artificial insemination was successful, experimental females did not lay as many fertile eggs or for the same duration as captive females inseminated naturally via copulation (Luer and Gilbert 1985).

Since artificial insemination bypasses physical involvement of the male, the natural insemination process could include fluids and/or secretions of male origin that might influence the motility of sperm as well as the storage of viable sperm in the oviducal glands. Consequently, fluids and/or secretions that might accompany seminal fluid during copulation, including seminal plasma, alkaline gland fluid, and clasper gland secretion, were examined for their chemical compositions and their effects on sperm viability and motility. Preliminary biochemical analyses of these fluids or secretions are summarized in Table 1. Both seminal plasma and clasper gland secretion were relatively high in protein and free amino acids, while clasper gland secretion was also high in fatty acids. By comparison, alkaline gland fluid contained only trace amounts of protein and fatty acids, and only one detectable amino acid.

The effects of different combinations of these same fluids or secretions on sperm viability and motility during in vitro exposures are compared in Table 2. Sperm viability was high in all prepara-

Table 1 Preliminary biochemical analysis of selected fluids/secretions from male clearnose skates, *Raja eglanteria*

Biochemical parameter	Seminal plasma ($n = 3$)	Clasper gland secretion ($n = 2$)	Alkaline gland fluid ($n = 4$)
Total protein, mg ml^{-1}	150–200	200–300	0–0.1
Total number of free amino acids (detected at ≥0.05 mM)	7	20	1
Fatty acid content, mg ml^{-1}	0.1–0.2	3.5–6.0	0–0.1
Urea concentration, mM	~400	~100	~115

tions. Sperm motility, assessed by relative movement of sperm tails, varied considerably. Sperm in undiluted seminal fluid displayed very little tail movement. Addition of elasmobranch-modified semen extender or 0.45 micron-filtered seawater resulted in sperm whose tails were visibly more active, and maintained a constant but gentle undulating motion. Maximum motility was observed whenever alkaline gland fluid was included in the mixture, inducing such rapid movement that tails were perceived as a blur rather than as individual tails. Addition of clasper gland secretion had no obvious visible effect on sperm motility.

Table 2 Viabilities and relative motilities of skate sperm in fresh seminal fluid combined with equal volumes of seawater, semen extender, clasper gland secretion, and/or alkaline gland fluid

Treatment, $n = 4$	Sperm viability (% Live)	Relative motility
Seminal fluid	85%	+
Seminal fluid + seawater	91%	++
Seminal fluid + semen extender	91%	++
Seminal fluid + clasper gland secretion	90%	+
Seminal fluid + alkaline gland fluid	90%	+++
Seminal fluid + semen extender + clasper gland secretion	88%	++
Seminal fluid + alkaline gland fluid + clasper gland secretion	91%	+++

Relative sperm motility scale: minimal movement of tails (+); gentle undulating motion of tails (++); rapid beating of tails (+++)

Discussion

Embryonic development

Although developmental series are highly desirable for studies in comparative experimental embryology, very few representatives of class Chondrichthyes exist in the literature. One of the major reasons that cartilaginous fishes are underrepresented in the vertebrate developmental series literature is the difficulty in obtaining complete sets of embryological material. Because the clearnose skate, *R. eglanteria*, is seasonally abundant along much of the Atlantic coast of the United States and is easily maintained in captivity, a reasonable supply of embryos at various stages of development is available during the egg-laying season. This paper describes the first complete description of embryonic development for a batoid elasmobranch including a timetable.

As in the staging of *S. canicula* described by Ballard et al. (1993), each stage in the development of *R. eglanteria* from blastodisc to hatching is characterized by the appearance of one or more conspicuous structures or developmental events. Unlike *S. canicula* where advanced cleavages and considerable blastomere division occur while eggs are still in the uterus, *R. eglanteria* eggs are laid while the blastodisc is undergoing early cleavage. With egg pairs being laid at intervals of four to five days (Luer and Gilbert 1985), the time that *R. eglanteria* eggs reside in the uterus between fertilization and oviposition is relatively short, justifying the rationale for designating the day of laying as the initial reference point for developmental ages in the timeline for *R. eglanteria* embryogenesis.

The timeline presented here is based on an incubation temperature of 20°C. Within a tolerance of ±1°C, the appearance, progression, and completion of embryonic events are remarkably reproducible from year to year. The mean incubation time of 85 ± 6 days at 20°C is considerably shorter than that of skates residing in colder environments. For example, estimates of incubation times for *Leucoraja erinacea* (formerly *Raja erinacea*) range from six to nine months for eggs maintained in Connecticut (Richards et al. 1963) to more than one year for *L. erinacea* eggs maintained in Maine (Perkins 1965). *R. eglanteria* developmental rates can be increased or decreased by raising or lowering the incubation temperature, respectively, although eggs fail to develop or die during development at temperatures lower than about 12°C or in excess of 25°C (Luer personal observation).

For the first 4 weeks, or approximately the first trimester of development, skate ontogeny is remarkably similar to that of sharks. Until the pectoral fins begin to expand into the characteristic batoid disc and the tail begins to lengthen, shark and skate development are nearly indistinguishable. Also after about 4 weeks of incubation (25–28 days after oviposition; Luer and Gilbert 1985), the respiratory canals in the egg case become unplugged and seawater freely circulates through the case propelled by the constant beating of the embryo tail. As a result, embryos can be removed from their protective egg case environment any time during the latter two-thirds of development and, provided proper seawater conditions are maintained, embryos will continue to grow and develop normally until the yolk sac is completely resorbed. This provides approximately 8 weeks for direct observation of embryonic growth, with the advantage that experiments and manipulations can be performed without sterile environments or sophisticated experimental chambers. Studies on the development of the ocular lens and cornea, for example, have demonstrated the value of skate embryos as models for vertebrate sensory system ontogeny (Sivak and Luer 1991; Conrad et al. 1994). Also, since eggs are laid in pairs, one embryo can be maintained as a control, while its same-aged sibling can be manipulated experimentally (Pardue et al. 1995).

Artificial insemination

The artificial insemination studies reported here are the first known attempts to achieve fertilization in any elasmobranch without copulation as a prerequisite behavior. Since captive breeding of elasmobranch fishes is limited to a small number of species (Uchida et al. 1990; Pratt and Carrier 2001; Carrier et al. 2004), methods for successful artificial insemination of females actively ovulating or producing mature ova would provide the opportunity to produce offspring from captively maintained specimens. Experimental procedures were based on success in the poultry industry, where semen is accompanied by a semen extender solution as a way to prolong sperm viability. Attempts to inseminate *R. eglanteria* utilized a semen extender developed for turkeys (Sexton 1982) and modified to approximate elasmobranch osmotic conditions.

Insemination was successful in two of the three females used in the experimental procedure. In the successfully inseminated females (females A and B), the first pair of eggs laid after the insemination was not fertile, but could have been in the process of encapsulation at the time of the procedure and not available for fertilization. In female A, eight pairs of fertile eggs resulted from depositing seminal fluid into the common cloacal receptacle, providing evidence that viable sperm were able to move up both uterine horns to the oviducal glands. In female B, the egg produced from the side in which seminal fluid was introduced was fertile in 17 pairs of eggs, although in seven of those pairs, the egg produced from the opposite side of the reproductive tract was also fertile. While this observation is with only one animal and certainly not conclusive, the possibility exists that sperm might be able to move between oviducal glands.

Microscopic examination of seminal fluid reveals that *R. eglanteria* sperm are extremely elongated, measuring approximately 250 μm in total length. Individual sperm consist of a head containing the nucleus and acrosome, a midpiece containing mitochondria, and a tail or flagellum that consumes nearly two-thirds of the total length of the sperm (Figs. 5a, b). The exceptional length and extended midpiece packed with active mitochondria are features that are consistent with *R. eglanteria* sperm being highly motile (Cardullo and Baltz 1991). Examination of seminal fluid aspirated from male seminal vesicles reveals that skate sperm also form aggregates (Figs. 5c, d) with heads clustered in the center and free clumps of tails on the outside (Pratt and Tanaka 1994). Video recording of seminal fluid combined with semen extender confirms that sperm tails exhibit

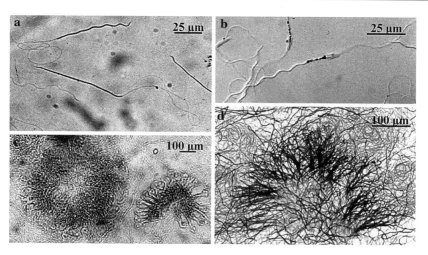

Fig. 5 Skate sperm are extremely elongated (**a**) and consist of a head containing the nucleus and acrosome, a midpiece containing mitochondria (**b**), and a tail or flagellum that consumes nearly two-thirds of the total length of the sperm. Sperm are observed to form aggregates with heads clustered in the center and free clumps of tails on the outside (**c, d**). Stained with methylene blue (**a, d**), MTT (**b**), or unstained wet mount (**c**)

a rhythmic beating motion that is more active than in the absence of semen extender, and that tails are active in both individual sperm and as part of the aggregate.

Although artificial insemination was successful, the combination of seminal fluid and semen extender did not result in as many fertile eggs (30–35 pairs per individual) or for the same duration (up to 6 months) as captive females inseminated naturally via copulation (Luer and Gilbert 1985). Seminal plasma contains protein and free amino acids, taurine being the most abundant, but only trace amounts of fatty acids that might be used as a nutritional source for sperm longevity (Table 1). While it is possible that the quantity of sperm utilized in the initial insemination was insufficient to sustain continued fertilization, it is also possible that the prolonged storage of sperm typical of female skates requires male-derived fluids or secretions that potentially accompany sperm into the female during copulation. Based on gross anatomy, male tissues that could logically contribute fluids or secretions during copulation are the alkaline gland and clasper gland.

The alkaline gland is a paired organ located on the ventral posterior surface of each kidney in male skates and some rays (Maren et al. 1963; Maren 1967; Grabowski et al. 1999). Its name is derived from its fluid, which maintains a pH of approximately 9.2. In skates, the fluid is clear and is composed of trace amounts of protein and fatty acids, with valine being the only free amino acid detected in amounts greater than 0.05 mM (Table 1). Earlier studies established that the electrolyte content of alkaline gland fluid is relatively high, with anions OH^-, Cl^-, and SO_4^{-2} actively concentrated against the electrochemical gradient and Na^+ and K^+ concentrations about twice those of plasma (Maren 1963). Conversely, the urea content is lower than in other elasmobranch fluids, resulting in alkaline gland fluid being isosmotic with plasma (Maren 1967). Relatively low levels of urea were also found in alkaline gland fluid from *R. eglanteria* (Table 1). While no function for this gland has been clearly established, speculation has been that alkaline gland fluid serves to protect sperm from contact with acidic urine, since the alkaline gland opens into the same chamber in which both seminal fluid and urine empty (Maren et al. 1963). Based on video documentation of seminal fluid wet mounts, however, the dramatic increase in sperm tail activity caused by addition of alkaline gland fluid strongly supports its role in enhancing sperm motility (Table 2).

Clasper glands are located on the ventral side of each clasper near their proximal end (Leigh-Sharpe 1920). Exposure of the gland through the muscular membrane of the clasper gland sac

Biology of Skates

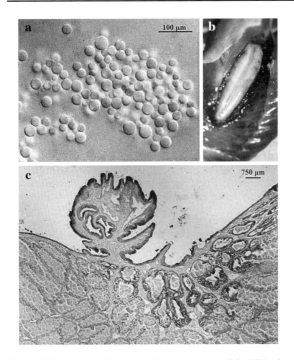

Fig. 6 Wet mount of clasper gland secretion (a). With the skin and muscle layer removed from its ventral surface, the clasper gland can be viewed as an elongated tissue with a longitudinal groove extending the entire length of the gland (b). A histological cross section of the clasper gland reveals that it is composed of a series of tubules, each surrounded by a simple columnar epithelium (c). The tubules release their viscous secretion through a row of papillae lining the middle of the clasper groove

reveals a longitudinal groove into which the clasper gland secretion is discharged (Figs. 6a, b). The gland is composed of a series of tubules, each surrounded by a simple columnar epithelium (LaMarca 1964), that release their contents through a row of papillae lining the middle of the clasper groove (Fig. 6c). The clasper gland secretion is rich in protein, free amino acids, and fatty acids (Table 1). Of the 20 amino acids detected, the most abundant is taurine, an amino acid whose metabolic roles include bile acid conjugation, detoxification, membrane stabilization, osmoregulation, and cell volume regulation (Huxtable 1992). Speculation of functions for clasper gland secretion include clasper lubricant, sealant for the clasper groove, and a protective medium for sperm suspension and transport (Leigh-Sharpe 1920; LaMarca 1964).

Because of its appearance on freshly prepared wet mounts (Fig. 6a), its relatively high protein and fatty acid content (Table 1), and its relative inability to affect sperm motility (Table 2), clasper gland secretion may serve as a nutritive substance to enhance the storage of viable sperm in the oviducal gland. In support of this speculation, the fatty acid profile for clasper gland secretion contains several of the same fatty acids and in similar proportions as mammalian and elasmobranch secretions known to have nutritive functions (Table 3). Specifically, palmitic, stearic,

Table 3 Percent composition of fatty acids from mammalian milk (domestic cow, *Bos taurus*, and human, *Homo sapiens*, Jensen 1995), histotroph (butterfly ray, *Gymnura micrura*, Wourms and Bodine 1983 and cownose ray, *Rhinoptera bonasus*, Luer et al. 1994) and clasper gland secretion (clearnose skate, *Raja eglanteria*)

Common Name	Chain length	Mammalian milk		Histotroph		Clasper gland secretion, n = 2
		B. taurus	H. sapiens	G. micrura	R. bonasus	R. eglanteria
Short-chain	C_{4-10}	9	2	19		
Lauric acid	12:0	4	6			
Myristic acid	14:0	11	7	15		3
Palmitic acid	16:0	26	27	26	27	23
Palmitoleic acid	16:1	3	4	8	4	3
Stearic acid	18:0	11	10	7	10	17
Oleic acid	cis 18:1	28	35	17	9	12
Elaidic acid	trans 18:1				2	5
Linoleic Acid	18:2	2	7	7	1	2
Linolenic acid	18:3	2	1			
Arachidonic acid	20:4			2	8	3
Other long-chain	C_{20-24}				30	21
Other fatty acids		2			9	11
		98	99	101	100	100

and oleic acids, that make up 65–72% of the fatty acids in cow and human milk, respectively (Jensen 1995), are also the predominant fatty acids in the nutritive uterine milk (histotroph) of *G. micrura* and *R. bonasus* where they comprise 50% and 46% of their respective secretions (Wourms and Bodine 1983; Luer et al. 1994). By comparison, the same fatty acids make up 52% of *R. eglanteria* clasper gland secretion.

Future attempts to improve the experimental insemination procedures described here will address issues such as quantity of sperm introduced per insemination, value of repeat inseminations, inclusion of alkaline gland fluid and/or clasper gland secretion in the mixture of seminal fluid and semen extender, and development of methods to cryopreserve *R. eglanteria* sperm. Although achieved with limited numbers, the successful insemination of *R. eglanteria* is encouraging and should pave the way for biologists to attempt insemination with other captive elasmobranchs.

Acknowledgements The authors gratefully acknowledge the use of facilities at Mote Marine Laboratory, Sarasota, FL and the Department of Animal and Veterinary Science, Clemson University, Clemson, SC. The authors appreciate the photographic skills of Jose Castro and Patricia Blum for their contributions to some of the figures, and thank Evanne Thies and Rebecca Rodgers for their assistance with the biochemical analyses.

References

Aristotle (343 BC) Thompson translation (1949) Historia animalium. In: The works of Aristotle, vol 4. Clarendon Press, Oxford, pp 485–633

Balfour FM (1878) A monograph on the development of elasmobranch fishes. Macmillan, London

Ballard WW, Mellinger J, Lechenault H (1993) A series of normal stages for development of *Scyliorhinus canicula*, the lesser spotted dogfish (Chondrichthyes: Scyliorhinidae). J Exp Zool 267:318–336

Beaver BV, Reed W, Leary S et al (2001) 2000 Report of the American Veterinary Medical Association Panel on Euthanasia. J Am Vet Med Assoc 218:669–696

Bilgili SF, Renden JA (1984) Fluorometric determination of avian sperm viability and concentration. Poult Sci 63:2275–2277

Cardullo RA, Baltz JM (1991). Metabolic regulation in mammalian sperm: mitochondrial volume determines sperm length and flagellar beat frequency. Cell Motil Cytoskeleton 19:180–188

Carrier JC, Pratt HL, Castro JI (2004) Reproductive biology of elasmobranchs. In: Carrier JC, Musick JA, Heithaus MR (eds) Biology of sharks and their relatives. CRC Press, Boca Raton

Clark RS (1922) Rays and skates (Raiae). No. 1. Eggcapsules and young. J Mar Biol Assoc UK 12(4):577–643

Clark RS (1927) Rays and skates. No. 2. Description of embryos. J Mar Biol Assoc UK 14(3):661–683

Conrad GW, Paulsen AQ, Luer CA (1994) Embryonic development of the cornea in the eye of the clearnose skate, *Raja eglanteria*. I. Stromal development in the absence of an endothelium. J Exp Zool 269:263–276

Dean B (1906) Chimaeroid fishes and their development. Publication number 32. Carnegie Institution of Washington, Wilkens-Sheiry Printing Co, Washington, DC

Didier DA, Leclair EE, Vanbuskirk DR (1998) Embryonic staging and external features of development of the chimaeroid fish, *Callorhinchus milii* (Holocephali, Callorhinchidae). J Morph 236:25–47

Fitz ES Jr, Daiber FC (1963) An Introduction to the biology of *Raja eglanteria* Bosc 1802 and *Raja erinacea* Mitchell 1825 as they occur in Delaware Bay. Bull Bing Oceanogr Coll 18:69–96

Gilbert PW, Wood FG Jr (1957) Method of anesthetizing large sharks and rays safely and rapidly. Science 126:212–213

Grabowski GM, Blackburn JG, Lacy ER (1999) Morphology and epithelial ion transport of the alkaline gland in the Atlantic stingray (*Dasyatis sabina*). Biol Bull 197:82–93

Hamlett WC, Knight DP, Koob T, Jezior M, Luong T, Rozycki T, Brunette N, Hysell MK (1998) Survey of oviducal gland structure and function in elasmobranchs. J Exp Zool 282:399–420

Hamlett WC, Musick JA, Hysell CK, Sever DM (2002) Uterine epithelial-sperm interaction, endometrial cycle and sperm storage in the terminal zone of the oviducal gland in the placental smoothhound, *Mustelus canis*. J Exp Zool 292:129–144

Huxtable RJ (1992) Physiological actions of taurine. Physiol Rev 72:101–163

Jensen RG (ed) (1995) Handbook of milk composition. Academic Press, San Diego

LaMarca JJ (1964) The functional anatomy of the clasper and clasper gland of the yellow stingray, *Urolophus jamaicensis* (Cuvier). J Morph 114:303–323

Leigh-Sharpe WH (1920) The comparative morphology of the secondary sexual characters of elasmobranch fishes. Mem I J Morph 34:245–265

Luer CA (1989) Elasmobranchs (sharks, skates and rays) as animal models for biomedical research. In:Woodhead AD (ed) Nonmammalian animal models for biomedical research. CRC Press, Inc., Boca Raton, FL

Luer CA, Gilbert PW (1985) Mating behavior, egg deposition, incubation period, and hatching in the clearnose skate, *Raja eglanteria*. Env Biol Fish 13(3):161–171

Luer CA, Walsh CJ, Bodine AB, Rodgers RS, Wyffels J (1994) Preliminary biochemical analysis of histotroph secretions from the cownose ray *Rhinoptera bonasus* and the Atlantic stingray *Dasyatis sabina*, with observations on the uterine villi from *R. bonasus*. Paper presented at the 10th Annual Meeting of the American Elasmobranch Society, University of Southern California, Los Angeles, 2–8 June 1994

Maren TH (1967) Special body fluids of the elasmobranchs. In: Gilbert PW, Mathewson RF, Rall DP (eds) Sharks, skates, and rays. The Johns Hopkins Press, Baltimore

Maren TH, Rawls JA, Burger JW, Myers AC (1963) The alkaline (Marshall's) gland of the skate. Comp Biochem Physiol 10:1–16

Melouk MA (1949) The external features in the development of the Rhinobatidae. Pub Mar Biol Sta Ghardaqa 7:3–98

Metten H (1939) Studies on the reproduction of the dogfish. Phil Trans Roy Soc London 230:217–238

Mosmann TJ (1983) Rapid colorimetric assay for cellular growth: application to proliferation and cytotoxicity assays. J Immunol Meth 65:55–63

Pardue MT, Luer CA, Callendar MG, Chou BR, Sivak JG (1995) The absence of a photopic influence on the refractive development of the embryonic eye of the clearnose skate, *Raja eglanteria*. Vision Res 35:1675–1678

Perkins FE (1965) Incubation of fall-spawned eggs of the little skate, *Raja erinacea* (Mitchell). Copeia 1965:114–115

Perry EJ (ed) (1968) The artificial insemination of farm animals. 4th edn., Rutgers University Press, New Brunswick, NJ

Pratt HL Jr (1993) The storage of spermatozoa in the oviducal glands of western North Atlantic sharks. Env Biol Fish 38:139–149

Pratt HL Jr, Tanaka S (1994) Sperm storage in male elasmobranchs: A description and survey. J Morph 219:297–308

Pratt HL Jr, Carrier JC (2001) A review of elasmobranch reproductive behavior with a case study on the nurse shark, *Ginglymostoma cirratum*. Env Biol Fish 60:157–188

Rasmussen LEL, Hess DL, Luer CA (1999). Alterations in serum steroid concentrations in the clearnose skate, *Raja eglanteria*: correlations with season and reproductive status. J Exp Zool 284:575–585

Richards SW, Merriman D, Calhoun LH (1963) Studies on the marine resources of southern New England. IX. The biology of the little skate, *Raja erinacea* Mitchell. Bull Bing Oceanogr Coll 18:5–67

Scammon RE (1911) Normal plates of the development of *Squalus acanthias*. In: Deibel F (ed) Normentafeln Entwicklungsgeschichte der Wirbeltiere. Gustav Fischer, Jena

Sexton TJ (1979) Preservation of poultry semen—a review. In: Hawk HW (ed) Animal reproduction, Beltsville Symposia in agricultural research, no. 3. Allanheld, Osmun & Co., Montclair NJ, p 159

Sexton T (1982) Beltsville poultry semen extender. 6. Holding turkey semen for six hours at 15°C. Poult Sci 61:1202–1208

Sivak JG, Luer CA (1991) Optical development of the ocular lens of an elasmobranch, *Raja eglanteria*. Vision Res 31:373-382

Spackman DH, Stein WH, Moore S (1958) Automatic recording apparatus for use in the chromatography of amino acids. Anal Biochem 30:1190–1206

Sukhija PS, Palmquist DL (1988) Rapid method for determination of total fatty acid content and composition of feedstuffs and feces. J Agric Food Chem 36:1202–1206

Uchida S, Toda M, Kamei Y (1990) Reproduction of elasmobranchs in captivity. In: Pratt HL, Gruber SH, Taniuchi T (eds) Elasmobranchs as living resources: Advances in the biology, ecology, systematics, and the status of the fisheries. NOAA Tech. Rep. NMFS 90:211–237

Walsh CJ, Luer CA (2004) Elasmobranch hematology: Identification of cell types and practical applications. In: Smith M, Warmolts D, Thoney D, Hueter R (eds) Elasmobranch husbandry manual: proceedings of the first international Elasmobranch husbandry symposium, 2001, Special publication of the Ohio Biological Survey, Number 16, Columbus, OH

Wourms JP, Bodine AB (1983) Biochemical analysis and cellular origin of uterine histotrophe during early gestation of the viviparous butterfly ray. Amer Zool 23(4):1018

Ziegler HE, Ziegler F (1892) Beitrage zur Entwicklungsgeschichte von *Torpedo*. Arch Mikr Anat 39:56–102

Endocrinological investigation into the reproductive cycles of two sympatric skate species, *Malacoraja senta* and *Amblyraja radiata*, in the western Gulf of Maine

Jeff Kneebone · Darren E. Ferguson · James A. Sulikowski · Paul C. W. Tsang

Originally published in the journal Environmental Biology of Fishes, Volume 80, Nos 2–3, 257–265.
DOI 10.1007/s10641-007-9215-8 © Springer Science+Business Media B.V. 2007

Abstract The smooth skate, *Malacoraja senta*, and thorny skate, *Amblyraja radiata*, are two commercially exploited batoids found within the Gulf of Maine. During the past five years, we conducted a large study to accurately describe important biological life history parameters previously lacking for these species. As part of that project, the current study reports our findings on the hormonal profiles associated with the reproductive cycles of *M. senta* and *A. radiata*. Blood samples were obtained from mature *M. senta* and *A. radiata* of both sexes from all months of the year, and plasma testosterone (T), estradiol (E_2) and progesterone (P_4) concentrations were determined using radioimmunoassay (RIA). In female *M. senta* and *A. radiata*, monthly T concentrations ranged from 4,522 pg ml^{-1} to 1,373 pg ml^{-1} and 31,940 pg ml^{-1} to 14,428 pg ml^{-1}, E_2 concentrations from 831 pg ml^{-1} to 60 pg ml^{-1} and 8,515 pg ml^{-1} to 2,902 pg ml^{-1}, and P_4 concentrations from 3,027 pg ml^{-1} to 20 pg ml^{-1} and 3,264 pg ml^{-1} to 331 pg ml^{-1}, respectively. No statistical differences were detected between any months for any hormone. Estradiol concentrations were not correlated with ovary weight, shell gland weight, or diameter of the largest follicles in either species. Monthly T concentrations in male *M. senta* and *A. radiata* ranged from 23,146 to 12,660 pg ml^{-1} and from 57,500 pg ml^{-1} to 24,737 pg ml^{-1}, while E_2 concentrations ranged from 7.5 pg ml^{-1} to undetectable and 103 to 30 pg ml^{-1}, respectively. No statistical differences were observed between months for either steroid. Testosterone concentrations were weakly correlated with testes weight and percent of stage VI spermatocysts in *A. radiata*, however, no correlation was detected between T and stage VI spermatocysts in *M. senta*. Collectively, these data support the previous conclusion that *M. senta* and *A. radiata* of both sexes are capable of reproducing year round in the western Gulf of Maine.

Keywords *Malacoraja senta* · *Amblyraja radiata* · Radioimmunoassay · Testosterone · Estradiol · Progesterone

J. Kneebone (✉) · D. E. Ferguson · P. C. W. Tsang
Department of Animal and Nutritional Sciences, University of New Hampshire, Durham, NH 03824, USA
e-mail: jkneebone@hotmail.com

J. A. Sulikowski
Marine Science Center, University of New England, Biddeford, ME 04005, USA

Introduction

Traditionally, investigations into the reproductive biology of elasmobranch fish were accomplished through visual observations of reproductive organs during necropsy. Currently, as increasing fishing pressure continues to cause declines in many elasmobranch populations, the need for extensive data on reproductive biology is crucial for the development of proper management plans. To complicate matters, these data are greatly lacking for many of the exploited species.

The reproductive endocrinology of elasmobranchs has received increasing attention in recent years. Several studies have successfully correlated plasma steroid hormone concentrations with visual observations of gonadal status for both viviparous and oviparous species. For example, in females, elevated estradiol (E_2) concentrations are associated with egg development during the follicular phase in the oviparous species: *Scyliorhinus canicula*, (Sumpter and Dodd 1979), *Leucoraja erinacea* (Koob et al. 1986), *Raja eglanteria* (Rasmussen et al. 1999), *Hemiscyllium ocellatum* (Heupel et al. 1999), and *Leucoraja ocellata* (Sulikowski et al. 2004). Similarly, E_2 titers are correlated with ovarian follicle development in the viviparous species, *Squalus acanthias* (Tsang and Callard 1987), *Sphyrna tiburo* (Manire et al. 1995), and *Dasyatis sabina* (Tricas et al. 2000). In males, testosterone (T) concentrations are correlated with the presence of mature spermatocysts in the *H. ocellatum* (Heupel et al. 1999), *D. sabina* (Tricas et al. 2000), and *L. ocellata* (Sulikowski et al. 2004). As a result of these observations, it appears as though the reproductive status of an individual could potentially be assessed simply by measuring blood steroid hormone concentrations over time. Such a non-lethal approach avoids the need to collect valuable information on reproduction through necropsy, especially from heavily exploited species.

The smooth, *Malacoraja senta*, and thorny, *Amblyraja radiata*, skates are two sympatric oviparous species commonly found within the Gulf of Maine. Currently, both species are listed as overfished, with *A. radiata* recently being identified as a prohibited species (NEFMC 2005). Fortunately, some life history information is now available for these species, based on several timely studies by Sulikowski et al. (2005a, b, 2007) and Natanson et al. (this volume). Monthly morphological assessments of mature *M. senta* and *A. radiata* reproductive tracts revealed no distinct annual patterns, suggesting that both species are capable of reproducing year round in the western Gulf of Maine (Sulikowski et al. 2005b, 2007). These conclusions were based upon the persistence of large shell glands (also known as oviducal or nidamental gland) and ovary weights as well as the presence of pre-ovulatory follicles in females throughout the year. Further, the persistence of mature stage VI spermatocysts, mature sperm that are organized into packets around the inner periphery of the spermatocysts (Maruska et al. 1996), suggested that males are capable of producing sperm year round. In the present study, the biochemical assessment of plasma T, E_2 and P_4 profiles was conducted to determine their association with these known morphological reproductive parameters in *M. senta* and *A. radiata*.

Materials and methods

Sampling techniques

Malacoraja senta and *A. radiata* specimens were captured by otter trawl in a 2300 km^2 area centered at 42°50′ N and 70°15′ W in the Gulf of Maine. These locations varied from 30 km to 40 km off the coast of New Hampshire, USA. Approximate depths at these locations ranged from 100 m to 120 m. Collection of skates occurred between the 10th and 20th of each month, beginning in May 2001 and ending in May 2003. Immediately after capture, a 5–10 ml aliquot of blood was collected by cardiac puncture. Samples were centrifuged at 1,300 × *g* for 5 min, and the separated plasma placed in a shipboard cooler (4°C) for 4–8 h before storage at –20°C in the laboratory. The dissection and processing of skates followed protocols previously described in Sulikowski et al. (2005b).

All specimens utilized in this study were deemed to be mature based on morphological criteria described in Sulikowski et al. (2005b) for *A. radiata* and Sulikowski et al. (2007) for *M. senta*. A subset of the morphological reproductive parameters from these studies was used in the data analysis of the present study.

Analysis of steroid hormones

Thin layer chromatography

Stock solutions of radiolabeled testosterone and estradiol (Amersham Biosciences, Piscataway, New Jersey) but not progesterone (Perkin Elmer Life Sciences, Boston, MA) were purified by thin layer chromatography (TLC) on silica gel plates (HLF, scored 20 × 20 cm, 250 μm; Analtech, Inc., Newark, Delaware). The solvent systems and specific procedures for TLC were identical to those described in Sulikowski et al. (2004).

Plasma steroid hormone extraction

Approximately 1,000 counts min^{-1} (cpm) of an appropriate radiolabeled steroid were added to plasma samples to account for procedural losses. Each sample was extracted twice with 10 volumes of diethyl ether (testosterone and estradiol), or petroleum ether (progesterone) before snap freezing in an acetone/dry ice bath. The ether was then decanted into a fresh test tube, evaporated to dryness under a stream of nitrogen and reconstituted in phosphate-buffered saline with 0.1% gelatin (PBSG). Buffer blanks, charcoal stripped-plasma, and plasma spiked with steroids were processed in an identical manner. They were stored at −20°C until assay. The overall mean recoveries for T, E$_2$ and P$_4$, were 74%, 76% and 68%, respectively for *A. radiata*, and 79%, 88%, 77%, respectively for *M. senta*. Each sample was corrected for recovery.

Radioimmunoassay

Plasma concentrations of testosterone, estradiol, and progesterone were determined by radioimmunoassay modified from procedures of Tsang and Callard (1987) and Goldberg et al. (1996). All nonradioactive steroids were obtained from Steraloids, Inc. (Wilton, New Hampshire). The specifics of the radiolabeled steroids, the antibody characteristics and titers, and scintillation counting are found in Sulikowski et al. (2004). The intra-assay coefficients of variance were 6.6% for T, 8.1% for E$_2$, and 10.4% for P$_4$. The inter-assay coefficients of variance were 10.0% for T, 11.0% for E$_2$, and 10.2% for P$_4$. Detection limits for RIA's were 5.0 pg for E$_2$ and 12.5 pg for T and P$_4$.

Statistics

Monthly hormone concentrations are presented as mean standard error of the mean (SEM) and evaluated by Kruskal–Wallis analysis of variance (ANOVA) followed by a Dunn's post-hoc test. Statistical significance was accepted at $P < 0.05$. To determine whether a relationship exists between plasma steroid concentrations and measured morphological reproductive parameters, a Pearson correlation analysis was performed. Progesterone concentrations in females containing no egg cases, or egg cases at all stages of development, were compared by a two-tailed *t* test.

Results

Malacoraja senta

Mature female skates ($n = 70$) ranged in size from 508 to 630 mm total length (570 ± 1.8; mean ± SEM) and from 0.6 to 1.2 kg (0.8 ± 0.05) in total body mass. Mature male skates ($n = 54$) ranged from 550 mm to 660 mm total length (607.6 ± 3.6) and from 0.7 to 1.3 kg (1.1 ± 0.02) in total body mass.

In females, T, E$_2$ and P$_4$ concentrations remained relatively constant throughout the year and no differences were observed between months ($P > 0.05$; Fig. 1). In addition, no correlations were found between E$_2$ concentrations and shell gland weight, gonadosomatic index (GSI), ovary weight or diameter of the largest follicles or between T concentrations and diameter of the largest follicles. Of note, E$_2$ concentrations remained high, in comparison with immature specimens (Sulikowski, unpublished observation), throughout the year along with the presence of large pre-ovulatory follicles within the ovaries (Fig. 1). Females containing no egg cases had higher P$_4$ concentrations than females containing egg cases in various stages of development ($P = 0.02$; Table 1).

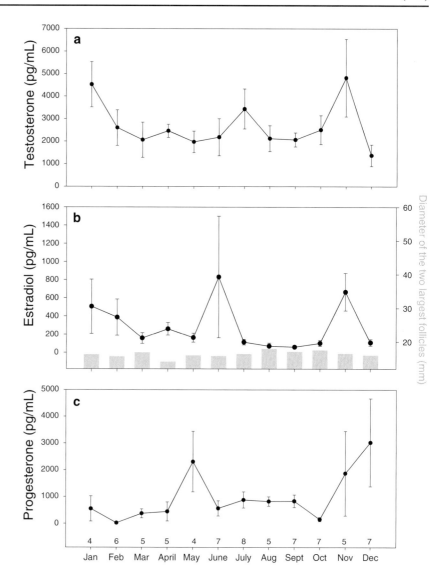

Fig. 1 Mean plasma steroid hormone concentrations in female *Malacoraja senta*. Testosterone (**a**), estradiol (**b**) and progesterone (**c**) concentrations expressed in picograms per milliliter ± standard error of the mean (pg ml^{-1} ± SEM). Average diameters of the largest follicles are presented along with the estradiol profile, demonstrating the presence of high estradiol concentrations and mature follicles year round. The number of skates for each month is indicated above the *x*-axis

In males, T concentrations remained consistent throughout the year, and no differences were observed between months ($P > 0.05$; Fig. 2). Estradiol concentrations were low, and undetectable in many samples (Fig. 2). Testes weight was weakly correlated with T concentrations ($r = 0.266$) and GSI ($r = 0.233$), and no association was found between T and percent of stage VI spermatocysts. Average T concentrations remained relatively high throughout the year in comparison with immature specimens (Sulikowski, unpublished observation), in the presence of a high percentage of mature stage VI spermatocysts within the testes (Fig. 2).

Amblyraja radiata

Mature female skates ($n = 47$) ranged from 820 to 1050 mm TL (922.8 ± 7.8) and from 4.4 kg to 10.2 kg (7.7 ± 0.2) in total mass. Mature male skates ($n = 52$) ranged from 800 to 1040 mm TL (945.9 ± 8.3) and from 5.4 kg to 10.8 kg (8.4 ± 0.3).

In females, average T, E_2 and P_4 concentrations were robust throughout the year but no differences were observed ($P > 0.05$; Fig. 3). No correlations were found between E_2 concentrations and shell gland weight, GSI, ovary weight or diameter of the largest follicles or between T and

Table 1 Progesterone concentrations (pg ml^{-1} ± SEM) and egg case dynamics in females skates

	Developing[a]	Formed[a]	All stages[b]	No egg case[a]
M. senta				
Number of skates	2	5	7	62
Mean P$_4$ concentration	262	41 ± 17	104 ± 71	1,075 ± 255
A. radiata				
Number of skates	5	11	16	29
Mean P$_4$ concentration	4,183 ± 1594	1,853 ± 404	2,581 ± 604	2,021 ± 546

[a] Upon necropsy, uteri were instpected for the presence or absence of developing (partially formed) or fully formed egg cases

[b] All stages of egg case development are combined

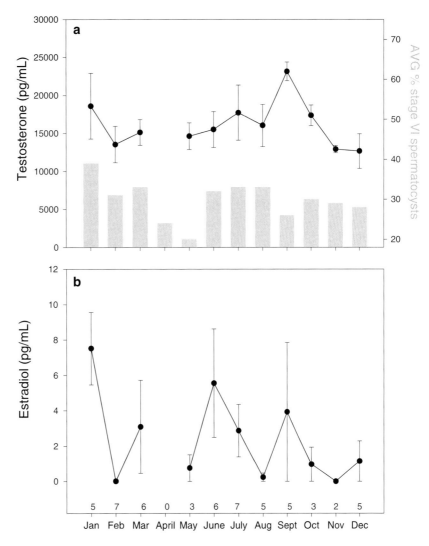

Fig. 2 Mean plasma steroid hormone concentrations in male *Malacoraja senta*. Testosterone (**a**) and estradiol (**b**) concentrations expressed in picograms per milliliter ± standard error of the mean (pg ml^{-1} ± SEM). Males persistently exhibit stage VI spermatocysts within the testes coupled with high testosterone concentrations. The number of skates for each month is indicated above the *x*-axis

diameter of the largest follicles. However, high E$_2$ concentrations, in comparison with immature specimens (Sulikowski, unpublished observation), persist throughout the year along with the presence of large pre-ovulatory follicles within the ovaries (Fig. 3). Progesterone concentrations were no different for females with and without egg cases ($P > 0.05$; Table 1).

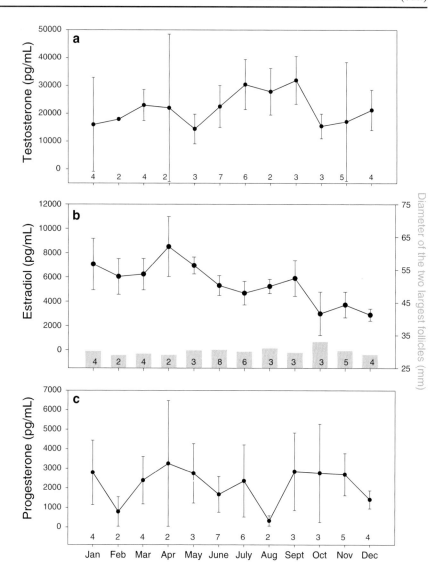

Fig. 3 Mean plasma steroid hormone concentrations in female *Amblyraja radiata*. Testosterone (**a**), estradiol (**b**) and progesterone (**c**) concentrations expressed in picograms per milliliter ± standard error of the mean (pg ml^{-1} ± SEM). Average diameters of the largest follicles are presented along with the estradiol profile, demonstrating the presence of high estradiol concentrations and mature follicles year round. The number of skates for each month is indicated above the *x*-axis

In males, T and E$_2$ did not differ throughout the year ($P > 0.05$; Fig. 4). Testosterone concentrations were weakly correlated with testes weight ($r = 0.342$), GSI ($r = 0.133$) and percent of stage VI spermatocysts ($r = 0.163$). Average T concentrations remained relatively high throughout the year in comparison with immature specimens (Sulikowski, unpublished observation), while a high percentage of mature stage VI spermatocysts was also present within the testes (Fig. 4).

Discussion

The results from the present study support and strengthen our previous conclusion that *M. senta* and *A. radiata* are capable of reproducing year round. In both species, no distinct patterns were revealed for any hormone in either sex. While visual inspection of T, E$_2$ and P$_4$ profiles for female *M. senta* suggest that two peaks may exist during May–July and again in October–December, the lack of sample homogeneity preclude any statistical significance. Such variations between samples within each month may represent individuals of widely different reproductive status and steroid producing potential, further supporting the presence of reproductively capable skates at all times of the year.

The overall hormone concentrations were much lower in *M. senta* than *A. radiata*. While

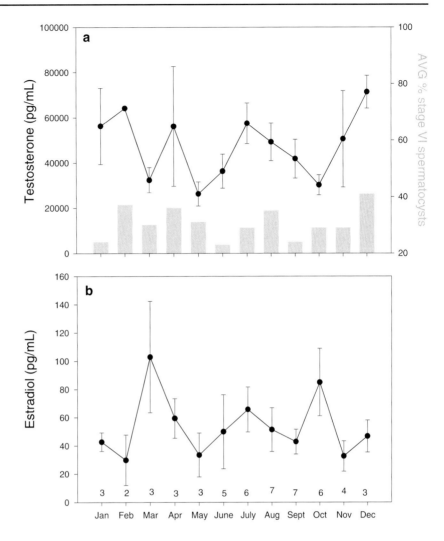

Fig. 4 Mean plasma steroid hormone concentrations in male *Amblyraja radiata*. Testosterone (**a**) and estradiol (**b**) concentrations expressed in picograms per milliliter ± standard error of the mean (pg ml^{-1} ± SEM). Males persistently exhibit stage VI spermatocysts within the testes coupled with high testosterone concentrations. The number of skates for each month is indicated above the *x*-axis

such a large difference may be attributable to the size discrepancy between the species, their steroid hormone concentrations were comparable to other oviparous elasmobranchs of similar size (Koob et al. 1986; Rasmussen et al. 1999; Tricas et al. 2000; Sulikowski et al. 2004), with the exception of E$_2$ in male *M. senta*. In these skates, E$_2$ concentrations did not reach above 20 pg ml^{-1}, with a majority of the specimens having no detectable E$_2$ levels. However, E$_2$ concentrations have typically been reported to be relatively low in males of oviparous species (Heupel et al. 1999; Rasmussen et al. 1999; Sulikowski et al. 2004).

The lack of correlation between E$_2$ concentrations and various morphological reproductive parameters in female *M. senta* and *A. radiata* is likely a consequence of the year round reproductive capability in these species. Previous studies on female *L. erinacea* and *L. ocellata* skates have demonstrated strong correlations between E$_2$ concentration and shell gland weight, GSI, ovary weight and diameter of the largest follicle (Koob et al.1986; Tsang and Callard 1987; Sulikowski et al. 2004). Elevated T concentrations have also been documented during the egg-laying stages of the reproductive cycle in the *R. eglanteria* (Rasmussen et al. 1999) and *L. ocellata* (Sulikowski et al. 2004). Several studies have also associated elevated E$_2$ concentrations with egg development during the follicular phase (Sumpter and Dodd 1979; Rasmussen and Gruber 1993; Manire et al. 1995; Snelson et al. 1997; Heupel et al. 1999; Tricas et al. 2000), while Tsang and Callard (1982) demonstrated that follicular cells synthesize E_2. Collectively, these findings suggest that correlations should have been detected in *M. senta* and *A. radiata*, however, closer examination of the reproductive biology of these sympatric species may provide insight into the apparent lack of association between hormone concentrations and gonadal/reproductive tract status. Since female *M. senta* and *A. radiata* are capable of reproducing year round, we would therefore expect to catch mature females in various stages of one single reproductive event (e.g. one egg-laying cycle) at any one time. Such a high degree of variability among individuals, coupled with the overall lack of synchrony within the *M. senta* and *A. radiata* populations, will likely preclude the ability to detect correlations between hormone concentrations and gonadal/reproductive tract status in species lacking a defined reproductive cycle. Interestingly, in studies where a strong correlation was identified for an oviparous female elasmobranch, these associations were observed during a single egg-laying cycle or for a species with a more defined annual reproductive cycle (Sumpter and Dodd 1979; Koob et al. 1986; Heupel et al. 1999; Rasmussen et al. 1999; Sulikowski et al. 2004). Thus, the lack of statistical correlation between gonadal/reproductive tract status and hormone concentrations in the present study is likely due to the reproductive asynchrony within the two skate populations, and does not suggest that steroid hormones are not associated with various events in the female reproductive cycle.

The persistence of high T concentrations throughout the year suggests that male *M. senta* and *A. radiata* are capable of producing sperm year round. Several studies have suggested a close link between T concentration and stages of spermatogenesis (Heupel et al. 1999; Tricas et al. 2000; Sulikowski et al. 2004), with heightened T production typically associated with increasing numbers of mature spermatocysts. Although weak correlations were found between T and percent of stage VI spermatocysts in *A. radiata*, none existed in *M. senta* males. In addition, T concentrations were also found to be weakly correlated with testes weight and GSI in both species. Thus, these data, at best, suggest a tenuous association between T and male reproductive readiness for both *M. senta* and *A. radiata*. However, as with females, the present findings may be indicative of the individual variability that accompanies the lack of synchrony within the population of reproductively capable males.

The role of P_4 in oviparous species is somewhat of an enigma, but its function has been speculated in several oviparous species. Koob et al. (1986) and Sulikowski et al. (2004) reported a strong relationship between elevated P_4 concentrations and egg retention and oviposition in *L. erinacea* and *L. ocellata*, respectively, while P_4 concentrations are elevated immediately after oviposition in *H. ocellatum* and *R. eglanteria* (Heupel et al. 1999; Rasmussen et al. 1999). In the present study, no statistical differences in P_4 concentration were found between female *A. radiata* bearing developing, fully formed or no egg cases (Table 1). As a result of the relatively few *M. senta* bearing developing egg cases, a similar analysis, as performed for *A. radiata*, was not feasible. However, when the data were reconfigured by combining females at all stages of egg case development into one group and compared to those without egg cases, an unexpected outcome was revealed. Contrary to what has been found in *A. radiata*, *L. ocellata* and *L. erinacea*, *M. senta* females without any egg cases had higher P_4 concentrations than those carrying egg cases of any developmental stage. A more detailed analysis is needed before we can conclude whether or not P_4 is associated with egg case development and dynamics. For example, an in-depth study on P_4 concentrations during a single egg-laying cycle should be conducted in order to draw any conclusions about the roles of P_4 in these two skate species.

In summary, the present data on hormonal profiles in *M. senta* and *A. radiata* reinforce our previous data on gross reproductive morphology (Sulikowski et al. 2005b, 2007) that both species are capable of reproducing year round in the western Gulf of Maine. In female *M. senta* and *A. radiata*, T, E_2 and P_4 concentrations remained relatively constant throughout the year, with no differences found between any months. Estradiol concentrations were not correlated with shell gland weight, GSI, ovary weight or diameter of the largest follicles in either species. In male *M. senta* and *A. radiata*, monthly T concentrations remained high, with no differences detected between any months of the year. Testosterone concentrations were weakly correlated with testes weight and percent of stage VI spermatocysts in *A. radiata*, although no correlation between T and stage VI spermatocysts was present in *M. senta*. Collectively, these data suggest that the maintenance of high hormone concentrations year round most likely indicate continuous reproductive capability in these two skate species. In the future, a detailed study monitoring hormone concentrations during a single egg-laying cycle is necessary in order to decipher the particular role that each steroid hormone plays in coordinating specific reproductive events in these species.

Acknowledgements We thank Joe Jurek of the F/V Mystique Lady for all skate collections. Further thanks are extended to Scott Elzey and Suzie Biron (UNH Department of Zoology) for their help in processing of skates. This project was supported by a grant from the Northeast Consortium (NA16FL1324) to PCWT and JAS.

References

Goldberg MJ, Moses M, Tsang PCW (1996) Identification of matrix metalloproteinases and metalloproteinase inhibitors in bovine corpora lutea and their variation during the estrous cycle. J Anim Sci 74:849–857

Heupel MR, Whittier JM, Bennett MB (1999) Plasma steroid hormone profiles and the reproductive biology of the epaulette shark, *Hemiscyllium ocellatum*. J Exp Zool 284:586–594

Koob TJ, Tsang PCW, Callard IP (1986) Plasma estradiol, testosterone and progesterone levels during the ovulatory cycle of the little skate, *Raja erinacea*. Biol Reprod 35:267–275

Manire CA, Rasmussen LEL, Hess DL, Hueter RE (1995) Serum steroid hormones and the reproductive cycle of the female bonnet head shark, *Sphyrna tiburo*. Gen Comp Endo 97:366–376

Maruska KP, Cowie EG, Tricas TC (1996) Periodic gonadal activity and protracted mating in elasmobranch fishes. J Exp Zool 276:219–232

Natanson LJ, Sulikowski JA, Kneebone J, Tsang PCW (this volume) Age and growth estimates for the smooth skate (*Malacoraja senta*) in the western Gulf of Maine. Envir Biol Fish

New England Fisheries Management Council (NEFMC) (2005) Skate Annual Review. National Marine Fisheries Service, February 2005

Rasmussen LEL, Gruber SH (1993) Serum concentrations of reproductively related circulating steroid hormones in the free ranging lemon shark, *Negaprion brevirostris*. Envir Biol Fish 38:167–174

Rasmussen LEL, Hess DL, Luer CA (1999) Alterations in serum steroid concentrations in the clearnose skate, *Raja eglanteria*: correlations with season and reproductive status. J Exp Zool 284:575–585

Snelson FF Jr, Rasmussen LEL, Johnson MR, Hess DL (1997) Serum concentrations of steroid hormones during reproduction in the Atlantic stingray, *Dasyatis sabina*. Gen Comp Endo 108:67–79

Sulikowski JA, Tsang PCW, Howell WH (2004) An annual cycle of steroid hormone concentrations and gonad development in the winter skate, *Leucoraja ocellata*, from the western Gulf of Maine. Mar Biol 144:845–853

Sulikowski JA, Kneebone J, Elzey S et al (2005a) Age and growth estimates for the thorny skate (*Amblyraja radiata*) in the western Gulf of Maine. Fish Bull 103:161–168

Sulikowski JA, Kneebone J, Elzey S et al (2005b) The reproductive cycle of the thorny skate (*Amblyraja radiata*) in the western Gulf of Maine. Fish Bull 103:536–543

Sulikowski JA, Elzey S, Kneebone J et al (2007) The reproductive cycle of the smooth skate (*Malacoraja senta*) in the western Gulf of Maine. J Mar FW Res 58:98–103

Sumpter JP, Dodd JM (1979) The annual reproductive cycle of the female lesser spotted dogfish, *Scyliorhinus canicula*, and its endocrine control. J Fish Biol 15:687–695

Tricas TC, Maruska KP, Rasmussen LEL (2000) Annual cycles of steroid hormone production, gonad development, and reproductive behavior in the Atlantic stingray. Gen Comp Endo 118:209–225

Tsang PCW, Callard IP (1982) Steroid production by isolated skate ovarian follicular cells. Bull Mt Desert Island Biol Lab 22:96–97

Tsang PCW, Callard P (1987) Morphological and endocrine correlates of the reproductive cycle of the aplacental viviparous dogfish, *Squalus acanthias*. Gen Comp Endo 66:182–89

Morphological variation in the electric organ of the little skate (*Leucoraja erinacea*) and its possible role in communication during courtship

Jason M. Morson · John F. Morrissey

Originally published in the journal Environmental Biology of Fishes, Volume 80, Nos 2–3, 267–275.
DOI 10.1007/s10641-007-9221-x © Springer Science+Business Media B.V. 2007

Abstract Skates discharge an electrical current too weak to be used for predation or defense, and too infrequent and irregular to be used for electrolocation. Additionally, skates possess a specialized sensory system that can detect electrical stimuli at the same strength at which they discharge their organs. These two factors are suggestive of a communicative role for the electric organ in skates, a role that has been demonstrated in similarly weakly electric teleosts (e.g., mormyrids and gymnotiforms). There is evidence that the sexual and ontogenetic variations in the electric organ discharge (EOD) in these other weakly electric fishes are linked to morphological variations in electric organs and the electrogenerating cells of the organs, the electrocytes. Little work has been done to examine possible sexual and ontogenetic variations in skate EODs or variations in the electrocytes responsible for those discharges. Electric organs and electrocyte morphology of male and female, and mature and immature little skates, *Leucoraja erinacea*, are characterized here. Female electric organs were bigger than male electric organs. This is suggestive of a sexually dimorphic EOD waveform or amplitude, which might be used as a sex-specific identification signal during courtship. The shapes of electrocytes that make up the organ were found to be significantly different between mature and immature individuals and, in some cases, posterior membrane surface area of the electrocytes increased at the onset of maturity due to the formation of membrane surface invaginations and papillae. This is evidence that the EOD of skates may differ in its waveform or amplitude or frequency between mature and immature skates, and act as a signal for readiness to mate. This study supports a communicative role during courtship for the weak electric organs of little skates, but studies that characterize skate EOD dimorphisms are needed to corroborate this speculation before conclusions can be drawn about the role the electric organ plays in communication during courtship.

Keywords Electrocyte · EOD · Electric fishes

Introduction

Skates are small, dorsoventrally flattened relatives of sharks and rays. The electric organs of skates are spindle shaped and found bilaterally within the musculature of the tail (Ewart 1892). Each electric organ is made up of individual electrogenerating cells called electrocytes, which collectively make up the skate electric organ

J. M. Morson (✉) · J. F. Morrissey
Biology Department, 114 Hofstra University,
Hempstead, NY 11549-1140, USA
e-mail: jmorso1@pride.hofstra.edu

discharge or EOD (Bennett 1971). Skates differ from strongly electric fishes, such as *Torpedo*, in that their EODs are too weak to be used for predation or defense (Bennett 1971; New 1994). Additionally, infrequent and irregular discharges make an electrolocation function for the organs highly unlikely (Bennett 1971; Bratton and Ayers 1987). On the other hand, evidence continues to support the notion that skate electric organs play some role in communication. Skate EODs are species-specific (Bratton and Ayers 1987; Bratton and Williamson 1993; Bratton et al. 1993; Hayes and Bratton 1996), indicating that their discharge may be used for species identification during courtship. One study offered evidence that there may be a seasonality to the skate EOD that coincides with their reproductive season (Mikhailenko 1971, in New 1994), whereas Mortenson and Whitaker (1973) recorded an increase in peak frequency for skates swimming in pairs as compared to skates swimming alone. Similarly, Bratton and Ayers (1987) recorded an increase in the number of times individuals discharge their organs when in contact with conspecifics. In addition, the peak discharge frequency detectable by skates adjusts at the onset of maturity to match the peak frequency of EODs generated by conspecifics (New 1990; Sisneros et al. 1998; Sisneros and Tricas 2002) and reproductively active skates discharge their organs at similar peak frequencies to those detectable by other skates (Sisneros and Tricas 2002). Although this evidence is suggestive of a communicative role for skate electric organs, studies identifying sexual and ontogenetic dimorphisms in the skate EOD are needed to identify its role, if any, in gender-specific or readiness-to-mate signaling during courtship.

Sexually, ontogenetically, and seasonally dimorphic EODs of similarly weakly electric freshwater teleosts (e.g., mormyrids and gymnotiforms) already have been identified and shown to play an important role in communication during mating (for review see Bass 1986a). These differences in EODs are strongly correlated with variations in fish electric organs and electrocytes. For instance, dimorphism in electric organ size is correlated with differences in EOD waveform and amplitude (Hopkins et al. 1990). Furthermore, increased electrocyte size, and the formation of papillae and surface invaginations on the electrocyte membrane have been correlated with variations in EOD waveform (Bass 1986b; Freedman et al. 1989; Hopkins et al. 1990) and pulse duration (Freedman et al. 1989; Hopkins et al. 1990). The correlation between dimorphic EODs and dimorphic electric organs and electrocytes in weakly electric freshwater teleosts suggests similar correlative relationships between organ morphology and organ discharges in skates. The skate electric organ must be characterized sexually, ontogenetically, and seasonally to identify such relationships in skates. If differences are found, it then will be necessary to identify correlative differences in EODs. Finally, behavioral studies will need to record skate interactions in reproductively active individuals and electrophysiological studies will need to identify the mechanisms by which organ morphology shapes EODs.

There is already some evidence that skate electric organs are dimorphic. For instance, Ishiyama (1958) found evidence that some skate electric organs are sexually dimorphic, Lyons (1990) discovered that female *Leucoraja ocellata* have bigger electrocytes than males, and Jacob et al. (1994) noted that in some skate species the electrocytes were not fully developed until individuals reached sexual maturity. Here, we elaborate on these morphological studies by presenting a comprehensive examination of the electric organ of one skate species, *Leucoraja erincaea*. We identify sexual and ontogenetic dimorphisms in electric organ morphology. These differences then are discussed in the context of possible correlative effects they might have on skate EODs, and therefore, their potential role in signaling during courtship or mating.

Materials and methods

We obtained 30 male and 45 female *Leucoraja erinacea* by otter trawl or gill net from the waters of Long Island, NY, between 27 April 2005 and 16 August 2005. *Leucoraja erinacea* was distinguished from *L. ocellata* according to Collette and Klein-MacPhee (2002). Skates smaller than 35 mm total length with between 52 and 58 tooth rows were not used in this study because they were impossible to distinguish from *L. ocellata*.

Maturity was assessed according to Richards et al. (1963). Males with long, calcified, rotating claspers, spurs, and alar spines present, and well-developed ductus deferens and clasper glands were considered reproductively mature. Females with well-developed oviducal glands and yolky follicles in the ovaries were considered reproductively mature.

We severed tails directly posterior to the cloaca and removed right and left electric organs from the tail under a dissecting microscope. The electric organs were measured for their length and width, weighed, and fixed in 10% neutral-buffered formalin for a minimum of 2 weeks. After fixation, we removed a 10-mm section from each, 25% posterior to the anterior tip of the tail. This tail segment then was placed in histo-cassettes, rinsed in running water for 2 h, and put through the following ascending ethanol dehydration series: 25, 50, 75, 95, and 100% ethanol for 2 h each, then 100% ethanol for 12 h. This long, gradual dehydration series was used in order to limit the effects of tissue shrinkage. Immediately following dehydration we placed tissue in Citrisolv for three 1-h intervals and then into a vacuum oven at 57°C, where samples were embedded using an ascending Paraplast/Citrisolv series at 2-h intervals of 25%:75%, 50%:50%, and 75%:25%, and then twice in 100% Paraplast. Following embedding frontal sections were cut 10 μm thick using a Spencer 820 microtome and then placed in a water bath. Samples were then placed on microscope slides, dried, and stained with hematoxylin and eosin according to common staining procedures.

We photographed each of the organs at 4× magnification and five representative electrocytes from each organ at 10× magnification. Representative electrocytes were chosen based on their having stalks. If no stalks were present then the largest electrocytes were photographed (Jacob et al. 1994). We measured the width, thickness, and area of each electrocyte photographed. The contours of anterior and posterior membranes were traced and the lengths of those lines were used as estimators of membrane surface area (Fig. 1).

We employed principle components (PC) analyses to reduce the number of variables to scores

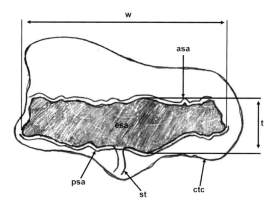

Fig. 1 Sagittal section schematic of an individual skate elelctrocyte. Abbreviations: electrocyte width (w), electrocyte thickness (t), electrocyte area (esa), anterior membrane surface area (asa), posterior membrane surface area (psa), stem (st), and connective tissue compartment (ctc)

for skate size, electric organ size, electrocyte size, and electrocyte shape. Skate size scores were calculated using skate disc width and skate total length (Table 1). Electric organ size scores were calculated using electric organ weight, width, and length (Table 2). Electrocyte size and shape scores were calculated using electrocyte width, thickness, and area (Table 3).

To examine whether or not gender or maturity affected electric organ size we performed a two-way analysis of variance (ANOVA) with skate size as a covariate. By using skate size as a covariate we were able to examine the effects on organ size that were due only to gender and maturity and not to differences in the size of the animal. For similar reasons, to examine the effects of gender or maturity on electrocyte size we performed a two-way repeated measures ANOVA with skate size as a covariate. Because five electrocytes were measured per individual we used repeated measures, which enabled us to examine within-individual variation, whereas comparing mean values from each individual would not have. A two-way repeated measures ANOVA was used to examine the effects of sex and maturity on electrocyte shape with skate size as a covariate, however skate size was determined to be an inefficient covariate and was removed from the model. Finally, to examine the effect of sex and maturity on surface area of the electrocytes' anterior and posterior membranes,

Table 1 Correlation matrix, eigenvalues, and eigenvectors for principle components created from the variables skate total length and disc width

	Total length		Disc width	
Correlation matrix				
Total Length	1.000		0.965	
Disc width	0.965		1.000	
Eigenvalues of the correlation matrix				
Principle component	Eigenvalue	Difference	Proportion	Cumulative
PC1	1.965	1.931	0.982	0.982
PC2	0.034		0.017	1.000
Eigenvectors				
Variable	PC1		PC2	
Total length	0.707		0.707	
Disc width	0.707		-0.707	

Table 2 Correlation matrix, eigenvalues, and eigenvectors for principle components created from the variables electric organ weight, length, and width

	Weight	Length	Width	
Correlation matrix				
Weight	1.000	0.962	0.975	
Length	0.962	1.000	0.977	
Width	0.975	0.977	1.000	
Eigenvalues of the correlation matrix				
Principle Component	Eigenvalue	Difference	Proportion	Cumulative
PC1	2.943	2.905	0.981	0.981
PC2	0.037	0.019	0.012	0.993
PC3	0.018		0.006	1.000
Eigenvectors				
Variable	PC1	PC2	PC3	
Weight	0.576	0.731	0.364	
Length	0.576	–0.680	0.452	
Width	0.579	–0.050	–0.813	

Table 3 Correlation matrix, eigenvalues, and eigenvectors for principle components created from the variables electrocyte width, thickness, and area

	Width	Thickness	Area	
Correlation matrix				
Width	1.000	0.017	0.516	
Thickness	0.017	1.000	0.569	
Area	0.516	0.569	1.000	
Eigenvalues of the correlation matrix				
Principle component	Eigenvalue	Difference	Proportion	Cumulative
PC1	1.776	0.793	0.592	0.592
PC2	0.983	0.742	0.328	0.920
PC3	0.240		0.080	1.000
Eigenvectors				
Variable	PC1	PC2	PC3	
Width	0.478	0.740	0.471	
Thickness	0.525	–0.671	0.523	
Area	0.703	–0.002		
			–0.711	

we performed two-way repeated measures ANOVAs with electrocyte width as a covariate. Membrane surface area increases with increasing electrocyte width. Therefore, to isolate differences in surface area due to changes in membrane contour (surface invaginations and the formation of papillae), and not due to increases in electrocyte width, we used electrocyte width as a covariate.

Results

Scores obtained from PC analyses were accurate representations of the variables from which they were compiled. Skate size scores accounted for 98% of the variation in skate disc width and skate total length (Table 1). Electric organ size scores accounted for 98% of the variation in electric organ weight, width, and length (Table 2). Electrocyte size scores accounted for 59%, while electrocyte shape scores accounted for 33%, of the variation in electrocyte width, thickness, and area (Table 3).

In no case was there a within-individual effect for any parameter measured on the five electrocytes in each organ. In other words, within an individual organ, representative electrocytes were similar in their size ($F_{2,146} = 0.262$, $P > 0.05$), shape ($F_{2,146} = 0.751$, $P > 0.05$), and anterior ($F_{2,146} = 0.591$, $P > 0.05$) and posterior ($F_{2,146} = 0.946$, $P > 0.05$) surface areas. However, for the purposes of our study electrocytes were chosen with a size bias, whereby the largest electrocytes were always selected over relatively smaller or medium-sized electrocytes. In fact, qualitative observations showed that there was a large amount of variation in electrocyte size parameters within a single electric organ, with larger electrocytes sometimes being twice the size of surrounding smaller electrocytes.

After eliminating the effects of skate size, maturity did not have a significant effect on electric organ size ($F_{1,73} = 0.003$, $P > 0.05$). While the overall maturity effect was not significant, it is, however, important to note that some mature skates did in fact have much larger electric organs than some immature skates of the same size (Fig. 2a). The gender effect on organ size was significant ($F_{1,73} = 6.261$, $P < 0.05$) after eliminating the effect of skate size. Female organs were larger than male organs in skates of the same size (Fig. 2b).

After eliminating the effects of skate size, electrocyte size was not affected by gender

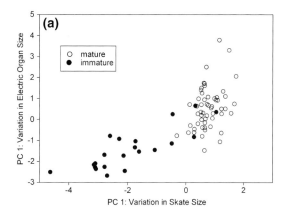

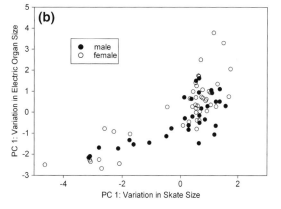

Fig. 2 (a) Effect of maturity on the relationship between skate size and electric organ size. Skate size accounts for 98% of the variation in skate disk width and skate total length. Electric organ size accounts for 98% of the variation in electric organ weight, width, and length; (b) Effect of gender on the relationship between skate size and electric organ size. A two-way ANOVA with skate size as a covariate tested the effect of maturity and gender on these relationships. Maturity did not have a significant effect ($F_{1,73} = 0.003$, $P > 0.05$), but gender did ($F_{1,73} = 6.261$, $P < 0.05$), with females having larger electric organs than males of the same size

($F_{1,73} = 0.455$, $P > 0.05$) or maturity ($F_{1,73} = 0.854$, $P > 0.05$; Fig. 3). Electrocyte shape was not affected by gender ($F_{1,73} = 0.143$, $P > 0.05$), but was significantly affected by maturity ($F_{1,73} = 27.525$, $P < 0.001$; Fig. 4). Immature skates had electrocytes that were thick and narrow whereas mature skates had electrocytes that were thin and wide (see Fig. 4 and Table 3). Dramatic differences in electrocyte shape can be seen in similar-sized skates that differ in their maturity status (Fig. 5).

After eliminating the effects of electrocyte width, electrocyte anterior surface area was not

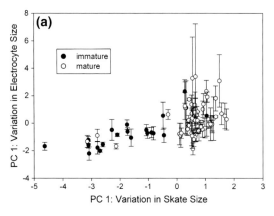

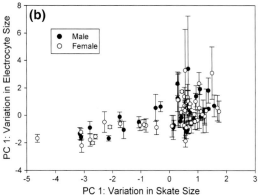

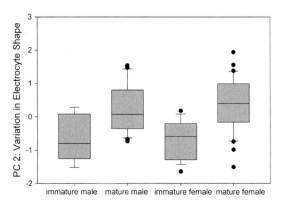

Fig. 3 (a) Effect of maturity on the relationship between skate size and electrocyte size. Skate size accounts for 98% of the variation in skate disk width and skate total length. Electrocyte size accounts for 59% of the variation in electrocyte width, thickness, and area; (b) Effect of gender on the relationship between skate size and electrocyte size. Mean electrocyte size values were calculated from five representative electrocytes in each individual. Error bars represent standard deviation. A two-way repeated measures ANOVA with skate size as a covariate tested the effect of maturity and gender on these relationships. Neither gender ($F_{1,73} = 0.455$, $P > 0.05$), nor maturity ($F_{1,73} = 0.854$, $P > 0.05$), had a significant effect

affected by gender ($F_{1,73} = 2.783$, $P > 0.05$) or maturity ($F_{1,73} = 1.371$, $P > 0.05$). The contours of the anterior membrane surface were smooth in immature male and female skates and remained smooth after the onset of sexual maturity (Fig. 5). Electrocyte posterior surface area, however, was affected by maturity ($F_{1,73} = 16.116$, $P < 0.001$), but not by gender ($F_{1,73} = 1.703$, $P < 0.05$). Increased electrocyte posterior membrane surface area resulted from surface invaginations and the formation of papillae along the membrane (Fig. 5a, c). This phenomenon was not present in

Fig. 4 Effect of gender and maturity on electrocyte shape. Electrocyte shape accounts for 33% of the variation in electrocyte width, thickness, and area. High shape scores indicate electrocytes that are thin and wide, while low shape scores represent electrocytes that are thick and narrow (see Table 3 for details). Mean shape score is represented by horizontal lines, standard error by boxes, and standard deviation by crosshairs. Outliers are represented by solid dots. A two-way repeated measures ANOVA tested the effect of gender and maturity on electrocyte shape. Electrocyte shape was not affected by gender ($F_{1,73} = 0.143$, $P > 0.05$), but was affected by maturity ($F_{1,73} = 27.525$, $P < 0.001$). Immature skates had thick, narrow electrocytes, while mature skates had thin, wide electrocytes

all mature individuals. In some sexually mature skates the posterior membrane was as smooth as the anterior membrane. However, contoured posterior membrane surface and papillae were never present in immature individuals (Fig. 5b, d).

Discussion

The objective of this study was to determine if sexual or ontogenetic variation exists in the electric organ or the electrocytes of little skates. If present, our secondary goal was to identify these differences and suggest what they might tell us about skate EODs and their role in communication during courtship. We found that electric organs are bigger in females than in males. In addition, although organ size was not statistically different between mature and immature individuals, many mature individuals had much larger electric organs than immature individuals of the same size. Furthermore, at the onset of maturity, electrocytes undergo a dramatic shape change, becoming wider and thinner, and in some indi-

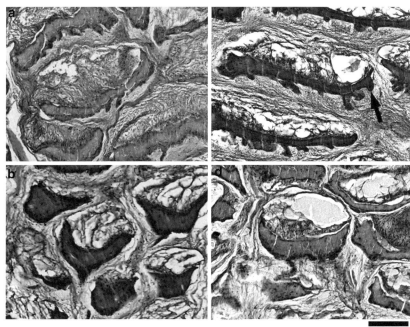

Fig. 5 Micrographs of sagittally sectioned electrocytes. (**a**) Electrocyte from a mature male, 39.0 cm total length (TL); (**b**) electrocyte from an immature male, 37.7 cm TL; (**c**) electrocyte from a mature female, 40.0 cm TL; (**d**) electrocyte from an immature female, 38.1 cm TL. Black arrow indicates the location of papillae along the membrane surface

viduals, gain surface area due to increased invaginations and the formation of papillae on their posterior membrane surface. Here we interpret these results in the context of the effects they may have on EODs and consider if these effects are used to facilitate communication during courtship or mating.

The differences found in overall size of male and female electric organs corroborate gender differences found in some Japanese skate species (Ishiyama 1958). This suggests the possibility of a sexually dimorphic skate EOD, which could be used in gender-specific signaling during courtship or mating. However, when both male and female *Leucoraja erinacea* discharges were previously measured, and their waveform and duration characterized, no gender differences were reported (Bratton and Ayers 1987; New 1994). Additionally, although organ size was found to differ in the two sexes, no correlative sexual dimorphism could be found at the individual elctrocyte level. Electrocyte size, shape, and anterior and posterior surface area were similar for male and female skates. It may be possible then that females simply have more electrocytes that make up the organ than do males, however, total number of electrocytes per organ was not determined in this study. It is difficult to assess what the functional advantage may be of a larger female electric organ, if that function is not to produce a different EOD from that of a male because no substantial evidence exists for functions other than communication for the organ. Sexually dimorphic organs result in sexually dimorphic EODs in other weakly electric fishes (Hopkins et al. 1990). As Koester (2003) suggests, perhaps the lack of evidence for sexually dimorphic skate EODs is due to the paucity of attention they have received and not necessarily because differences do not exist.

Ewart (1888) and Lyons (1990) found that large increases in organ weight to small increments of overall skate growth occurred around the onset of maturity. A similar trend was found in this study, although not in every instance. Organ weight increased at a fairly steady rate with skate size, but at the onset of maturity many skates had robust gains in organ size. No studies have compared EODs of mature and immature skates, but the large increase in organ size noted at the onset of maturity in some skates in this study suggests a correlative change in EOD structure. An interesting finding here was that organs were not always larger in mature individuals. In other words, some mature skates had electric organs that were similar in size to those of immature skates. This trend of variation in mature skates was noted as well in electrocyte size and in the presence of contoured posterior membrane surfaces. The seemingly dynamic nature of these dimorphisms in mature skates hints at a possible seasonal variation to the skate EOD, as was suggested by Mikhailenko (1971), in New (1994). However, skates in this study were collected in only one season, the summer. Furthermore, little skates are known to be reproductively active all year round (Richards et al. 1963). It is particularly during the breeding season that many of the characteristically dimorphic electric organs and EODs are seen in freshwater teleosts (Hopkins 1986). In fact, during the non-breeding season some species have indistinguishable male and female EODs. The seasonal, sexual, and ontogenetic variations in electric organ and electrocyte morphology of these fishes is under hormonal control (Zakon 1996). Assuming the mechanism is similar in skates, and that little skates are reproductively active year round, it would certainly make sense that some mature individuals have temporarily dimorphic electric organs, electrocytes, and possibly EODs. More confident conclusions could be drawn about the differences in organ and electrocyte morphology among mature individuals if a species with a defined reproductive season, such as *Raja eglanteria*, was sampled year round.

The most notable differences in morphology were at the individual electrocyte level. Although gender differences were lacking, easily notable were significant changes in shape at the onset of maturity, with immature individuals having electrocytes that were thick and narrow and mature individuals having electrocytes that were thin and wide. Skate electrocytes are categorized as being either disc shaped or cup shaped (Ewart 1888) and shape is thought to be species specific (Jacob et al. 1994). However, the skate electrocytes studied here seem to change shape from a cup-type to a disc-type at the onset of maturity. This obvious metamorphosis in electrocytes of maturing skates suggests a change in the EODs they produce. Furthermore, mature skates not only have wider electrocytes, but in some the posterior membrane becomes highly invaginated and papillae form along its surface. This phenomenon is well documented in other weakly electric fishes (Bass 1986b; Freedman et al. 1989; Hopkins et al. 1990; Mills et al. 1992) and its correlative relationship with changes in EOD waveform (Bass 1986b; Hopkins et al. 1990) and pulse duration (Freedman et al. 1989; Hopkins et al. 1990) has been demonstrated. The mechanisms by which electroctye membranes shape the EOD waveform in skates are explained by Bennett (1961) and Bass (1986a). Skate electrocytes are innervated on their anterior face and when that face is depolarized the current flows in an anterior–posterior direction and a post-synaptic potential (PSP) is generated. This PSP has the potential to turn on the conductance of the posterior face. If it does not, then only one peak is observed in the external EOD waveform, but if it does, a second peak arises. The potential of a PSP to increase the conductance of the electrocyte posterior face is surface-area dependent, whereby the conductance of a larger surface area is more difficult to turn on. Therefore, the combined maturity effects of electrocyte shape change and the formation of papilla and invaginations on the posterior surface, collectively resulting in a much larger posterior membrane with a much larger surface area, indicate that there may be a change in the waveform of the EOD at the onset of maturity in little skates. However, these electrophysiological implications for increased membrane surface area have yet to be examined in skates.

Of course, differences in electrocyte morphology do not necessarily correspond to differences in EODs. Even though many studies reveal correlations between EODs and electrocyte morphology (Bass 1986a), Mills et al. (1992) found that there was no correlation between the morphology of electrocytes and EODs in *Sternopygus macrurus*, a fish with a sexually dimorphic EOD. This suggests that the morphology, whether it be surface area or size or shape of the electrocyte, or even size of the electric organ, is not the only factor that influences EODs. On the contrary, hormones may modulate the EOD by changing membrane properties without affecting size or area at all (Mills et al. 1992). Bass (1986a) suggests the use of immunohistochemistry to quantify the density of sodium and potassium ion channels along the surface of the membrane. Although much evidence exists that EODs are controlled by electrocyte morphology there is evidence that in some gymnotiform fishes they are not. For this reason it would be important to consider the biochemical properties of electrocyte membranes in skates, as well, when considering mechanisms for dimorphic EODs.

Acknowledgements We thank James Bricca, Todd Gardner, John Maniscalco, and Mark Wuenschel for their assistance in obtaining skates. We thank Donna McLaughlin for performing maturity assessments and for expert technical assistance. We thank Julie Heath for statistical advice. Helpful comments were made by David Koester throughout manuscript preparation. This research was funded in part by a Hofstra University Graduate Research Fellowship.

References

Bass AH (1986a) Electric organs revisited: evolution of a vertebrate communication and orientation organ. In: Bullock TH, Heiligenberg W (eds) Electroreception. Wiley, New York, pp 13–70

Bass AH (1986b) A hormone sensitive communication system in an electric fish. J Neurosci 10(8):2660–2671

Bennett MVL (1961) Modes of operation of electric organs. Ann NY Acad Sci 94:458–509

Bennett MVL (1971) Electric organs. In: Hoar WS, Randall DJ (eds) Fish physiology, vol 5. Academic Press, New York, pp 347–491

Bratton BO, Ayers JL (1987) Observations on the electric organ discharge of two skate species (Chondrichthyes: Rajidae) and its relationship to behavior. Environ Biol Fishes 20:241–254

Bratton BO, Williamson R (1993) Waveform characteristics of the skate electric organ discharge. J Comp Physiol A 173:174

Bratton BO, Christiano A, McClennen N, Murray M, O'Neil E, Ritzen K (1993) The electric organ discharge of two skate species (Chondrichthyes: Rajidae): waveform, occurrence and behavior relationships. Soc Neurosci Abstr 19:374

Collette BC, Klein-MacPhee G (2002) Fishes of the Gulf of Maine, 3rd edn. Smithsonian Institute Press, Washington, DC

Ewart JC (1888) The electric organ of the skate: on the development of the electric organ of *Raja batis*. Phil Trans R Soc Lond 179B:410–416

Ewart JC (1892) The electric organ of the skate: observations on the structure, relations, progressive development, and growth, of the electric organ of the skate. Phil Trans R Soc Lond 183B:389–420

Freedman EG, Olyarchuk J, Marchaterre MA, Bass AH (1989) A temporal analysis of testosterone-induced changes in the electric organs and electric organ discharges of mormyrid fishes. J Neurobiol 20(7):619–634

Hayes LD, Bratton B (1996) Pulse train characteristics of the electric organ discharge (EOD) during the communicative behavior of two skate species (*Raja ocellata* and *Raja erinacea*). Soc Neurosci Abstr 22(Part 1):449

Hopkins CD (1986) Behavior of Mormyridae. In: Bullock TH, Heiligenberg W (eds) Electroreception. Wiley, New York, pp 527–576

Hopkins CD, Comfort NC, Bastian J, Bass AH (1990) Functional analysis of sexual dimorphism in an electric fish, *Hypopomus pinnicaudatus*, order Gymnotiformes. Brain Behav Evol 35:350–367

Ishiyama R (1958) Studies on the rajid fishes (Rajidae) found in the waters around Japan. J Shimonoseki Coll Fish 7:189–394

Jacob BA, McEachran JD, Lyons PL (1994) Electric organs in skates: variation and phylogenetic significance (Chondrichthyes: Rajoidei). J Morphol 221:45–63

Koester DM (2003) Anatomy and motor pathways of the electric organ of skates. Anat Rec A 273A:648–662

Lyons PL (1990) Morphology of the electric organ in two species of skate, *Raja erinacea* Mitchell and *Raja ocellata* Mitchell. Masters Thesis, University of Massachusetts

Mills AH, Zakon H, Marchaterre MA, Bass AH (1992) Electric organ morphology of *Sternopygus macrurus*, a wave-type, weakly electric fish with a sexually dimorphic EOD. J Neurobiol 23(7):920–932

Mortenson J, Whitaker RH (1973) Electric discharge of free swimming female winter skates (*Raja ocellata*). Amer Zool 13:1266

New JG (1990) Medullary electrosensory processing in the little skate. J Comp Physiol A 167:285–294

New JG (1994) Electric organ discharge and electrosensory reafference in skates. Biol Bull 187:64–75

Richards SW, Merriman D, Calhoun LH (1963) The biology of the little skate *Raja erinacea* Mitchell. In: Studies on the marine resources of southern New England IX. Bull Bingham Oceanogr Collect, Yale University, 18(3):5–68

Sisneros JA, Tricas TC, Luer CA (1998) Response properties and biological function of the skate electrosensory system during ontogeny. J Comp Physiol A 183:87–99

Sisneros JA, Tricas TC (2002) Neuroethology and life history adaptations of the elasmobranch electric sense. J Physiol (Paris) 96:379–389

Zakon HH (1996) Hormonal modulation of communication signals in electric fish. Dev Neurosci 18:115–123

Reproductive biology of *Rioraja agassizi* from the coastal southwestern Atlantic ecosystem between northern Uruguay (34°S) and northern Argentina (42°S)

Jorge H. Colonello · Mirta L. García · Carlos A. Lasta

Originally published in the journal Environmental Biology of Fishes, Volume 80, Nos 2–3, 277–284.
DOI 10.1007/s10641-007-9239-0 © Springer Science+Business Media B.V. 2007

Abstract A total of 552 individuals of *Rioraja agassizi* (257 females and 295 males) were collected by bottom-trawl during research cruises. Sexual dimorphism was observed with females are heavier than males for a given total length (TL). Using logistic regression, it was determined that TL at 50% maturity of males was 475 mm TL and of females 520 mm TL. This estimation agrees with the morphological parameters measured. Although not statistically significant, a peak in reproductive activity was observed in males during late spring and summer. The monthly variation in the gonadosomatic index and oviducal gland width, together with the largest diameter of ovarian follicles, suggests that *R. agassizi* females have a partially defined annual reproductive cycle with two peaks, one from November (spring) to February (summer) and another in July (winter).

Keywords Rajidae · Maturity · Reproductive cycle · Life history

J. H. Colonello (✉) · C. A. Lasta
Instituto Nacional de Investigación y Desarrollo Pesquero, Paseo Victoria Ocampo No 1, Escollera Norte, Mar del Plata 7600, Argentina
e-mail: jcolonello@inidep.edu.ar

J. H. Colonello · M. L. García
Consejo Nacional de Investigaciones Científicas y Técnicas, Rivadavia 1917, Buenos Aires, Argentina

M. L. García
Departamento de Zoología de Vertebrados, Facultad de Ciencias Naturales y Museo, Universidad Nacional de La Plata, Paseo del Bosque S/N, La Plata 900, Argentina

Introduction

Four genera of skates occur in the coastal (<50 m depth) southwestern Atlantic between 34°S and 42°S: *Atlantoraja*, *Psammobatis*, *Sympterygia* and *Rioraja* (Menni and Stehmann 2000). The monotypic genus *Rioraja* is represented by the Rio skate *Rioraja agassizi* Müller and Henle 1841, a common endemic skate from coastal waters between southern Brazil to northern Argentina (42°S) (Menni 1973; Menni and Lopez 1984; Menni and Stehmann 2000). This species grows to a maximum total length (TL) of 632 mm and matures at approximately 420–520 mm TL for males and 526–545 mm TL for females (Massa et al. 2001).

Skates, like other elasmobranchs, are particularly sensitive to fishery pressure due to their life history traits and ecological characteristics (Hoeing and Gruber 1990; Walker and Hislop 1998; Dulvy et al. 2000; Dulvy and Reynolds 2002). In the southwestern Atlantic coastal ecosystem between 34°S and 42°S, a multi-species fishery

(Lasta et al. 1999) has historically produced sustainable captures of skates. However, since 1994 an increase in fishing effort, mainly due to larger fishing vessels, has resulted in a decline in the abundance of skates in this region (Massa et al. 2004).

Rioraja agassizi, although not directly targeted by the fisheries, is one of the most abundant skates taken as by-catch in this region (Massa et al. 2004). Depending on their size, *R. agassizi* captures are discarded or kept by artisanal fisheries in this region (Tamini et al. 2006). This indirect fishing pressure has resulted in significant declines of this species. This problem is further magnified due to a lack of pertinent life history information that is necessary for proper management. Thus, the aim of this study was to estimate the size at maturity and to examine seasonal variation in its reproductive condition in order to define critical breeding periods needed for their successful management.

Materials and methods

The study area consists of two distinct coastal (<50 m depth) systems in the southwestern Atlantic between 34°S and 43°S, a homogeneous coastal zone primarily consisting of an estuarine system generated by the discharge of the Negro and Colorado rivers (south of 37°S) and a stratified coastal zone formed by the discharge of the Río de La Plata river (north of 37°S) (Lucas et al. 2005, Fig. 1). In the homogeneous zone the bottom temperature varies from 13 to 15°C from November to March, and may reach a low of 10°C from July to September (Jaureguizar et al. 2006). In the stratified zone the mean bottom water temperature is 20°C during the period December-March (late spring-summer) and 10-12°C during the period June-September (winter) (Guerrero et al. 1997).

Samples of *R. agassizi* were collected from 14 bottom-trawl surveys carried out by RVs "Dr Eduardo L. Holmberg" and "Cap. Oca Balda" of the Instituto Nacional de Investigación y Desarrollo Pesquero (National Institute of Fisheries Research and Development, Argentina). The survey area included the coastal (<50 m depth)

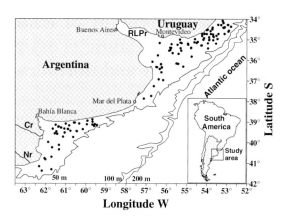

Fig. 1 Map of the study area showing the location of trawl stations. *Black dots* show the position of trawls where *Rioraja agassizi* were collected. The *inset* shows the location of study area in South America. *RLPr*, *Cr* and *Nr* are the Río de La Plata, Colorado and Negro rivers, respectively

southwestern Atlantic between 34°S and 42°S (Fig. 1). Cruises were conducted in December (2003 and 2005), February (2004), March (2002), May (2003), July (2004), August (2002 and 2004) and November (2005). A comparison of samples taken from the same month between different years showed no variability. Therefore, the data from the same month were grouped.

A total of 552 *R. agassizi* (257 females and 295 males) were obtained and used to examine sexual development and reproduction. Each individual skate was measured, weighed and sexed after capture. Total length (TL, mm), total weight (TW, g) and liver weight (LW, g) were recorded. TL-TW and TW-LW relationships were calculated for each sex, the data were ln-transformed for regressions, and the null hypothesis of no differences between slopes was tested using ANCOVA (Zar 1984).

Maturity was assessed macroscopically for females and males. The oviducal gland[1] width (OGW, mm), uterus width (UW, mm), number and diameter of the largest ovarian follicles (DLOF, mm) and ovaries weight (OW, g) were recorded for females. Ovary weight was

---

[1] Oviducal gland refers to the portion of oviduct that produces the rigid egg capsule, the thin pliable transient egg candle case and egg jelly that surrounds the fertilized egg (Hamlett et al. 1998).

expressed as gonadosomatic index (GSI = OW/TW × 100). Symmetry and functional parity of the ovaries were analyzed. For this purpose, the null hypothesis of no differences between mean weight of ovaries and number and DLOF of each ovary were tested with a paired-sampled t-test (Zar 1984). The inner claspers length (CL, mm) of males, as measured from the apex of the cloaca to claspers tip, number of alar thorns rows and testes weight (TEW, g) were recorded for each fish. TEW was expressed as gonadosomatic index (GSI = TEW/TW × 100).

Maturity was assessed by visual inspection of reproductive organs following guidelines published by Walmsley-Hart et al. (1999) and Stehmann (2002). Females were considered to be juvenile when ovarian follicles were undifferentiated, oviducal glands indistinct from the anterior oviduct, and the uterus was uniformly thin. Subadult females includes those whose ovarian follicles were translucent with some of them accumulating yolk, oviducal glands still undifferentiated and the uterus was narrow and constricted. Adult females had ovaries with yellow ovarian follicles ≥3 mm, oviducal glands that were distinctly differentiated and the uterus was pendulous. Males were assessed as juveniles when testes were thin, the efferent ducts were straight and the claspers were shorter than pelvic fins and uncalcified. Subadult males had lobular testes, efferent ducts beginning to coil, elongate and flexible but not calcified claspers. Mature males had enlarged testis, highly coiled efferent ducts, large and calcified claspers. A logistic ogive was fitted to the data using maximum-likelihood approach in order to estimate size at 50% maturity (TL_{50}). Differences in TL_{50} between sexes were evaluated through log-likelihood test.

To asses temporal changes in female reproductive cycles the hepatosomatic index (HSI = LW/TW × 100) and GSI (OW/TW × 100), OGW and DLOF were used. Male HSI and GSI (GSI = TEW/TW × 100) were calculated to assess monthly changes in their reproductive condition. Monthly differences were tested using analysis of variance (ANOVA, $P < 0.05$) followed for post-hoc comparisons by Tukey's test (Zar 1984).

Results

Females ranged from 176 to 698 mm TL and males from 134 to 620 mm TL (Fig. 2). There was a significant difference between the TL-TW relationship of females (TW = $4e^{-07} \times TL^{3.42}$; $r^2 = 0.97$) and males (TW = $1e^{-06} \times TL^{3.26}$; $r^2 = 0.93$) (ANCOVA, $t = 6.82$, $df = 531$, $P = 0.009$), with females heavier than males at a given TL (Fig. 3a). Liver size at a given weight was also sexually dimorphic (ANCOVA, $t = 25.04$, $df = 414$, $P = 8.3 \times 10^{-07}$) (Fig. 3b).

Based on an abrupt change in OGW, UW and ovary weight female maturity was determined to occurred between 500 and 550 mm TL (Fig. 4a, b, c). The smallest matured female was 495 mm and the largest immature female 577 mm. The TL_{50} was estimated at 520 mm TL (78% of the maximum TL) (Fig. 4d). The number and DLOF between ovaries

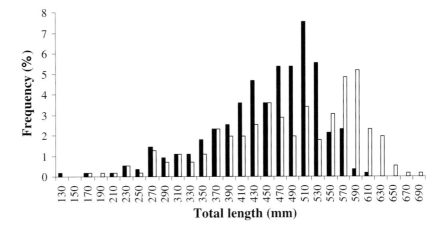

Fig. 2 *Rioraja agassizi*. Length frequency distribution of individuals analyzed. *Empty bars* represent females ($n = 257$); *filled bars* represent males ($n = 295$)

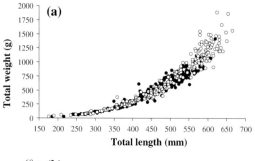

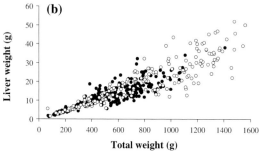

Fig. 3 *Rioraja agassizi.* (**a**) Relationship between total length and total weight. (**b**) Relationship between total weight and liver weight. *Empty circles* represent females; *filled circles* represent males

did not differ (Number: $t = 1.53$, $df = 62$, $P = 0.12$, Diameter: $t = 5.52$, $df = 66$, $P = 0.59$). However, significant differences between left and right ovaries were observed in ovarian masses ($t = 2.51$, $df = 88$, $P = 0.01$).

Males matured between 400 and 500 mm TL (Fig. 5a, b, c) as an abrupt change in the CL was observed along with an increase in TEW and the development of alar thorns rows occurred within this size range. The largest immature male measured 522 mm and the smallest mature male was 418 mm. The TL_{50} was estimated at 475 mm TL (76% of the maximum TL) (Fig. 5d), which was significantly smaller than TL_{50} for females ($t = 19.30$, $df = 1$, $P = 1.11 \times 10^{-5}$).

A significant difference in female HSI was observed throughout the year ($F = 11.71$, $df = 93$, $P = 9.63 \times 10^{-10}$) with highest values in July (winter) and lowest values in February (summer) (Fig. 6a). Significant differences were also observed in GSI ($F = 19.68$, $df = 80$, $P = 1.84 \times 10^{-13}$) (Fig. 6b), OGW ($F = 9.31$, $df = 86$, $P = 7.34 \times 10^{-8}$) (Fig. 6c) and DOLF ($F = 8.35$, $df = 65$, $P = 1 \times 10^{-6}$) (Fig. 6d). In general, the highest values of these reproductive parameters were observed in November, December and February. Interestingly, the DLOF of females collected in July were >25 mm (presumptive preovulatory) (Fig. 6d).

Males showed no statistical differences in the GSI throughout the year (ANOVA, $F = 1.28$, $df = 95$, $P = 0.27$), although higher values of GSI were observed in December (late spring) and February (summer) and lower in May (fall) and July (winter) (Fig. 7a). Conversely, statistical differences were found in HSI (ANOVA, $F = 7.62$, $df = 108$, $P = 7.74 \times 10^{-7}$) (Fig. 7b), but without a clear trend.

Fig. 4 Female *Rioraja agassizi.* Relationship between total length and (**a**) oviducal gland width, (**b**) uterus width, (**c**) gonadosomatic index. *Black circles* represent juvenile; *empty circles* represent mature; *triangles* represent egg-laying, (**d**) proportion of mature individuals per 20 mm total length intervals. *Dashed lines* mark the total length at which 50% of the individuals are mature

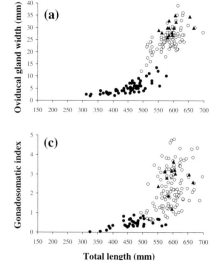
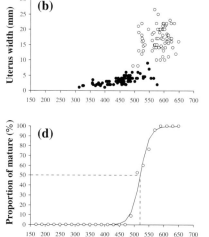

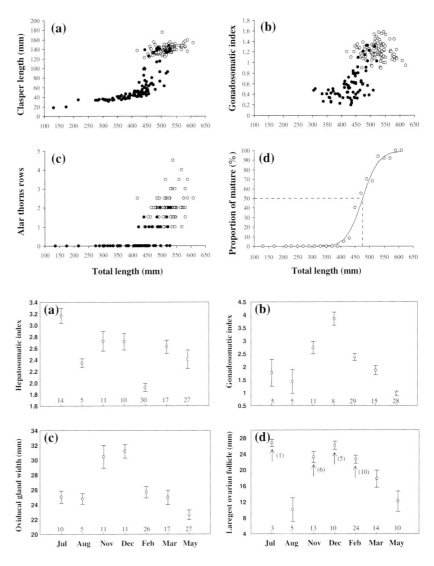

Fig. 5 Male *Rioraja agassizi*. Relationship between total length and (a) clasper length, (b) gonadosomatic index, (c) number of alar thorn rows. *Black circles* represent juvenile; *empty circles* represent mature, (d) proportion of mature individuals per 20 mm total length intervals. *Dashed lines* mark the total length at which 50% of the individuals are mature

Fig. 6 Mature female *Rioraja agassizi*. Monthly variation in: (a) hepatosomatic index, (b) gonadosomatic index, (c) oviducal gland width and (d) diameter of the largest ovarian follicles. Values are expressed as means ± error standard. Sample size is given above each month. The presence (*arrows*) and number of egg-laying females are indicating in (d)

Discussion

In Chondrichthyans, asymmetry between ovaries are common, but in skates symmetry appears to be a common characteristic (Walmsley-Hart et al. 1999; Braccini and Chiaramonte 2002; Mabragaña et al. 2002; Ebert 2005; Oddone and Vooren 2005; San Martín et al. 2005; Ruocco et al. 2006). However, as observed in *Sympterygia bonapartii* (Mabragaña et al. 2002), differences in weight between ovaries were found in *R. agassizi*. This may suggest that development of the epigonal organ is different between ovaries.

The use of OW and TEW to determine the onset of sexual maturity was not good in *R. agassizi*, which may be associated with the seasonal changes in the reproductive condition of mature individuals. As was observed in other skates, the use of CL and oviducal glands measurements provides good information to estimate the onset of sexual maturity. In agreement with these morphological parameters, the estimated size at maturity was 520 mm TL for females and 475 mm TL for males. These estimations agree with the values proposed by Massa et al. (2001). However, sexual dimorphism in maximum size and size at maturity was found in our study, with females larger and maturing at a larger size than males. This dimorphism, common in viviparous elasmobranchs, is quite variable in skates. Some

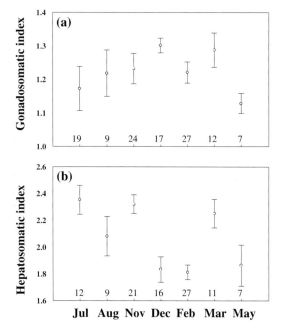

Fig. 7 Mature male *Rioraja agassizi*. Monthly variation in: (a) gonadosomatic index and (b) hepatosomatic index. Values are expressed as means ± error standard. Sample size is indicated above each month

females matured at a larger size (Ebert 2005; Oddone and Vooren 2005; Oddone et al. 2005; Sulikowski et al. 2005a; Whittamore and McCarthy 2005; Coelho and Erzini 2006), smaller size (Walmsley-Hart et al. 1999; Braccini and Chiaramonte 2002; Mabragaña et al. 2002; Mabragaña and Cousseau 2004; San Martín et al. 2005), or at the same size males (Ebert 2005; Ruocco et al. 2006). The energetic demand could be different between the reproductive modes and selection pressure for a larger size at maturity in females is not as strong in oviparous as among viviparous elasmobranchs (Klimley 1987; Ebert 2005). Local ecological characteristics may affect the size at maturity also, which may result in intraspecific differences in the sexual dimorphism of skates (e.g. *Psammobatis extenta*; Martins et al. 2005).

Sexual dimorphism in TL-TW relationship was observed in *R. agassizi*, with adult females heavier than males for a given TL. This appear to be common in skates (Mabragaña et al. 2002; Mabragaña and Cousseau 2004; Ruocco et al. 2006) and could be related the production of eggs (Ruocco et al. 2006). Also, females have heavier livers than males for a given TW suggesting that the importance of the liver as energy source is different between sexes. The role of the liver in reproduction is still unclear in males. It has been suggested that changes in liver reserves of Atlantic sharpnose shark, *Rhizoprionodon terraenovae*, males are related with feeding success and migration (Hoffmayer et al. 2006), although no similar study has been conducted for skates. In females, a significant proportion of lipid is transferred to the growing oocytes from the liver (Craik 1978). The lack of samples for all months in our study makes it difficult to determine if the HSI variation in females is due to seasonal reproductive cycling or to differential lipid deposition. It is possible that the monthly variation observed in *R. agassizi* is an indication of a lack of seasonality, and that it could be related to lipid deposition at different times over the course of a reproductive cycle. This is in contrast to other skates, capable of reproducing throughout the year, where reserves are stored and metabolized continuously (Sulikowski et al. 2004, 2005b).

In general, the reproductive parameters analyzed for females (GSI, OGW and DLOF) displayed a steadily increasing trend during November, December and February. This suggests that *R. agassizi* females have a partially defined annual reproductive cycle with at least one peak from November (spring) to February (late summer). The presence of egg-laying females and females with presumptive pre-ovulatory follicles in July (winter) could indicate a second peak in reproductive activity. Although mean OGW in this month was smaller than November and December, this could be indicative of reproductive activity.

A peak in reproductive activity was observed in *R. agassizi* males between late spring and summer. This pattern of GSI increasing seasonally has been observed in other skates (Braccini and Chiaramonte 2002; San Martín et al. 2005; Coelho and Erzini 2006). However, recent studies suggest that GSI and reproductive readiness may not be positively correlated (Sulikowski et al. 2004, 2005b). Changes in the spermatogenesis stage would likely be undetected with GSI.

The occurrence of peaks within the reproductive activity between species and areas appear to be variable. Research on other coastal (<70 m depth) skates in the region, such as *P. extenta* (Braccini and Chiaramonte 2002), *S. bonapartii* (Mabragaña et al. 2002), *S. acuta* (Mabragaña et al. 2004) and *P. bergi* (San Martín et al. 2005) suggest a distinct seasonal reproductive cycle consisting of one peak. In contrast, in a few deep water (>70 m depth) southwestern Atlantic skate species, no peak in reproductive condition has been observed (Oddone and Vooren 2005; Ruocco et al. 2006). These differences could be explained by an environmental cue, less variable in deeper waters, which might stimulate reproductive activity. Holden et al. (1971) stated that the onset and rate of egg-laying is apparently temperature-dependent. It is possible that the peak in reproductive activity of *R. agassizi* females is related to an abrupt change of temperature in coastal waters.

In summary, *R. agassizi* females reach a larger maximum size and mature at a larger size than males. Mature females have a reproductive cycle with two peaks, one from November (spring) to February (summer) and another in July (winter). This conclusion was based on monthly changes in ovary weight, OGW and diameter of ovarian follicles. For mature males, although not statistically significant, the highest values of TEW were observed during late spring and summer. Currently, histological analyses of the testes are in progress for this species which may provide a precise determination of sexual maturity and reproductive season.

Acknowledgements This paper is part of J. Colonello PhD thesis and was supported by CONICET and INIDEP. JHC would like to thank the American Elasmobranch Society for travel funds to attend this meeting where this paper was presented as part of a symposium on the biology of skates. We wish to express our gratitude to crews of the INIDEP and to C. Carozza and A. Massa. We also thank R. Menni, A. Milessi, A. Vagelli, G. Ortí, D. Ebert and J. Sulikowski for comments and suggestions. INIDEP contribution Number 1440.

References

Braccini JM, Chiaramonte GE (2002) Reproductive biology of *Psammobatis extenta*. J Fish Biol 61:272–288

Coelho R, Erzini K (2006) Reproductive biology of the undulate ray, *Raja undulata*, from the south coast of Portugal. Fish Res 81:80–85

Craik JCA (1978) An annual cycle of vitellogenesis in the elasmobranch *Scyliorhinus canicula*. J Mar Biol Assoc UK 58:719–334

Dulvy NK, Reynolds JD (2002) Predicting extinction vulnerability in skates. Conserv Biol 16:440–450

Dulvy NK, Metcalfe K, Glanville JD, Pawson J, Reynolds MG (2000) Fishery stability, local extinction, and shift in community structure in skates. Conserv Biol 14:283–293

Ebert DA (2005) Reproductive biology of skates, *Bathyraja* (Ishiyama), along the eastern Bering Sea continental slope. J Fish Biol 66:618–649

Guerrero RA, Acha EM, Framiñan MB, Lasta CA (1997) Physical oceanography of the Río de la Plata estuary, Argentina. Cont Shelf Res 17:727–742

Hamlett WC, Knight DP, Koob TJ, Jezior M, Luong T, Rozycki T, Brunette N, Hysell MK (1998) Survey of oviducal gland structure and function in elasmobranchs. J Exp Zool 282:399–420

Hoeing JM, Gruber SH (1990) Life-history patterns in elasmobranchs: implications in fisheries management. NOAA Technical Report, NMFS-90, 16 pp

Hoffmayer ER, Parson GR, Horton J (2006) Seasonal and interannual variation in the energetic condition of adult male Atlantic sharpnose shark *Rhioprionodon terranovae* in the northern Gulf of Mexico. J Fish Biol 68:645–653

Holden MJ, Rout DW, Humphreys CN (1971) The rate of egg laying by three species of ray. J Cons Int Explor Mer 33:335–339

Jaureguizar AJ, Menni RC, Lasta CA, Guerrero R (2006) Fish assemblages of the northern Argentine coastal spatial patterns and their temporal variations. Fish Oceanogr 15:326–344

Klimley AP (1987) The determinants of sexual segregation in the scalloped hammerhead shark, *Sphyrna lewini*. Environ Biol Fish 18:27–40

Lasta CA, Carozza CR, Ruarte CO, Jaureguizar AJ (1999) Ordenamiento pesquero en el ecosistema costero bonaerense (Fishery classification of Buenos Aires coastal ecosystem). Informe Técnico Interno INIDEP (INIDEP technical report, Argentina) N° 25, 20 pp

Lucas AJ, Guerrero R, Mianzán H, Acha EM, Lasta CA (2005) Coastal oceanographic regimen of the Northern Argentina Continental Shelf (34°–43°S). Estua Coast Shelf Sci 65:405–420

Mabragaña E, Cousseau MB (2004) Reproductive biology of two sympatric skates in the south-west Atlantic: *Psammobatis rudis* and *Psammobatis normani*. J Fish Biol 65:559–573

Mabragaña E, Lucifora LO, Massa AM (2002) The reproductive ecology and abundance of *Sympterygia bonapartii* endemic to the south-west Atlantic. J Fish Biol 60:951–967

Mabragaña E, Lucifora LO, Colonello JH, Saicha V, Bernatene F (2004) Biología reproductiva de *Sympterygia acuta* (Chondrichthyes, Rajidae) en el Atlántico sudoccidental (Reproductive biology of *Sympterygia bonapartii* (Chondrichthyes, Rajidae) in southwestern Atlantic). XXIV Congreso Ciencias del Mar (XXIV Meeting of Science of the Sea), Chile

Martins IA, Martins CL, Leme MHA (2005) Biological parameters and population structure of *Psammobatis extenta* in Ubatuba region, north coast of the State of Sao Paulo, Brazil. J Mar Biol Assoc UK 85:1113–1118

Massa AM, Mabragaña E, Lucifora LO (2001) Reproductive ecology of the rio skate, *Rioraja agassizi*, from northern Argentina. IX COLACMAR (IX Latin American meeting of science of the Sea), Colombia

Massa AM, Lucifora LO, Hozbor NM (2004) Condrictios de la región costera bonaerense y uruguaya (Condrichthyans of Buenos Aires and Uruguayan coast). In: Boschi EE (ed) El Mar Argentino y sus recursos pesqueros. Los peces marinos de interés pesquero. Caracterización biológica y evaluación del estado del estado de explotación. INIDEP Press, Argentina, pp 85–99

Menni RC (1973) Rajidae del litoral bonaerense. I. Especies de los géneros *Raja*, *Bathyraja* y *Sympterygia* (Chondrichthyes) (Species of the genera *Raja*, *Bathyraja* and *Sympterygia* (Chndrychthyes)). Physis (Argentina) A 32:413–439

Menni RC, Lopez HL (1984) Distributional patterns of argentine marine fishes. Physis (Argentina) A 42:71–85

Menni RC, Stehmann MFW (2000) Distribution, environment and biology of batoid fishes off Argentina, Uruguay and Brazil. A review Rev Mus Argentino Ciencias Nat (J Argent Nat Hist Mus) 2:69–109

Oddone MC, Vooren CM (2005) Reproductive biology of *Atlantoraja cyclophora* (Regan 1903) (Elasmobranchii: Rajidae) off southern Brazil. ICES J Mar Sci 62:1095–1103

Oddone MC, Paesh L, Norbis W (2005) Size at sexual maturity of two species of rajoi skates, genera *Atlantoraja* and *Dipturus* (Pisces, Elasmobranchii, Rajidae), from the south-western Atlantic Ocean. J Appl Ichthyol 21:70–72

Ruocco NL, Lucifora LO, Díaz de Astarloa JM, Wöhler O (2006) Reproductive biology and abundance of the white-dotted skate, *Bathyraja albomaculata*, in the Southwest Atlantic. ICES J Mar Sci 63:105–116

San Martín MJ, Perez JE, Chiaramonte GE (2005) Reproductive biology of the South West Atlantic marble sand skate *Psammobatis bergi* Marini, 1932 (Elasmobranchii, Rajidae). J Appl Ichthyol 21:504–510

Stehmann MFW (2002) Proposal of a maturity stages scale for oviparous and viviparous cartilaginous fishes (Pisces, Chondrichthyes). Arch Fish Mar Res 50:23–48

Sulikowski JA, Tsang PCW, Howell WH (2004) An annual cycle of steroid hormone concentrations and gonad development in the winter skate, *Leucoraja ocellata*, from the western Gulf of Maine. Mar Biol 144:845–853

Sulikowski JA, Tsang PCW, Howell WH (2005a) Age and size at sexual maturity for the winter skate, *Leucoraja ocellata*, in the western Gulf of Maine based on morphological, histological and steroid hormone analyses. Environ Biol Fish 72:429–441

Sulikowski JA, Kneebone J, Elzey S (2005b) The reproductive cycle of the thorny skate (*Amblyraja radiata*) in the western Gulf of Maine. Fish Bull 103:536–543

Tamini LL, Chiaramonte GE, Perez JE, Cappozzo HL (2006) Batoids in a coastal fishery of Argentina. Fish Res 77:326–332

Templeman W (1987) Differences in sexual maturity and related characteristics between populations of thorny skate (*Raja radiata*) from the Northwest Atlantic. J Northw Atl Fish Sci 7:155–167

Walker PA, Hislop JRG (1998) Sensitive skates or resilient rays? Spatial and temporal shifts in ray species composition in the central and north-western North Sea between 1930 and the present day. ICES J Mar Sci 55:392–402

Walmsley-Hart SA, Sauer WHH, Buxton CD (1999) The biology of the skates *Raja wallacei* and *R. pullopunctata* (Batoidea: Rajidae) on the Agulhas Bank, South Africa. S Afr J Mar Sci 21:165–179

Whittamore JM, McCarthy ID (2005) The population biology of the thornback ray *Raja clavata* in Caernarfon Bay, north Wales. J Mar Biol Assoc UK 85:1089–1094

Zar H (1984) Biostatistical analysis, 2nd edn. Englewood Cliffs, Prentice Hall, 718 pp

Profiling plasma steroid hormones: a non-lethal approach for the study of skate reproductive biology and its potential use in conservation management

James A. Sulikowski · William B. Driggers III · G. Walter Ingram Jr. · Jeff Kneebone · Darren E. Ferguson · Paul C. W. Tsang

Originally published in the journal Environmental Biology of Fishes, Volume 80, Nos 2–3, 285–292.
DOI 10.1007/s10641-007-9257-y © Springer Science+Business Media B.V. 2007

Abstract Information regarding sexual maturity and reproductive cycles in skates has largely been based on gross morphological changes within the reproductive tract. While this information has proved valuable in obtaining life history information, it also necessitates sacrificing the skates to obtain this data. In contrast, few studies have used circulating steroid hormones to establish when these batoids become reproductively capable or for the determination of reproductive cyclicity. This study summarizes our current knowledge of hormonal analyses in determining skate reproductive status and offers information that suggests analysis of circulating steroid hormone concentrations provide a means to determine size at sexual maturity and asses reproductive cycles without the need to sacrifice the skate.

Keywords Skate · Sexual maturity · Reproduction · Testosterone · Estradiol · Non-lethal technique

J. A. Sulikowski (✉)
Marine Science Center, University of New England, Biddeford, ME 04005, USA
e-mail: jsulikowski@une.edu

W. B. Driggers III · G. W. Ingram Jr.
National Marine Fisheries Service, Southeast Fisheries Science Center, Mississippi Laboratories, P.O. Drawer 1207, Pascagoula, MS 39568, USA

J. Kneebone · D. E. Ferguson · P. C. W. Tsang
Department of Animal and Nutritional Sciences, University of New Hampshire, Durham, NH 03824, USA

Introduction

Existing information regarding sexual maturity and reproductive cycles in skates is largely based on gross examination of morphological changes associated with reproductive organs and structures. For example, past studies have used structural changes to claspers in males and ovary weight in females, while others have utilized the gonadosomatic index (GSI) to help assess reproductive status in batoids (e.g. Zeiner and Wolf 1993; Walmsley-Hart et al. 1999; Francis et al. 2001; Ebert 2005; Ruocco et al. 2006). While this information has proven valuable in obtaining life history information, collecting pertinent data requires that specimens be sacrificed. This can be problematic, especially for species that have been classified as endangered or threatened. Thus, unless biomarkers can be identified through non-lethal sampling techniques, related information regarding reproductive biology for prohibited species will be difficult to obtain.

In recent years, circulating concentrations of steroid hormones, such as 17-β-estradiol (E$_2$) and testosterone (T), have been used with gross morphological changes to evaluate events associated with the reproductive cycles in a limited number of sharks and rays (Manire et al. 1995; Snelson et al. 1997; Heupel et al. 1999; Tricas et al. 2000) and three species of skates (Rasmussen et al. 1999; Sulikowski et al. 2004; Kneebone et al. 2007). To date, only three studies have used a comprehensive approach of combining morphological and biochemical variables to assess developmental changes in elasmobranchs as these mature. They include puberty in male bonnethead sharks, *Sphyrna tiburo* (Gelsleichter et al. 2002), sexual maturity in the winter skate, *Leucoraja ocellata* (Sulikowski et al. 2005a), and an investigation of sexual maturity in the thorny skate, *Amblyraja radiata* (Sulikowski et al. 2006). Although circulating concentrations of hormones have been measured in only a few elasmobranch species, elevated titers of T and E_2, as well as other reproductive steroid hormones, are correlated with specific morphological events during the reproductive cycle and during sexual maturation (e.g. Manire et al. 1995; Snelson et al. 1997; Heupel et al. 1999; Tricas et al. 2000; Gelsleichter et al. 2002; Sulikowski et al. 2004, 2005a, 2006). This link between elevated E_2, T, and morphological events in the reproductive tissues suggests that analyses of these steroid hormone concentrations alone could provide the necessary information to gauge sexual maturity and reproductive cycle status of elasmobranchs species. If this is true, then profiling these hormones from blood samples is a non-lethal approach that can be used to assess the reproductive status of elasmobranchs. Thus, the goal of the current study was to perform a quantitative analysis of the morphological, histological, and plasma steroid hormone data obtained from our previous investigations (Sulikowski et al. 2004, 2005a, 2006; Kneebone et al. 2007) in order to determine whether circulating concentrations of T and E_2 can indeed be used as a non-lethal alternative to study reproductive biology of female and male skates, respectively.

Materials and methods

All sampling protocols and procedural details are in Sulikowski et al. (2004, 2005a, 2005b, 2006) and Kneebone et al. (2007). Briefly, skates were captured by otter trawl in a 5800 km^2 area, centered at about 42°15′ N and 70°15′ W, in the Gulf of Maine. Collection of skates occurred between the 10th and 20th of each month, beginning in November 2000 and ending in June 2003. For each skate, blood (5–10 ml) was collected by cardiac puncture using chilled, heparinized syringes with a 21 gauge needle, followed by centrifugation at 1,300 g for 5 min. The separated plasma was then placed in a cooler (4°C) for 4–8 h before storage at −20°C in the laboratory. After blood was drawn, the skates were euthanized (in a bath of 0.05 MS222 g l^{-1}) before taking external morphological measurements.

Gross morphological examination of female reproductive tracts included removal and weighing of ovaries, shell glands, and uteri to the nearest gram. Follicle development was assessed using calipers to measure all eggs that were greater than or equal to 1 mm in diameter. Testes were removed from male specimens, blotted dry, and weighed to the nearest gram. Testes were histologically processed following the protocol of Sulikowski et al. (e.g. 2004, 2005a) to assess spermatogenic development. Specifically, the mean proportion of mature spermatocysts was measured along a straight-line distance across a representative full lobe cross-section of a testis (Sulikowski et al. 2004, 2005a), in accordance with the compound testis in rajids (Pratt 1988).

Criteria for determining sexual maturity and reproductive activity in skates

The criteria for determining sexual maturity and reproductive activity in winter and thorny skates are detailed in Sulikowski et al. (2004, 2005a, 2005b, 2006). For convenience, we have included these criteria in Table 1.

Hormone preparation and analysis

The full procedural details for plasma steroid extraction and radioimmunoassay are found in Tsang and Callard (1987) and Sulikowski et al. (2004). Briefly, thin layer chromatography (TLC) was used to purify stock solutions of tritiated E_2 and T. Each skate plasma sample was extracted twice with 10 volumes of diethyl ether (anhydrous), which was evaporated under a stream of nitrogen before the dried extracts were reconstituted in phosphate-buffered saline with 0.1% gelatin (PBSG) and stored at −20°C until assay. To assess linearity and parallelism, the R^2 for standard curve/samples was calculated for E_2 to be 0.97/0.98 and for T to be 0.98/0.97. Correction for procedural losses during the extraction process was accomplished by adding 1,000 counts per minute (cpm) of the appropriate tritiated steroid into each plasma sample. The intra-assay coefficients of

Table 1 Reproductive parameters for female (f) and male (m) *Leucoraja ocellata* and *Amblyraja radiata* previously published in Sulikowski et al. (2004, 2005a, 2005b, 2006)

Skate species	Follicle size (mm)	Shell gland mass (g)	Estradiol (pg/ml)	Clasper length (mm)	% mature spermatocysts	Testosterone (pg/ml)
L. ocellata (f)	20 ± 2	22 ± 4	1,000 ± 250			
L. ocellata (m)				180 ± 7	26 ± 4	30,000 ± 4,400
A. radiata (f)	26 ± 3	30 ± 8	3,000 ± 500			
A. radiata (m)				180 ± 2	26 ± 4	30,000 ± 5,200

The sizes of each reproductive parameter represents the minimum necessary for females and males to be considered sexually mature and capable of successful mating, egg encapsulation, and oviposition. Values given as mean ± SE

variance were 6.5% for E_2 and 8.1% for T. The inter-assay coefficients of variance were 10.1% for E_2 and 9.8% for T. Radioactivity was determined in a Beckman LS6000IC (Fullerton, California) liquid scintillation counter.

Statistics

Logistic maturity ogives based upon both length and age were used to test for potential differences between maturity estimates derived from morphological characters and those derived from E_2 and T concentration. This was accomplished using PROC LOGISTIC in SAS (v. 9.1.3) based on the following equation:

$$Y = \frac{1}{(1 + e^{-(a+bx+c)})}$$

where Y is the dependent binomial maturity variable (i.e. immature or mature); a is the intercept; b is the slope, which describes the relationship between Y

Table 2 Size and age at 50% maturity for winter skates, *L. ocellata*

Gender (Significance of covariance parameter)	Method of maturity assessment	Parameter	Estimate	S.E.	n	Total Length (mm)
Size at 50% maturity						
Female (*P*-value = 0.48)	Morphometric analyses	a	−50.17	12.96	85	774.17
		b	0.06	0.02		
	Estradiol	a	−41.05	10.20		768.71
		b	0.05	0.01		
Male (*P*-value = 0.54)	Morphometric analyses	a	−53.14	15.51	93	755.85
		b	0.07	0.02		
	Testosterone	a	−49.80	13.70		761.44
		b	0.07	0.02		
		Parameter	Estimate	S.E.	n	Age (years)
Age at 50% maturity						
Female (*P*-value = 0.47)	Morphometric analyses	a	−15.46	3.34	85	12.58
		b	1.23	0.26		
	Estradiol	a	−12.68	2.66		12.25
		b	1.03	0.21		
Male (*P*-value = 0.59)	Morphometric analyses	a	−24.06	6.89	93	11.49
		b	2.09	0.58		
	Testosterone	a	−22.98	6.57		11.62
		b	1.98	0.55		

There were no significant differences in parameter estimates for size and age at maturity for each sex based on type 3 analyses (p-values shown under gender) (maturity) and x (age or size); and c is the method of maturation assignment, included in the model as a covariate (Agresti 2002). Type 3 analysis of the model variables was used to determine the statistical significance of that covariate. Likewise, reproductive activity of females for both species was modeled using a logistic regression approach using PROC LOGISTIC in SAS (v. 9.1.3), due to the binomial nature of a skate's assignment as reproductively active or not. The method of activity assignment was included in the model as a covariate, and type 3 analysis was used to determine the statistical significance of that covariate. A temporal variable, which had two levels: warm season (June - November) and cool season (December - May), was included in the model in order to standardize any effects of seasonality on reproductive activity.

Results

Type 3 analyses (Tables 2–3) indicated that ogives developed from E_2 or T for female and male skates of each species (Figs. 1–2) did not differ significantly from those developed from morphological characters. The logistic models describing reproductive activity in female winter skates indicated that the season variable was highly significant and the type 3 analyses developed from E_2 did not differ significantly from that developed from morphological characters (Table 4). Logistic models for adult thorny skates would not converge, therefore, a comparison between reproductive activity and morphological characters could not be made.

Discussion

Profiling circulating steroid hormones has the potential to be used as a non lethal approach for assessing the reproductive status of elasmobranchs. This statement is based on the quantitative results of the current study, and research that suggests the physiological and morphological processes central to elasmobranch reproductive biology (as in other vertebrates) are likely regulated by endocrine factors (e.g. Koob and

Table 3 Size and age at 50% maturity for thorny skates *A. radiata*

Gender (Significance of covariance parameter)	Method of maturity assessment	Parameter	Estimate	S.E.	n	Total Length (mm)
Size at 50% maturity						
Female (*P*-value = 0.8182)	Morphometric analyses	a	−44.73	10.97	89	868.53
		b	0.05	0.01		
	Estradiol	a	−41.98	10.14		865.59
		b	0.05	0.01		
Male (*P*-value = 0.5002)	Morphometric analyses	a	−771.80	1148.40	78	879.14
		b	0.88	1.30		
	Testosterone	a	−63.61	21.35		874.94
		b	0.07	0.02		
		Parameter	Estimate	S.E.	n	Age (years)
Age at 50% maturity						
Female (*P*-value = 0.8236)	Morphometric analyses	a	−23.44	5.31	89	10.83
		b	2.16	0.49		
	Estradiol	a	−21.32	4.73		10.78
		b	1.98	0.43		
Male (*P*-value = 0.4311)	Morphometric analyses	a	−130.90	525.40	78	10.90
		b	12.00	47.76		
	Testosterone	a	−38.38	9.77		10.67
		b	3.60	0.92		

The last maximum likelihood iteration is reported. There were no significant differences in parameter estimates for size and age at maturity for each sex based on type 3 analyses (*P*-values shown under gender)

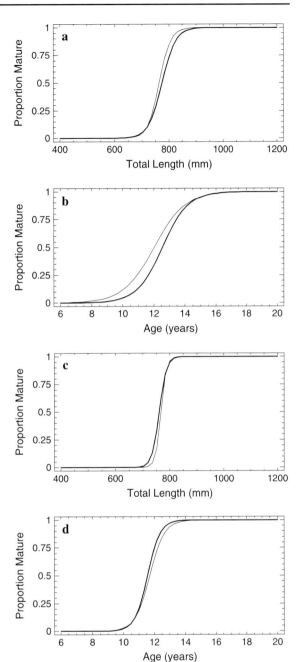

Fig. 1 (**a**) Size at 50% maturity ogives for female winter skates, *L. ocellata*. Thin line is based on estradiol concentration and solid line is based on overall maturity as assessed by examining ovarian weight, shell gland weight, and ovarian follicle diameter. (**b**) Age at 50% maturity ogives for female winter skates, *L. ocellata*. Dashed line is based on estradiol concentration and solid line is based on overall maturity as assessed by examining ovarian weight, shell gland weight, and ovarian follicle diameter. (**c**) Size at 50% maturity ogives for male winter skates, *L. ocellata*. Dashed line is based on testosterone concentration and solid line is based on overall maturity as assessed by examining clasper length, testis weight, and sperm production. (**d**) Age at 50% maturity ogives for male winter skates, *L. ocellata*. Dashed line is based on testosterone concentration and solid line is based on overall maturity as assessed by examining clasper length, testis weight, and sperm production.

Callard 1999). The results presented in this manuscript detail information obtained for T and E_2. Although these hormones have received the bulk of study, it is also important to consider the possible action of other steroids not presented in this study, such as dihydrotestosterone, 11-ketotestosterone, and progesterone which may be linked to key events in the reproductive cycle as well (e.g. Borg 1994; Manire et al. 1999; Sulikowski et al. 2004).

Follicular cells from large and small ovarian follicles of the little skate, *Leucoraja erinacea*, produce E_2 (Tsang and Callard 1982), and in the spiny dogfish, *Squalus acanthias*, the granulosa and theca cells of the follicle synthesize both T and E_2 (Tsang and Callard 1992). These endocrine hormones, in turn, are associated with physiological events that occur during the reproductive cycle. For example, Koob et al. (1986) and Tsang and Callard (1987) observed correlations between follicle size and E_2 concentrations for the little skate and spiny dogfish, respectively. Moreover, this association between elevated E_2 concentrations and egg development during the follicular phase has also been documented in several other female elasmobranch species (Sumpter and Dodd 1979; Manire et al. 1995; Snelson et al. 1997; Heupel et al. 1999; Tricas et al. 2000; Kneebone et al. 2007), strengthening the link between the ovarian follicle and E_2 concentrations. In male elasmobranchs, in vitro studies demonstrated that T produced by the Sertoli cells (Cuevas and Callard 1989; Du Bois et al. 1989; Callard and Cuevas 1992) is associated with spermatocyst development. Similar associations between T concentrations and distinct stages of spermatogenesis were also documented in the epaulette shark, *Hemiscyllium ocellatum* (Heupel et al. 1999), and the Atlantic stingray, *Dasyatis sabina*, (Tricas et al. 2000).

Steroid hormone concentrations are also associated with chronological changes in the reproductive tract as elasmobranchs mature. For example, as the bonnethead shark, *Sphyrna tiburo*, matures, changes in the development of claspers along with the

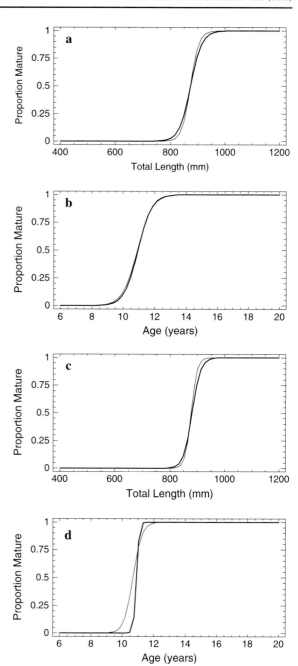

Fig. 2 (a) Size at 50% maturity ogives for female thorny skates, *A. radiata*. Thin line is based on testosterone concentration and solid line is based on overall maturity as assessed by examining ovarian weight, shell gland weight, and ovarian follicle diameter. (b) Age at 50% maturity ogives for female thorny skates, *A. radiata*. Dashed line is based on testosterone and solid line is based on overall maturity as assessed by examining ovarian weight, shell gland weight, and ovarian follicle diameter. (c) Size at 50% maturity ogives for male thorny skates, *A. radiata*. Dashed line is based on testosterone and solid line is based on overall maturity as assessed by examining clasper length, testis weight and sperm production. (d) Age at 50% maturity ogives for male thorny skates, *A. radiata*. Dashed line is based on testosterone concentration and solid line is based on overall maturity as assessed by examining clasper length, testis weight and sperm production

concomitant increases in spermatogenesis and circulating steroid hormone concentrations occur (Gelsleichter et al. 2002). These findings are consistent with the results of Rasmussen and Murru (1992), whose research found T and E_2 concentrations in immature sandbar sharks, *Carcharhinus plumbeus*, to be considerably lower ($P < 0.05$) than adults. Moreover, as immature bull sharks *Carcharhinus leucas* matured, T and E_2 concentrations increased to levels that were observed for specimens that were known to be sexually mature (Rasmussen and Murru 1992).

Previous research on the reproductive biology of winter and thorny skates (Sulikowski et al. 2004, 2005a, 2005b, 2006; Kneebone et al. 2007) supports and strengthens the findings of the aforementioned studies, both in adult elasmobranchs and in those that are maturing. Our results indicate that the coordinate examination of morphological structures of the reproductive system and histological variables, along with steroid hormone concentrations, is an accurate approach to determining age and size at sexual maturity and for discerning the reproductive cycles of adults. Moreover, the quantitative analyses performed in our current study indicate that steroid hormone profiles associated with maturation in both winter and thorny skates statistically different from those based on reproductive tract morphological parameters. Additionally, these findings could also be used for determining the reproductive cycle of adult female winter skates. Surprisingly, logistic models could not be developed to describe the reproductive cycle of adult female thorny skates. The lack of correlation between E_2 concentrations and various morphological reproductive parameters in female thorny skates is perplexing, especially since all other research on elasmobranch reproductive steroid hormones suggests that correlations should have been detected in thorny skates. However, a likely explanation may be due to the year round reproductive capability in this species (Sulikowski et al. 2005b; Kneebone et al. 2007). Since female thorny skates are capable of reproducing year round, we would expect to catch mature females in various stages of one single

Table 4 Logistic model parameters for determining differences in the method of reproductive activity assignment for the winter skate, *L. ocellata*

Method of reproductive activity assignment	Parameter	Estimate	S.E.	Parameter P-value	Significance of covariance parameter	N
Estradiol concentration	*Intercept*	−8.9070	5.0335	0.0768	P-value = 0.39	85
	TL	0.0117	0.00613	0.0570		
	season	−1.3245	0.3520	0.0002		
morphometric analyses	*Intercept*	−0.5923	5.1751	0.9089		
	TL	0.00134	0.00628	0.8308		
	season	−1.6277	0.3725	<.0001		

There were no significant differences in the method of reproductive activity assignment based on type 3 analyses (see Significance of covariance parameter)

reproductive event (e.g. one egg laying cycle) at any one time. Such a high degree of variability among individuals will likely limit the ability to detect correlations between hormone concentrations and reproductive tract status in a species lacking a defined reproductive cycle. The studies where strong correlations exist between hormone concentrations and female reproductive tissues were conducted on elasmobranch species that displayed a more defined annual reproductive cycle (Sumpter and Dodd 1979; Koob et al. 1986; Heupel et al. 1999; Rasmussen et al. 1999; Sulikowski et al. 2004). Based on this information, we suggest the lack of statistical correlation between gonadal/reproductive tissues status and hormone concentrations in thorny skates is likely due to reproductive asynchrony rather than a disassociation between steroid hormones and various events in the female reproductive cycle (Kneebone et al. 2007). In order to fully understand this phenomenon, a study monitoring hormone concentrations is necessary to determine the role that E_2 plays in coordinating specific reproductive events in thorny skates (Kneebone et al. 2007).

In summary, the present study reinforces the strong relationship that exists between biochemical and morphological events associated with the reproductive physiology of elasmobranchs. Further, our present study quantitatively demonstrates that analysis of circulating E_2 and T concentrations can serve as a non-lethal approach to assess sexual maturity and reproductive activity in skates. This capability could prove extremely useful for species that have been classified as endangered, threatened or prohibited when no other means of obtaining reproductive information is available.

Acknowledgements We thank Joe Jurek of the F/V Mystique Lady for all skate collections. Further thanks are extended to Scott Elzey and Suzie Biron (University of New Hampshire, Department of Zoology) for their help in processing of skates. This project was supported by a grant from the Northeast Consortium (NA16FL1324) to PCWT and JAS. This is contribution number #6 from University of New England's Marine Science Center.

References

Agresti A (2002) Categorical data analysis, 2nd edn. John Wiley and Sons, Inc, New York

Borg B (1994) Androgens in teleost fishes. Comp Biochem Physiol 109C:219–245

Callard G, Cuevas ME (1992) In vitro steroid secretion by staged spermatocysts (Sertoli/germ cell units) of dogfish (*Squalus acanthias*) testis. Gen Comp Endocrinol 88:151–165

Cuevas ME, Callard GV (1989) In vitro secretion by Sertoli/germ cell units (spermatocysts) derived from dogfish (*Squalus acanthias*) testis. Bull Mount Desert Island Biol Lab 28:30–31

Du Bois W, Mak P, Callard GV (1989) Sertoli cell function during spermatogenesis: the shark testis model. Fish Physiol Biochem 7:221–227

Ebert DA (2005) Reproductive biology of skates, *Bathyraja* (Ishiyama), along the eastern Bering Sea continental slope. J Fish Biol 66:618–649

Francis M, Maolagain CO, Stevens D (2001) Age, growth, and sexual maturity of two New Zealand endemic skates, *Dipturus nasutus* and *D. innominatus*. NZ J Mar Freshwater Res 35:831–842

Gelsleichter J, Rasmussen LEL, Manire CA, Tyminski J, Chang B, Lombardi-Carlson L (2002) Serum steroid concentrations and development of reproductive organs during puberty in male bonnethead sharks, *Sphyrna tiburo*. Fish Phys Biochem 26:389–401

Heupel MR, Whittier JM, Bennet MB (1999) Plasma steroid hormone profiles and reproductive biology of the epaulette shark, *Hemiscyllium ocellatum*. J Exp Zool 284:586–594

Kneebone J, Sulikowski JA, Ferguson DE, and Tsang PCW (2007) Endocrinological investigation into the reproductive cycles of two sympatric skate species, *Malacoraja senta* and *Amblyraja radiata*, in the western Gulf of Maine. Environ Bio Fish (This issue)

Koob TJ, Tsang P, Callard IP (1986) Plasma estradiol, testosterone and progesterone levels during the ovulatory cycle of the little skate, *Raja erinacea*. Biol Reprod 35:267–275

Koob TJ, Callard IP (1999) Reproductive endocrinology of female elasmobranchs: Lessons from the little skate (*Raja erinacea*) and spiny dogfish (*Squalus acanthias*). J Exp Zool 284:557–574

Manire CA, Rasmussen LEL, Hess DL, Hueter HE (1995) Serum steroid hormones and the reproductive cycle of the female bonnethead shark, *Sphyrna tiburo*. Gen Comp Endocrinol 97: 366–376

Manire CA, Rasmussen LEL, Gross TS (1999) Serum steroid hormones including 11-ketotestosterone, 11-ketoandrostenedione, and dihydroprogesterone in juvenile and adult bonnethead sharks, *Sphyrna tiburo*. J Exp Zool 284:595–603

Pratt HL (1988) Elasmobranch gonad structure—a description and survey. Copeia 1988:719–729

Rasmussen LEL, Murru FL (1992) Long term studies of serum concentrations of reproductively related steroid hormones in individual captive carcharhinids. Aus J Mar Freshwater Res 43:273–281

Rasmussen LEL, Hess DL, Luer CA (1999) Alterations in serum steroid concentrations in the clearnose skate, *Raja eglanteria:* correlations with season and reproductive status. J Exp Zool 284:575–585

Ruocco NL, Lucifora LO, Díaz de Astarloa JM, Wöhler O (2006) Reproductive biology and abundance of the white-dotted skate, *Bathyraja albomaculata*, in the southwest Atlantic. ICES J Mar Sci 63:105–116

Snelson FF Jr., Rasmussen LEL, Johnson MR, Hess DL (1997) Serum concentrations of steroid hormones during reproduction in the Atlantic stingray, *Dasyatis sabina*. Gen Comp Endocrinol 108:67–79

Sulikowski JA, Tsang PCW, Howell WH (2004) Annual changes in steroid hormone concentrations and gonad development in the winter skate, *Leucoraja ocellata*. Mar Biol 144:845–853

Sulikowski JA, Tsang PCW, Howell WH (2005a) Age and size at sexual maturity for the winter skate, *Leucoraja ocellata*, in the western Gulf of Maine based on morphological, histological and steroid hormone analyses. Environ Biol Fish 72:429–441

Sulikowski JA, Kneebone J, Elzey S, Danley P, Howell WH, Tsang PCW (2005b) The reproductive cycle of the thorny skate, *Amblyraja radiata*, in the Gulf of Maine. Fish Bull 103:536–543

Sulikowski JA, Kneebone J, Elzey S, Howell WH, Tsang PCW. Tsang (2006) Using the composite parameters of reproductive morphology, histology and steroid hormones to determine age and size at sexual maturity for the thorny skate, *Amblyraja radiata*, in the western Gulf of Maine. J Fish Biol 69:1449–1465

Sumpter JP, Dodd JM (1979) The annual reproductive cycle of the female lessor spotted dogfish, *Scyliorhinus canicula*, and its endocrine control. J Fish Biol 15:687–695

Tricas TC, Maruska KP, Rasmussen LEL (2000) Annual cycles of steroid hormone production, gonad development, and reproductive behavior in the Atlantic stingray. Gene Comp Endocrinol 118:209–225

Tsang P, Callard IP (1982) Steroid production by isolated skate ovarian follicular cells. Bull Mt Desert Isl Biol Lab 22:96–97

Tsang P, Callard IP (1987) Morphological and endocrine correlates of the reproductive cycle of the aplacental viviparous dogfish, *Squalus acanthias*. Gen Comp Endocrinol 66:182–189

Tsang PCW, Callard IP (1992) Regulation of ovarian steroidogenesis in vitro in the viviparous shark, *Squalus acanthias*. J Exp Zool 261:97–104

Walmsley-Hart SA, Sauer WHH, Buxton CD (1999) The biology of the skates *Raja wallacei* and *R. pullopunctata* (Batoidea: Rajidae) on the Agulhas Bank, South Africa. South African J Mar Sci 21:165–179

Zeiner SJ, Wolf PG (1993) Growth characteristics and estimates of age at maturity of two species of skates (*Raja binoculata* and *Raja rhina*) from Monterey Bay, California. Conserv. Biol. Elasmobr., NOAA Technical Report NMFS. 115 pp

Age and growth estimates for the smooth skate, *Malacoraja senta*, in the Gulf of Maine

Lisa J. Natanson · James A. Sulikowski · Jeff R. Kneebone · Paul C. Tsang

Originally published in the journal Environmental Biology of Fishes, Volume 80, Nos 2–3, 293–308.
DOI 10.1007/s10641-007-9220-y © Springer Science+Business Media B.V. 2007

Abstract Age and growth estimates for the smooth skate, *Malacoraja senta*, were derived from 306 vertebral centra from skates caught in the North Atlantic off the coast of New Hampshire and Massachusetts, USA. Males and females were aged to 15 and 14 years, respectively. Male and female growth diverged at both ends of the data range and the sexes required different growth functions to describe them. Males followed a traditional growth scenario and were best described by a von Bertalanffy curve with a set L_o (11 cm TL) where $L_{inf} = 75.4$ cm TL, $K = 0.12$. Females required the use of back-calculated values to account for a lack of small individuals, using these data they were best described by a von Bertalanffy curve where growth parameters derived from vertebral length-at-age data are $L_{inf} = 69.6$ cm TL, $K = 0.12$, and $L_o = 10$.

Keywords Vertebra · Skate · Age

L. J. Natanson (✉)
USDOC/NOAA/NMFS, 28 Tarzwell Drive, Narragansett, RI 02882, USA
e-mail: Lisa.Natanson@noaa.gov

J. A. Sulikowski
Marine Science Center, University of New England, Biddeford, ME 04005, USA

J. R. Kneebone · P. C. Tsang
Animal and Nutritional Sciences, University of New Hampshire, Durham, NH 03824, USA

Introduction

The smooth skate, *Malacoraja senta*, is one of the smallest (<70 cm total length; <2.0 kg wet weight) species of skate endemic to the western North Atlantic Ocean (McEachran 2002). *M. senta* has a relatively broad geographic distribution, ranging from Newfoundland and southern Gulf of St. Lawrence in Canada to New Jersey in the United States (Robins and Ray 1986; McEachran 2002), yet the only direct biological data for this species pertains to the reproductive cycle for specimens in the Gulf of Maine (Sulikowski et al. in press). Although no directed fisheries for this species exist in the Gulf of Maine, this skate is taken as bycatch (New England Fishery Management Council [NEFMC] 2001).

The vulnerability of sharks to exploitation as bycatch in commercial fisheries is well documented (Bonfil 1994; Musick 1999, 2004). However, recent population assessments of skates suggest these elasmobranchs are also at risk. For example, Dulvy and Reynolds (2002) recently confirmed the disappearance of the common skate, *Dipturus batis*, from the Irish Sea, and reported that the long nose skate, *D. oxyrhinchus* and the white skate, *Rostroraja alba*, were also absent from substantial parts of their ranges. In the western North Atlantic, there have been declines in barndoor, *Dipturus laevis* (Casey and

Meyers 1998), thorny, *Amblyraja radiata*, and smooth skate, stocks (NEFMC 2001, 2003). As a result of these circumstances, commercial landing of these three species is now prohibited in the Gulf of Maine (NEFMC 2003). However, these species continue to suffer the threat of population decline through the indirect fishing pressures of bycatch mortality.

Age information forms the basis for the calculations of growth rate, mortality rate and productivity, making it one of the most influential variables for estimating a population's status and assessing the risks associated with exploitation. Given the current predicament of *M. senta* stocks in the Gulf of Maine, the objective of the present study was to establish age and growth rates of *M. senta* based on annular bands counts within the vertebral centra.

Materials and methods

Collection

Smooth skates were captured 30–40 km off the coast of New Hampshire and Massachusetts, USA, by otter trawl in an approximate 2,300 square km area centered at 42° 50′ N and 70° 15′ W in the Gulf of Maine. Skates were maintained alive on board the F/V Mystique Lady until transport to the University of New Hampshire's Coastal Marine Laboratory (CML). There, individual fish were euthanized (0.3 g · l^{-1} bath of MS222) and morphological measurements, such as total length (TL in cm), disk width (DW in cm), and total wet weight (kg) were recorded. A block of 10 vertebrae from above the abdominal cavity was removed from smooth skates, labeled, and stored frozen. Vertebrae sent for histology were stored in ethanol for transport.

Histology—To ensure an adequate number of sections, two whole vertebrae from each sample were processed for histology. Standard techniques for calcified tissues were adjusted based on the size of the vertebrae (Humason 1972). As per Casey et al. (1985) *tert*-butyl alcohol was used in place of xylene during infiltration to eliminate hardening and distortion of the calcified tissue. Vertebrae were removed from ethanol, and placed in individual tissue capsules. Samples of similar size were decalcified together in a large beaker filled with 100% RDO® rapid decalcifying agent. Constant movement of the solution was provided by a mixing plate and magnetic stir-bar. The length of decalcification varied between a 30 min and 4 h depending on the size of the centrum. The stage of decalcification was determined by the ease with which a sharp scalpel could trim off a section of the vertebra. Once the blade could easily pass through the vertebra with no signs of calcification, decalcification was considered complete. Centra were then transferred to a beaker of running water for several hours. Decalcified centra were stored in 70% ethanol until processing.

Vertebrae were embedded using a nine-step process with each lasting for 2 h (Table 1a). At the completion of step nine, vertebrae were immediately placed in embedding molds filled with 100% Paraplast Plus®. One vertebra of the two processed from each specimen was sectioned, while the other was used as a spare. The embedded centra were sectioned with a sledge microtome to obtain 80–100 μm sections from the center of the centra. The focus of the section was determined by the presence of the notochordal remnant. One to eight sections were obtained from each sample. Sections were placed by sample into a tissue capsule and temporarily immersed in 100% xylene. Paraplast Plus® was progressively removed from the sections in preparation for staining (Table 1b). Once stained with Harris hematoxylin, the sections were brought into glycerin in preparation for mounting on microscope slides with Kaiser Glycerin Jelly (Table 1b). Lastly, the sections were sealed with clear nail polish (Humason 1972).

Each section was digitally photographed with an MTI CCD 72 video camera attached to a SZX9 Olympus stereomicroscope[1] using reflected light and a white background. Band pairs (consisting of one opaque and one translucent band) were counted and measured on the images using Image Pro 4 software. Measurements were made from the midpoint of the notochordal remnant (focus)

---

[1] Reference to trade names does not imply endorsement by the National Marine Fisheries Service.

Table 1 Embedding procedure and Staining procedure

Step	Percent alcohol	Formula	Time (h)	Temperature
(a) Embedding procedure				
1	70%	500 ml ethanol 300 ml distilled water 200 ml *tert*-butyl alcohol	2	Room
2	85%	500 ml ethanol 150 ml distilled water 350 ml *tert*-butyl alcohol	2	Room
3	100%	450 ml ethanol 550 ml *tert*-butyl alcohol	2	Room
4	100%	250 ml ethanol 750 ml *tert*-butyl alcohol	2	Room
5	100%	1000 ml *tert*-butyl alcohol	2	Room (warm enough to be liquid)
6	100%	1000 ml *tert*-butyl alcohol	2	Room (warm enough to be liquid)
7	50%	500 ml *tert*-butyl alcohol 500 ml Paraplast Plus®	2	60°F
8		1000 ml Paraplast Plus®	2	60°F
9		1000 ml Paraplast Plus®	2	60°F

Step	Formula	Time (min)	Notes
(b) Staining procedure			
1	100% xylene	3–10	Time in xylene depends on amount of Paraplast Plus® in tissue and size of tissue. Increase time for increased Paraplast Plus® and tissue size.
2	100% xylene	3–10	
3	100% xylene	3–10	
4	100% ethanol	5	
5	100% ethanol	5	
6	100% ethanol	5	
7	95% ethanol 5% distilled water	5	
8	80% ethanol 20% distilled water	5	
9	100% distilled water	5	Process can be stopped here for overnight
10	Harris Hematoxylin	10	Sections should be checked to ensure proper staining
11	Water rinse	until clear	

Table 1 continued

Step	Formula	Time (min)	Notes
(b) Staining procedure			
12	Acid alcohol*	2	This can be adjusted depending on staining strength
13	Water rinse	1	
14	Lithium carbonate**	5	Use agitation or until the tissue turns blue
15	Running water	10	
16	Distilled water	2	
17	25% glycerin	10	
18	50% glycerin	10	
19	75% glycerin	10	
20	100% glycerin	10	Tissues can be stored for longer periods at this step

* Acid alcohol—65% distilled water, 35% ethanol, 6 drops hydrochloric acid per 100 ml
** Lithium carbonate—400 ml distilled water, 0.2 gm lithium carbonate

of the full bow tie to the opaque growth bands at points along the internal corpus calcareum. The radius of each centrum (VR) was measured from the midpoint of the notochordal remnant to the distal margin of the intermedialia along the same diagonal as the band measurements.

The relationship between VR and TL was calculated to determine the best method for back-calculation of size at age data and to confirm the interpretation of the birth band. Regressions were fit to the male and female data and an ANCOVA was used to test for difference between the two relationships. The relationship between TL and VR was best described by a quadratic equation; therefore the data were ln-transformed before linear regression. The quadratic-modified Dahl-Lea was used for back calculation:

$$L_i = L_c * [(a + bVR_i + cVR_i^2)/(a + bVR_c + cVR_c^2)] \quad (1)$$

where a, b and c are the quadratic fit parameter estimates; VR_i = vertebral radius at band "i"; and VR_c = vertebral radius at capture. Due to a lack of small individuals back-calculations were performed on both sexes.

Vertebral centrum interpretation

A band pair consisted of one opaque and one translucent band. The criteria for designating a band pair were based on broad opaque and translucent bands, each of which was composed of layers of distinct thinner rings (sensu Cailliet et al. 1983; Martin and Cailliet 1988). In most cases the intermedialia was not present thus the bands were determined solely by their appearance on the corpus calcareum. Due to the consistency in counts between the corpus calcareum and intermedialia in those samples which had both we were confident that the counts from the corpus only were representative of the age.

Data analysis

Initial count comparison between co-authors was undertaken on the female sample. Ageing bias and precision of bands counts were examined using APE, D, (Beamish and Fournier 1981) and the coefficient of variation (CV) (Chang 1982). Additionally, a contingency table was made and Chi-square tests of symmetry (McNemar 1947; Bowker 1948; Hoeing et al. 1995; Evans and

Hoenig 1998) performed to determine whether differences between readers were biased or due to random error. An age-bias plot was also calculated (Campana et al. 1995).

Once the criteria for the bands were determined using the within reader comparison, reader 1 (JAS) counted the male sample set and reader 2 (JRK) counted the female sample set. Both sample sets were counted twice. Pairwise comparisons of precision and bias were conducted on the two counts for each sex. Samples that still did not agree were recounted, if a consensus was not reached at that time, the sample was discarded.

Von Bertalanffy growth functions (VBGF) were fit to observed total length-at-age data and back-calculated data using the original equation of von Bertalanffy (1938) with size at birth L_o rather than t_o:

$$L(t) = L_\infty - (L_\infty - L_o)exp^{-kt} \quad (2)$$

where $L(t)$ = predicted length at time t; L_∞ = mean asymptotic total length; k = a rate parameter (yr^{-1}); and L_o = total length at birth.

Two variations of the model were used: 3-parameter calculation estimated L_∞, k and L_o, 2-parameter method estimated L_∞ and k and incorporated a set L_o = 11 cm TL (Kulka and Sulikowski Unpub. data[2]).

As an alternative to the VBGF analysis, we used the Gompertz growth function (GGF) as described in Ricker (1979):

$$L(t) = L_0 e^{G(1-e^{(-kt)})} \quad (3)$$

where $L_\infty = L_o e^G$ is the mean maximum TL ($t = \infty$); k (= g in Ricker, 1975) is a rate constant (yr^{-1}), and L_o = total length at birth.

Two variations of this model were also fit to the data as above with unconstrained parameters and with L_o set using the same value as with the VBGF. All of these growth equations were fit using non-linear regression in Statgraphics (Manuguistics)® on both the observed length and back calculated at age data independently.

Additionally, we applied five cases of the Schnute (1981) growth model, as presented by Bishop et al. (2006), to the observed length at age data. The best fitting case was determined using Akaike's (1973) Information Criterion (AIC), and the Bayesian Information Criterions (BIC) using the following equations:

$$AIC = -2(LL) + 2p \quad (4)$$

where LL is the maximum log-likelihood for each model and p is the number of model parameters.

$$BIC = -2(LL) + p\ln(n) \quad (5)$$

where n = sample size.

Comparison of growth between the sexes

The Kimura (1980) method of using χ^2 tests on each likelihood ratio were used to compare the data between sexes for VBGF. Likelihood ratio and chi-square tests were performed in Microsoft Excel® as outlined by Haddon (2001). Additionally, the AIC and BIC of the best fitting Schnute (1981) case with sexes combined was compared the values for of sexes separate to determine if the sexes grew similarly.

Verification

The periodicity of band pair formation was investigated using marginal increment ratio (MIR) (Skomal 1990; Goldman 2004). The MIR was calculated as the ratio of the distance between the last band pair to the margin to the width of the last complete band pair. Average MIR was plotted by month of capture to identify trends in band formation, and a Kruskal-Wallis one-way analysis of variance on ranks was used to test for differences in marginal increment by month (Simpfendorfer et al. 2000).

Longevity

Three methods were used to estimate longevity. The oldest fish aged from the vertebral method provides an initial estimate of longevity; however, this value is likely to be underestimated in a

---

[2] Unpub. data. 2006. Kulka, D.W. Marine Fish Species at Risk, Fisheries and Oceans, Science, Aquatic Resources Northwest Atlantic Fisheries Centre, St Johns, Newfoundland & Labrador, A1C 5X1 Canada.

fished population as large individuals are removed from the fishery. Taylor (1958) defined the life span of a teleost species as the time required to attain 95% of L_∞. The estimated age at 95% of L_∞ (=longevity in years) was calculated by solving the VBGF and GGFs for t and replacing $L_{(t)}$ with $0.95L_\infty$. For the VBGF we obtained:

$$Longevity = (1/k) \ln \left[\frac{(L_\infty - L_0)}{L_\infty(1 - x)} \right] \qquad (6)$$

and for the Gompertz growth curves we obtained:

$$Longevity = (1/k) \ln \left[\frac{\ln(L_0/L_\infty)}{\ln(x)} \right] \qquad (7)$$

where $x = L(t)/L_\infty = 0.95$.

Results

Vertebral samples from 405 smooth skates (189 males, 216 females) were processed. Poor slide quality (90) and/or lack of length data (10) reduced the readable sample size to 306 (153 males, 153 females). Samples ranged in size from 30.0 to 68.0 cm TL for males and 32.5–62.5 cm TL for females.

Vertebral centrum interpretation

The TL-VR data was best described as a polynomial for both sexes (Fig. 1). The data were log-transformed and fit with a linear regression to compare between the sexes. There was no significant difference between the sexes for intercept ($p = 0.719$) however, there was a significant difference in the slope ($p = 0.0010$).

Therefore, we calculated the quadratic regressions for sexes separately. The regressions were:

$$TL = -5.60 + 33.75*VR$$
$$-4.39*VR^2 \quad R^2 = 0.73, \; n = 146, \text{ female;}$$
$$(8)$$

$$TL = -0.51 + 26.26*VR$$
$$-2.30*VR^2 \quad R^2 = 0.82, \; n = 144, \text{ male.}$$
$$(9)$$

Smooth skate vertebrae did not show consistent pre-birth marks; thus, the first distinct opaque band was defined as the birth band. The location of the birth band (BB) was just outside of the focus there was no angle change associated with this mark (Fig. 2). The mean BB value was not significantly different between males and

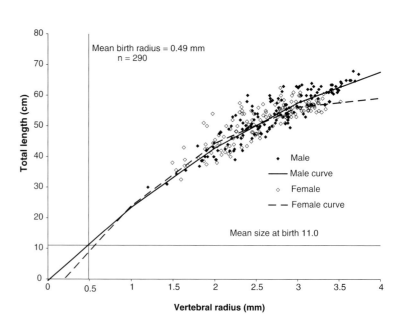

Fig. 1 Relationship between vertebral radius and total length for male and female smooth skates. The horizontal dotted line represents the size at birth and the vertical dotted line represents the mean radius of the birth mark

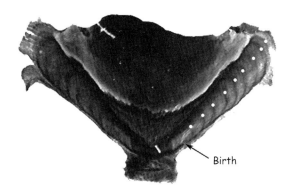

Fig. 2 Photograph of a vertebral section from a 50 cm TL male smooth skate estimated to be 9 years-old. The birth band (BB) is indicated and band pairs are marked with a light circle. Vertebral radius = 2.2 mm

females (Kolmogorov-Smirnov $p = 0.273$) and the sexes were combined for analysis. The mean BB value of the total sample was 0.49 mm ± 0.01 mm (±95% CI) ($N = 290$). Since the mesh size of the commercial trawl precluded the capture of young-of-the-year (YOY) to compare with this radius, we back calculated the lengths at BB for the entire sample using Eq. (1) to obtain an associated mean length at birth and ensure the BB was identified correctly. The calculated birth mean TL (10.84 cm ± 0.30 cm; $N = 290$) is very close to the 11 cm TL size at birth estimated for this species (Kulka and Sulikowski Unpub. data[2]).

Data analysis

Comparison of counts between readers indicated that all readers were identifying the same bands. The CV between Readers 1 and 2 ($n = 100$) varied about a mean of 4.3%. Age bias plots showed minimal variation around the 1:1 ratio and no systematic bias (Fig. 3a). Additionally, the Bowker (1948), McNemar (1947), Hoeing et al. (1995) Evans and Hoenig (1998) χ^2 tests of symmetry gave no indication that differences between readers were systematic rather than due to random error (χ^2 test, $p > 0.05$ in all cases).

Comparison of the first and second counts of each reader also indicated no systematic bias. The individual CV, APE and D were 1.52%, 1.16% and 1.08% for males (Reader 1; $N = 153$) and 5.54%, 5.88%, and 3.92% for females (Reader 2; $N = 153$) (Fig. 3b, c). In the absence of bias, the level of precision was considered acceptable.

Back-calculation

While the back-calculated values were slightly lower on average than the observed for both sexes (4.8% and 7.5% males and female, respectively), the majority of points overlapped and there was no evidence of Lee's Phenomenon (Fig. 4a, b). In the early portion of the female relationship the actual data are limited and do not fall to the known size at birth; the back-calculated data adequately fill in these size classes. Sizes estimated for birth (11.5 cm TL and 10.2 cm TL) were close to estimates of known values.

Growth rates

Male and female growth rates were significantly different (Kimura 1980, $\chi^2 = 36.75$; $p < 0.001$). Additionally, the AIC and BIC values from the best fitting Schnute case (2) were lower for the sexes separate than together, confirming a difference in growth rates between the sexes. Thus, growth models were fit separately by sex (Table 2). Using the assumption of annual band pair deposition, all of the growth curves were fitted to the observed length at age data, additionally the VBGF and the Gompertz models were fit to the back-calculated length at age data.

Male growth curves

All the growth curves fit the male observed data over the aged size ranges (Fig. 5a). The Schnute (1981) Case 2, which is basically a form of the GGF 3 parameter, was the best Case using the Schnute (1981) model. The Schnute (1981) Case 2 and GGF 3 curves were identical; and overestimated size at birth (Table 3 and Fig. 5a). The VBGF 2 and 3 parameter estimations were very similar and provided reasonable estimates of the parameters for males based on observed values of size at birth and maximum size. The residual sum of squares (RSS) (636.3) for the 2 parameter model indicate that this is a slightly better fit to the data.

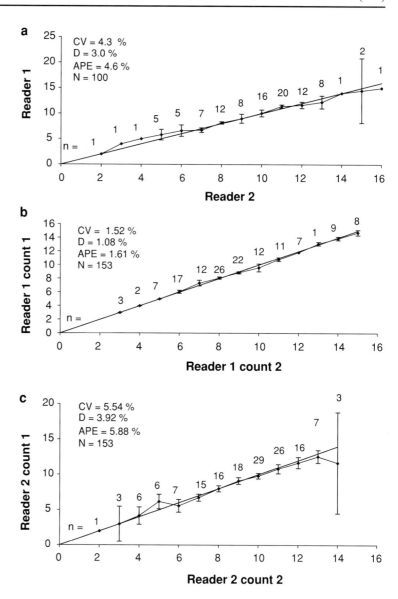

Fig. 3 Age bias graphs for pair-wise comparison of smooth skate vertebral counts from (a) two independent age readings by the one each by reader 1 and 2, n = 100; (b) two independent age readings by reader 1, n = 153; (c) two independent age readings by reader 2, n = 153. Each error bar represents the 95% confidence interval for the mean age assigned in reading 2 to all fish assigned a given age in reading 1. The one to one equivalence line is also presented as are the estimates of APE, D and CV

All the growth curves fit the male back-calculated data over the size range aged (Fig. 5b). The VBGF 3 parameter model provided the best fit (RSS = 9694.5) and provided reasonable estimates of the parameters for males based on observed values of size at birth and maximum size (Table 4). The fit of both the back-calculated VBGF 3 and the observed VBGF 2 are almost identical over the data (Fig. 5b, c). However, the L_∞ value is better represented in the observed VBGF 2 parameter model; thus, this model was chosen to represent the male growth in further analyses (Table 3).

Female growth curves

In contrast to the males, female observed growth was more difficult to define. As with the males, the Schnute (1981) Case 2 provided the best fit for the Schnute (1981) model and it was an identical curve to the GGF 3 parameter (Fig. 6a and Table 3). These two curves and the VBGF 3 parameter fit the data but overestimated the size at birth. These overestimates are probably due to a lack of small samples and the variability in those sampled. The VBGF and GGF models with the L_0 set provided good fits to the data at ages five to

Fig. 4 Smooth skate band pair counts versus total length based on back-calculated data (solid symbols) and observed data (opened diamonds) of (**a**) males and (**b**) females

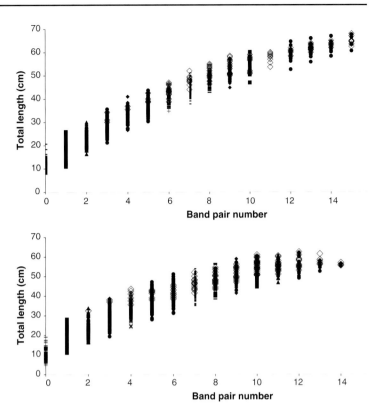

Table 2 Comparison of AIC and BIC for the Schnute case 2 model, for sexes combined and separate

Model	Sex	Growth function	N	RSS	Log likelihood	# params	AIC	BIC
1b	Combined	Schnute 2	305	2287	−740.02	4	1488.05	1502.93
2	Males	Schnute 2	153	629	−356.95			
2	Females	Schnute 2	152	1269	−354.62			
2	Separately	Schnute 2	305	1898	−711.57	7	1437.13	1463.18

14. Despite the high L_o, the comparison of RSS suggest that the GGF 3 model provided the best fit to the data.

All the growth curves fit the female back-calculated data over the size range aged (Fig. 6b and Table 4). The VBGF 3 parameter model provided the best fit (RSS = 18065.6) and reasonable estimates of the parameters for females based on observed values of size at birth and maximum size. The lower portion of the curve of the back-calculated VBGF 3 is more reasonable than that of the best fitting observed model (GGF 3 parameter, $R^2 = 80.2$) (based on observed values of size at birth and maximum size) and warrants the use of the back-calculated VBGF 3 model for the females. The L_∞ and K parameter estimations for both models are similar; the difference between the models lies in the estimations of L_0 which is improved by filling in the lower portion of the curve with the back-calculated values. The curves are similar after age 6, prior to this age they are effected by the different size at birth estimations (Fig. 6c). Due to the similarity between the back calculated and observed data, and the lack of any Lee's phenomenon, we feel that our results accurately represent growth rates of female smooth skates. Thus, the back-calculated VBGF 3 parameter estimation was chosen to represent the female growth rates in further analyses (Table 4).

Fig. 5 Male smooth skate growth curves based on (**a**) observed data; (**b**) back-calculated data; and (**c**) observed and back-calculated comparison

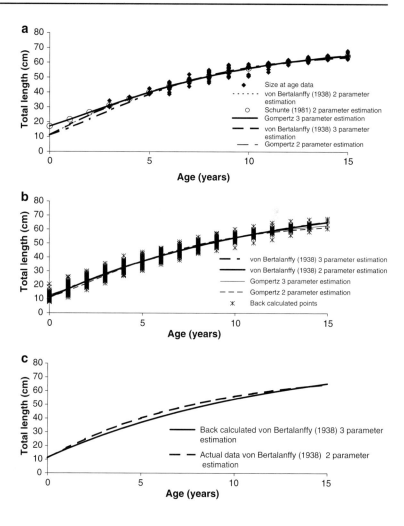

Verification

Marginal increment analyses were performed on 120 female and 131 male skates over the range of age and size classes. In males, marginal increments were significantly different between months (Kruskal-Wallis $p < 0.001$); with a distinct trend of increasing monthly increment growth peaking in July, followed by a decline then nadir in December (Fig. 7a). Based on this information, the increment analyses support the likelihood that a single opaque band may be formed annually on male vertebral centra during December. Although an increasing trend that peaked in October was present in female marginal increments, these values were not significantly different between months (Kruskal-Wallis $p = 0.081$; Fig. 7b). Thus, in contrast to males, the periodicity of annual band formation in females lacks statistical verification.

Longevity

The maximum age based on vertebral band pair counts was 15 and 14 years, for males and females, respectively. The calculated longevity estimates using the values appropriate for each sex are 24 years for males and 23 years for females.

Discussion

The histological processing of smooth skate vertebrae resulted in the elucidation of alternating opaque and translucent bands. Using this method,

Table 3 von Bertalanffy growth function parameters and 95% confidence intervals calculated by using observed vertebral counts

Method		L_{inf}	K	L_0	n	Longevity	Residual sum of squares
von Bertalanffy—3 parameter	Male	75.487	0.120	11.041	153	23.6	636.3
	CI±	4.523	0.025	5.069			
	Female	74.970	0.090	22.657	152	31.4	1289.4
	CI±	14.225	0.051	5.810			
von Bertalanffy—2 parameter	Male	75.456	0.120		153	23.6	636.3
	CI±	2.415	0.009				
	Female	63.762	0.170		152	16.5	1409.2
	CI±	2.614	0.021				
Gompertz—3 parameter	Male	70.857	0.182	17.080	153	18.3	629.1
	CI±		0.027	2.857			
	Female	68.402	0.145	24.228	152	20.7	1268.7
	CI±		0.054	4.323			
Gompertz—2 parameter	Male	66.888	0.240		153	14.9	710.3
	CI±		0.009				
	Female	59.184	0.298		152	11.7	1625.9
	CI±		0.022				

vertebral sections were easy to interpret, and false bands (those bands that did not completely encircle the centra) were easily distinguished from complete bands. Precision and lack of bias in our estimates suggest that our band interpretation methods represent a precise approach for ageing *M. senta*.

While we are confident in our ability to determine band pairs with consistency, we are lacking accuracy, as we have no validation of the periodicity of band pair formation. The Kruskal-Wallis test indicated that there was a significant difference in the monthly MIR in males, however, the female result, though very close to significant at ($p = 0.08$) was not. Yoccoz (1991) suggested that biological significance cannot always be related to statistical significance and in the current case we believe this argument is substantiated. In female smooth skates there are clear transitional peaks and numerical lows observed in the monthly MIR. Moreover, the 0.08 *p*-value was very close to statistical significant level set at 0.05, which suggests that, while not statistically significant, a biologically important event was occurring in female vertebrae.

Table 4 von Bertalanffy growth function parameters and 95% confidence intervals calculated by using back-calculated vertebral counts

Method		L_{inf}	K	L_0	n	Longevity	Residual sum of squares
von Bertalanffy—3 parameter	Male	88.037	0.081	11.478	1493	35.2	9694.5
	CI±	3.005	0.005	0.320			
	Female	69.606	0.125	10.021	1479	22.5	18065.6
	CI±	2.235	0.009	0.462			
von Bertalanffy—2 parameter	Male	85.756	0.086		1493	33.3	9750.4
	CI±	2.312	0.004				
	Female	72.156	0.114		1479	24.8	18280.4
	CI±	2.283	0.007				
Gompertz—3 parameter	Male	68.273	0.200	12.806	1493	17.4	10181.7
	CI±		0.006	0.278			
	Female	58.991	0.263	11.400	1479	13.2	18489.1
	CI±		0.011	0.397			
Gompertz—2 parameter	Male	64.809	0.232		1493	15.2	11305.8
	CI±		0.004				
	Female	58.460	0.272		1479	12.8	18539.6
	CI±		0.007				

Fig. 6 Female smooth skate growth curves based on (**a**) observed data; (**b**) back-calculated data; and (**c**) observed and back-calculated comparison

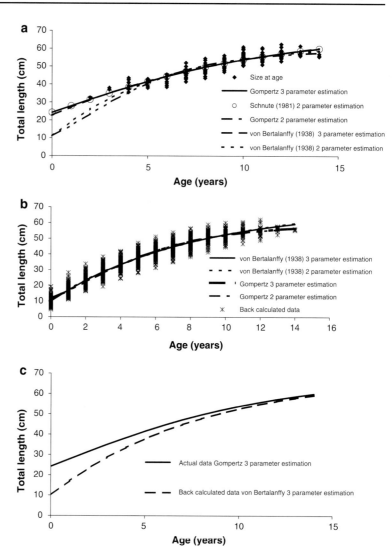

Another interesting aspect of female marginal increments concerns the lack of congruency between male and female increment growth. There was a distinct difference in the timing of the peaks between male and female mean MIR. Male marginal increments peaked 3 months earlier and reached numerical lows 2 months earlier than females suggesting a difference in the timing of band deposition of the band pairs. Since skates of all size classes were utilized for MIA, the disparity cannot be resolved due to differences in skate size. Since the reproductive cycles of the sexes are synchronous, it would be unlikely that reproductive events are responsible for the difference in band deposition cycles between the sexes (Sulikowski et al. in press). At this time the size and age at maturity for the smooth skate is unknown, but one possibility is that the females stop depositing band pairs on a regular basis once they are mature. This was directly observed in the little skate, *Raja erinacea*, in a laboratory study where two tetracycline injected females that were laying eggs did not deposit annual bands, whereas the six non-reproductively active females deposited clear band pairs (Natanson 1993). In the event that the mature smooth skate females deposit annual bands sporadically, or not at all, the timing would be unpredictable and the ages as related to size would be less predictable than the males. In fact in this study the females were much more difficult to age as is evidenced by the larger values for CV, APE and D in the females than

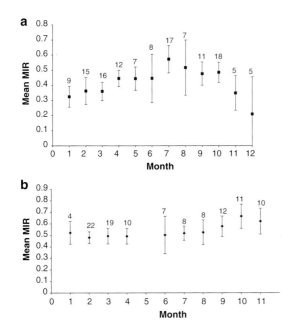

Fig. 7 Mean monthly marginal increments for (**a**) male and (**b**) female smooth skates, *Malacoraja senta*. Sample sizes are given above each corresponding month. Error bars represent 1 ± SD

males and there was greater variability in the observed length at age values. Additionally, the observed data was not as easily fit with a growth model.

Male and female growth diverged at both ends of the data range and thus the sexes required different growth functions to best describe them. While the fitting of growth models to the male smooth skate data was fairly straightforward, the fitting of models to the female data was challenging. The typical growth curve of the observed male data was able to be fit using a variety of models with the main difference being in the size at birth estimation. All of the models, which estimated all parameters, overestimated size at birth. Setting the size at birth and using the VBGF provides the best fit though in fact the curves for all of the models fit that data similarly past age three to four (Fig. 5a). The lack of individuals less than 6 years old is the most likely reason for the inability of the curve to adequately represent the birth size. The lack of small individuals for this study is cause for the use of back-calculation. However, the male growth curves produced using back-calculated values were not sufficiently different from those calculated using the observed values. The female observed data was not a typical growth curve and contained a depression at age six and did not completely reach an asymptote (Fig. 6a). This shape along with the lack of small individuals made it difficult for the models to adequately predict female growth. The best fit for the female observed data was the GGF 3 parameter estimation, though even that overestimated size at birth (Table 3). Despite the different growth parameters the curves for all of the models overlapped past age four; the difference before age four being a result of setting the size at birth on the two parameter estimations. In the case of the female data the use of the back-calculated data vastly improved the fit of all growth models and the traditional 3 parameter VBGF provided the best fit to these data. The difficulty in fitting VBGFs is not unprecedented in elasmobranchs in fact over estimation of the size at birth is quite common (Carlson et al. 2003; Bishop et al. 2006; Natanson et al. in press). The recent use of set sizes at birth indicates that this is problem needs to be addressed. Often as is probably the case in this study, the growth functions have difficulty due to low sample sizes in the younger ages. Additionally, the VBGF has difficulty with species that have rapid initial growth this had lead to the introduction of the use of different growth models to describe growth (Bishop et al. 2006; Natanson et al. 2006).

Females of other elasmobranch species have also been shown to have growth that is less easily defined than the males. Several studies on the shortfin mako, *Isurus oxyrinchus*, had similar findings with females such as no asymptote and high estimates of size at birth from their von Bertalanffy (1938) equation and felt that the Schnute (1981) four parameter function (Bishop et al. 2006) or the GGF 3 parameter model (Natanson et al. 2006) better described growth. While this is a very unrelated species, it is possible that the reproductive energy demands placed on females of a viviparous species with large young causes the female to divert calcium from the

Table 5 Comparison of von Bertalanffy growth parameters for several similar sized skate species

Scientific name	Sex	L_∞ (mm)	k	Max age (yr)	Source
Raja erinacea	♀♂	527 (TL)	0.35	8	Waring 1984
Raja wallacei	♀♂	422 (DW)	0.26	15	Walmsley-Hart et al. 1999
Raja montagui	♂	687 (TL)	0.19		Holden 1972
Raja montagui	♀	728 (TL)	0.18		Holden 1972
Malocoraja senta	♂	754 (TL)	0.12	15	This study
Malocoraja senta	♀	696 (TL)	0.12	14	This Study

DW = Disk width; TL = Total length

structural components to the young. This phenomenon needs further investigation before any conclusions can be drawn. The lack of comparison to other similar species also precludes a definitive answer at this time.

Growth rates and oldest aged specimens were similar for both sexes of smooth skates ($k = 0.12$; age = 14 for females and $k = 0.12$, age = 15 for males). These data are in agreement with the life history values from similar sized skates species (Table 5). For instance, the Antarctic skate *Amblyraja georgiana*, which reaches a total length of 71 cm, has been aged to 13 years and found to have a corresponding k value of 0.31 for combined sexes (Francis and Maolagáin 2005).

The current study of the smooth skate demonstrates the potential value of using alternative growth models to the commonly used VBGF. This is especially important as demographic analyses requires accurate determinations of the growth coefficient if stock assessments are to be successful in preventing overexploitation of a species (e.g., Cortés 1999; Goldman 2002; Neer and Thompson 2005). The basic age and growth parameters for the smooth skate provided in the present study support the hypothesis that *M. senta*, like other elasmobranchs, require conservative management because of their slow growth rate and susceptibility to over-exploitation (Brander 1981; Kusher et al. 1992; Zeiner and Wolf 1993; Frisk et al. 2001; Sulikowski et al. 2003, 2005).

Acknowledgements We would like to extend our appreciation to Malcolm Francis for running our Schnute growth functions and providing invaluable insights into that method. Additionally, both Malcolm and John Carlson provided much appreciated time saving spreadsheets, which made our analysis less painful. We would also like to thank Olivia Marcus for help in processing the histological samples and Joseph Deppen for sealing the slides. We appreciate the help provided by Karen Tougas in formatting this paper and the continued support of all members of the Apex Predators Investigation. This work was supported by New Hampshire Sea Grant Development grant #NA16RG1035.

References

Akaike H (1973) Information theory and the extension of the maximum likelihood principle. In: Petrov BN, Csaki F (eds) International symposium on information theory. Academiai Kaido, Budapest, pp 267–281

Beamish RJ, Fournier DL (1981) A method for comparing the precision of a set of age determinations. Can J Fish Aquat Sci 38:982–983

Bishop SDH, Francis MP, Duffy C, Montgomery JC (2006) Age, growth, maturity, longevity and natural mortality of the shortfin mako (Isurus oxyrinchus) in New Zealand waters. Mar Fresh Res 57:143–154

Bonfil R (1994) Overview of world elasmobranch fisheries. FAO Fisheries Technical Paper No. 341, 119 pp

Bowker AH (1948) A test for symmetry in contingency tables. J Am Stat Assoc 43:572–574

Brander K (1981) Disappearance of common skate Raia batis from Irish Sea. Nature 290:48–49

Cailliet GM, Martin LK, Kusher D, Wolf P, Welden BA (1983) Techniques for enhancing vertebral bands in age estimation of California elasmobranchs. In: Prince ED, Pulos LM (eds) Proceedings International Workshop on Age Determination of Oceanic Pelagic Fishes: Tunas, Billfishes, Sharks, NOAA Tech Rep NMFS 8:157-165

Campana SE, Annand MC, McMillan JI (1995) Graphical and statistical methods for determining consistency of age determinations. Trans Amer Fish Soc 124:131–138

Carlson JK, Cortés E, Bethea D (2003) Life history and population dynamics of the finetooth shark (Carcharhinus isodon) in the northeastern Gulf of Mexico. Fish Bull 101:281–292

Casey JG, Pratt HL Jr., Stillwell CE (1985) Age and growth of the sandbar shark(Carcharhinus plumbeus) from the western North Atlantic. Can J Fish Aquat Sci 42(5):963–975

Casey JM, Myers RA (1998) Near extinction of a large, widely distributed fish. Science 281:690–692

Chang WYB (1982) A statistical method for evaluating the reproducibility of age determination. Can J Fish Aquat Sci 39:1208–1210

Cortés E (1999) A stochastic stage-based population model of the sandbar shark in the western North Atlantic. In: Musick JA (ed) Ecology and conservation of long-lived marine animals. Amer Fish Soc Symp 23:115–136

Dulvy NK, Reynolds JD (2002) Predicting extinction vulnerability in skates. Conserv Biol 16:440–450

Evans GT, Hoenig JM (1998) Testing and viewing symmetry in contingency tables, with application to readers of fish ages. Biometrics 54:620–629

Francis MP, Ó Maolagáin C (2005) Age and growth of the Antarctic Skate (Amblyraja georgiana) in the Ross Sea. CCAMLR Sci 12:183–194

Frisk MG, Miller TJ, Fogarty MJ (2001) Estimation and analysis of biological parameters in elasmobranch fishes: a comparative life history study. Can J Fish Aquat Sci 58:969–981

Goldman KJ (2002) Aspects of Age, Growth, Demographics and Thermal Biology of Two Lamniform Shark Species. Ph.D. dissertation, College of William and Mary, School of Marine Science, Virginia Institute of Marine Science, 220 pp

Goldman KJ (2004) Age and growth of elasmobranch fishes. In: Management Techniques for Elasmobranch Fisheries. Musick JA, Bonfil R (eds) FAO Fisheries Technical Paper, No. 474. Rome, FAO. 251 p (electronic version available at http://www.flmnh.ufl.edu/fish/organizations/ssg/EFMT2004.htm)

Haddon M (2001) Growth of individuals In: Modelling and quantitative measures in fisheries, Chapman & Hall/CRC Press, Boca Raton, FL, pp 187–246

Hoeing JM, Morgan MJ, Brown CA (1995) Analyzing differences between two age determination methods by tests of symmetry. Can J Fish Aquat Sci 52:364–368

Holden MJ (1972) The growth rates of *Raja brachyura*, *R. clavata* and *R. montagui* as determined from tagging data. J Cons Int Explor Mer 34(2):161–168

Humason GL (1972) Animal tissue techniques, 4th edn. W.H. Freeman and Company, San Francisco, 661 pp

Kimura DE (1980) Likelihood methods for the von Bertalanffy growth curve. Fish Bull 77(4):765–776

Kusher D, Smith SE, Cailliet GM (1992) Validated age and growth of the leopard shark, Triakis semifasciata, with comments on reproduction. Environ Biol Fish 35:187–303

Martin LK, Cailliet GM (1988) Age and growth of the bat ray, Myliobatis californica, off central California. Copeia 1988(3):762–773

McEachran JD (2002) Skates: Family Rajidae. In: Collette B, Klein-MacPhee G (eds) Bigelow and Schroeder's Fishes of the Gulf of Maine, 3rd edn. Smithsonian Institution Press, pp 60–75

McNemar Q (1947) Note on the sampling error of the difference between correlated proportions or percentages. Psychometrika 12:153–157

Musick JA (1999) Ecology and conservation of long-lived marine animals. In: Musick JA (ed) Life in the slow lane: Ecology and conservation of long-lived marine animals. American Fisheries Society Symposium 23:1–10

Musick JA (2004) Introduction: management of sharks and their relatives (Elasmobranchii), pp 1–6. In: Management Techniques for Elasmobranch Fisheries. Musick JA, Bonfil R (eds) FAO Fisheries Technical Paper, No. 474. Rome, FAO. 251 p (electronic version available at http://www.flmnh.ufl.edu/fish/organizations/ssg/EFMT2004.htm)

Natanson LJ (1993) Effect of temperature on band deposition in the little skate, Raja erinacea. Copeia 1993(1):199–206

Natanson LJ, Ardizzone D, Cailliet GM, Wintner S, Mollet H (2006) Validated age and growth estimates for the shortfin mako, Isurus oxyrinchus, in the North Atlantic Ocean. Special volume from symposium of the American Elasmobranch Society, July 2005. Goldman KJ, Carlson JK (eds) Environ Biol Fish 77:367–383

Neer JA, Thompson BA (2005) Life history of the cownose ray, Rhinoptera bonasus, in the northern Gulf of Mexico, with comments on geographic variability in life history traits. Environ Biol Fish 73:321–331

New England Fishery Management Council (2001) 2000 Stock Assessment and Fishery Evaluation (SAFE) Report for the Northeast Skate Complex. Newburyport, MA, 179 pp

New England Fishery Management Council (2003) Skate Fisheries Management Plan. 50 Water Street, Mill 2 Newburyport, MA, 01950

Ricker WE (1975) Computations and interpretation of biological statisticsof fish populations. Fish Res Bd Canada Bull 191. 382 pp

Ricker WE (1979) Growth rates and models. In: Hoar WS, Randall DJ, Brett JR (eds) Fish physiology, vol VIII: Bioenergetics and growth. Academic Press, Florida, pp 677–743

Robins CR, Ray GC (1986) A Field Guide to Atlantic Coast Fishes of North America. Houghton Mifflin Company, Boston, USA, 354 pp

Schnute J (1981) A versatile growth model with statistically stable parameters. Can J Fish Aquat Sci 38:1128–1140

Simpfendorfer CA, Chidlow J, McAuley R, Unsworth P (2000). Age and growth of the whiskery shark, Furgaleus macki, from southwestern Australia. Environ Biol Fish 58:335–343

Skomal GB (1990) Age and growth of the blue shark, Prionace glauca, in the North Atlantic. Master's Thesis, University of Rhode Island, 82 pp

Sulikowski JA, Morin MD, Suk SH, Howell WH (2003) Age and growth of the winter skate (Leucoraja ocellata) in the western Gulf of Maine. Fish Bull 101:405–413

Sulikowski JA, Kneebone J, Elzey S, Danley P, Howell WH, Tsang PCW (2005) Age and growth estimates of the thorny skate, Amblyraja radiata, in the Gulf of Maine. Fish Bull 3:161–168

Sulikowski JA, Elzey S, Kneebone J, Howell WH, Tsang PCW (2007) The reproductive cycle of the smooth skate, Malacoraja senta, in the Gulf of Maine. J Mar Fresh Res 58:98–103

Taylor CC (1958) Cod growth and temperature. J Cons Int Explor Mer 23:366–370

von Bertalanffy L (1938) A quantitative theory of organic growth (inquiries on growth laws II). Hum Biol 10:181–213

Walmsley-Hart SA, Sauer WHH, Buxton CD (1999) The biology of the skates Raja wallacei and R. pullopunctata (Batoidea:Rajidae) on the Agulhas Bank, South Africa. S. Afr J Mar Sci 21:165–179

Waring GT (1984) Age, growth, and mortality of the little skate off the northeast coast of the United States. Trans Am Fish Soc 113:314–321

Yoccoz NG (1991) Use, overuse, and misuse of significance tests in evolutionary biology and ecology. Bull Ecol Soc Am 71:106–111

Zeiner SJ, Wolf P (1993) Growth characteristics and estimates of age at maturity of two species of skates (Raja binoculata and Raja rhina) from Monterey Bay, California. NOAA Tech Rep 115:87–99

Age, growth, maturity, and mortality of the Alaska skate, *Bathyraja parmifera*, in the eastern Bering Sea

Mary Elizabeth Matta · Donald R. Gunderson

Originally published in the journal Environmental Biology of Fishes, Volume 80, Nos 2–3, 309–323.
DOI 10.1007/s10641-007-9223-8 © Springer Science+Business Media B.V. 2007

Abstract The Alaska skate, *Bathyraja parmifera*, is the most abundant species of skate on the eastern Bering Sea shelf, accounting for over 90% of total skate biomass. However, little is known regarding the life history of this species despite its common occurrence as bycatch in several Bering Sea fisheries. This is the first study to focus on the age and growth of *B. parmifera*. From 2003 to 2005, more than one thousand specimens were collected by fisheries observers and on scientific groundfish surveys. Annual banding patterns in more than 500 thin sections of vertebral centra were examined for age determination. Caudal thorns were tested as a potentially non-lethal ageing structure. Annual band pair deposition was verified through edge and marginal increment analyses. A three-parameter von Bertalanffy growth function and a Gompertz growth function were fit to observed length-at-age data. Both models provided significant fits, although the Gompertz function best described the overall pattern of growth in both males and females, based upon statistical criteria and parameter estimates. Age and size at 50% maturity were 9 years and 92 cm TL for males and 10 years and 93 cm TL for females. The maximum observed ages for males and females were 15 years and 17 years, respectively. Estimates of natural mortality (M) ranged from 0.14 to 0.28, and were based on published relationships between M and longevity, age at maturity, and the von Bertalanffy growth coefficient. Due to these life history characteristics and a lack of long-term species-specific stock data, a conservative management approach would be appropriate for *B. parmifera*.

Keywords Elasmobranch · Rajidae · Life history · Natural mortality · Caudal thorns

Introduction

The Alaska skate, *Bathyraja parmifera*, is a large-bodied species (111 cm maximum total length (TL); Ebert 2005) that is widely distributed across the eastern Bering Sea (EBS) and typically found at depths from 50 m to 200 m (Stevenson 2004). While its range extends to the eastern Aleutian Islands and the Gulf of Alaska, *Bathyraja parmifera* is the dominant skate species across the EBS shelf, accounting for over 90% of total skate biomass there (Matta et al. 2006). Despite the abundance of *B. parmifera*, little is known regarding its life history. Additionally, no long-term species-specific catch data are available, since skate catch in the EBS is officially reported in aggregate, and fishery observers have only been

M. E. Matta (✉) · D. R. Gunderson
School of Aquatic and Fishery Sciences, University of Washington, Box 355020, Seattle, WA 98105, USA
e-mail: bmatta@u.washington.edu

identifying skates to species since 2003 (Matta et al. 2006).

While there is currently no directed fishery for *B. parmifera* in the EBS, the formation of one is possible due to its large body size and abundance. Skates already make up a large part of the observed incidental catch in several EBS groundfish fisheries, approximately 30% of which is retained (Matta et al. 2006). Declines of large-bodied skate species have been documented in other parts of the world and have often been attributed to overfishing (Brander 1981; Walker and Heessen 1996; Casey and Meyers 1998; Dulvy et al. 2000). Certain species of skate may have particularly low resilience to exploitation and increased vulnerability to extinction due to their life history characteristics (Walker and Hislop 1998; Dulvy and Reynolds 2002), highlighting the need for species-specific data when making management decisions.

Skates belong to the taxonomic group Elasmobranchii, which also includes sharks and rays. As a group, elasmobranchs can be classified as equilibrium (K-selected) strategists and typically share the biological characteristics of low fecundity, late maturity, high juvenile survivorship, slow body growth, and long life span (Hoenig and Gruber 1990; Winemiller and Rose 1992; Camhi et al. 1998; King and McFarlane 2003). Equilibrium strategists tend to have low reproductive rates and slow population growth and generally cannot sustain high levels of fishing pressure (Holden 1974; Adams 1980; Hoenig and Gruber 1990). King and McFarlane (2003) suggested that due to their life history characteristics, equilibrium strategists should only be harvested at low to moderate rates.

Life history studies are necessary for stock assessment, as the resulting parameters are often incorporated into fishery models. Age at maturity may be an especially good indicator of a population's resilience to fishing pressure, since it is negatively correlated with population growth rate (Cortés 2002). Furthermore, age at maturity, the von Bertalanffy growth coefficient, and longevity may be useful predictors of the instantaneous rate of natural mortality (Hoenig 1983; Jensen 1996). Such information is essential to understanding the biology and dynamics of a given population.

Ageing elasmobranchs is a complicated process, as there are few available calcified structures and difficulties in validation. However, in recent years researchers have successfully developed accurate ageing methods for skates (Cailliet and Goldman 2004). Skates are generally aged by counting vertebral annuli, although caudal thorns are semi-calcified structures that have been aged successfully in several species (Gallagher and Nolan 1999). Caudal thorns are external modified spines that are thought to aid in protection; the benefit of using caudal thorns as ageing structures is that they can presumably be obtained non-lethally.

The aims of this study were to: (1) determine whether vertebral thin sections and caudal thorns are appropriate ageing structures for *B. parmifera*; (2) describe the pattern of growth for both sexes; (3) verify age estimates through edge analysis and marginal increment analysis (MIA); (4) estimate size and age at maturity; and (5) estimate the instantaneous rate of natural mortality (M) from empirical longevity, von Bertalanffy growth coefficients, and age at maturity estimates. This study is the first to provide estimates of age and growth, age at maturity, and natural mortality for *B. parmifera* in the eastern Bering Sea.

Methods

Specimen collection and preparation

Skates were collected from NOAA Fisheries EBS groundfish trawl surveys during the summers of 2003 and 2004, and throughout 2004 and 2005 by the North Pacific Groundfish Observer Program (NPGOP) on flatfish trawlers and Pacific cod longline vessels. Due to morphological similarities between Bering Sea skates, fishery observers were specially trained in species identification prior to sampling. Samples were obtained every month of the year excluding March, April, and May. Each skate was sexed, and weighed when possible, and TL and disc width (DW) measurements were recorded to the nearest centimeter. Maturity stage was determined at sea using the modified criteria of Zeiner and Wolf (1993; Table 1) and later confirmed in the laboratory through visual inspection of the gonads. A length of vertebral column was removed from the thoracic (mid-dorsal) region of the body and frozen until further processing was possible. Caudal thorns were removed from the base of the tail where it joins the disc and stored frozen. Three to five vertebrae from each specimen were individually separated with a sharp blade, rinsed in tap water, and allowed to soak in 70–95% ethanol for at least 24 h. After air-drying, vertebrae were cut through the centrum with a double-bladed low-speed saw to create sagittal thin sections approximately 0.3 mm thick, which were then mounted on glass slides. Caudal thorns were cleaned by boiling in tap water for 10–20 min, and excess tissue was thereafter gently wiped away using a piece of tissue-paper and forceps.

Age estimates and verification

Vertebral band pairs, or annuli, were viewed against a dark background with reflected light using a dissecting microscope. Mineral oil was applied to the thin section surface to clarify the banding pattern. A band pair consisted of one wide opaque band of growth that appeared lightly colored and one narrow translucent band that appeared darkly colored (Fig. 1). Age estimates were generated by counting translucent bands. Each translucent band was considered to represent one year of growth if it accompanied an opaque band and if it could be detected across the corpus calcareum. The birthmark increment, assumed to correspond to the event of hatching from the egg case, was evident from a change in the angle of the corpus calcareum relative to the intermedialia (Goldman 2004) and was not included in the age estimate. In some cases, faint checks were present within the larger bands. These checks were distinguished from growth zones by their lack of continuity throughout the ageing structure, irregular spacing, and faint appearance.

The annuli in caudal thorns were also viewed with a dissecting microscope, using reflected light and a dark background (Fig. 2). As in vertebral thin sections, each band pair consisted of one translucent band and one opaque band, together assumed to represent one year of growth. Age estimates were generated by counting each translucent band with reference to a protothorn (birthmark increment) located at the apex of each caudal thorn (Gallagher and Nolan 1999). The protothorn was evident by an angle change relative to the next fully formed translucent band. Each translucent band had a ridge-like appearance and was considered to represent one year if it accompanied an opaque band and extended around the circumference of the thorn. Age estimates generated from caudal thorns and vertebrae were compared by means of a paired-sample t test and an age-bias plot.

Vertebral thin sections were aged three independent times by a primary reader, and a subset of 72 samples was aged three times by a second reader with experience in ageing skates. Several

Table 1 Criteria used to define maturity stages in male and female *B. parmifera*, adapted from Zeiner and Wolf (1993)

Stage	Male characteristics	Female characteristics
1 Juvenile (Immature)	No coiling of vas deferens Testes small and undeveloped Claspers do not extend past posterior edge of pelvic fin No alar thorns present	Small undeveloped ovaries and shell glands No differentiated ova present
2 Adolescent (Maturing)	Some coiling of vas deferens Testes enlarging Claspers extend past pelvic fin edge but are only somewhat calcified Alar thorns may be present	Ovaries beginning to enlarge and differentiate Ova white, distinguishable but smal Shell glands widening but still gelatinous in appearance
3 Adult (Mature)	Complete coiling of vas deferens Large developed testes Calcified claspers Prominent alar thorns	Large yellow (yolked) ova, greater than 2.0 cm in diameter Large, solid, functional shell glands

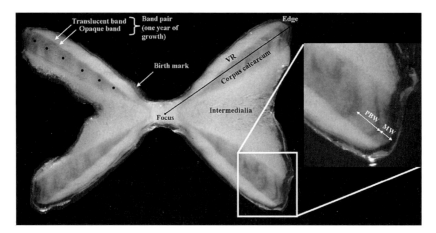

Fig. 1 Digital photograph of a vertebral thin section from a 75 cm TL female skate, viewed with reflected light against a dark background. The birthmark increment is indicated by the angle change in the corpus calcareum relative to the intermedialia. VR is vertebral radius. Inset shows measurements used for marginal increment analysis (MIA), where PBW is penultimate band pair width and MW is margin width. Translucent bands, indicated by solid black circles, were counted to give an age estimate of 6 years old for this specimen. Photographs were not digitally enhanced

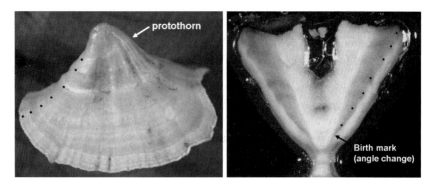

Fig. 2 Caudal thorn (left) and vertebral thin section (right) from a 7 year old, 82 cm TL skate. Translucent bands are indicated by solid circles

measures of precision were used to evaluate the consistency of inter- and intra-reader age estimates, including average percent error (APE; Beamish and Fournier 1981), the coefficient of variation (CV), and percent reader agreement (PA). Additionally, bias between readers was visually assessed using an age-bias plot.

Quality, or readability, was recorded by the primary reader on a scale of one to five, based on criteria described by the Committee of Age Reading Experts (CARE 2000). A quality rating of '1' was considered poor, with considerable variation expected between readings due to inaccurate band visualization. A quality of '5' corresponded to a very clear growth pattern, with an excellent chance of reproducing the same age in subsequent readings. By recording quality, samples with vague or 'checky' patterns could be omitted from the analysis if necessary, reducing the possibility of inaccurate age estimation. Only samples with APE less than 15% (Sulikowski et al. 2003) and a quality rating greater than '2' were used to estimate growth parameters.

TL was plotted against vertebral radius (VR) and against caudal thorn base length (TBL) to ensure that the growth of each ageing structure was proportional to the growth of the skate. In vertebral thin sections, this relationship was used to generate a back-calculation function:

$$TL_i = \left(\frac{VR_i}{VR_c}\right)^v TL_c$$

where TL_i = total length (cm) at time of formation of growth increment i, VR_i = vertebral radius (mm) at time of increment formation, TL_c = total length at time of capture, VR_c = vertebral radius at time of capture, and v = the slope parameter derived from the allometric TL–VR relationship (Francis 1990). The back-calculation function was used to determine the average length at time of birthmark increment formation (TL_{BM}). Length at birth was estimated independently from late-stage embryos and from the smallest free-swimming individual observed during this study; the actual length at birth was assumed to fall within this range of values. By comparing TL_{BM} with the independent estimates of length at birth, it was possible to verify the accurate identification of the birthmark when generating age estimates from vertebral thin sections.

Edge analysis and MIA were performed to verify the annual nature of band deposition. Edge type was assessed without knowledge of collection date by using digital photographs of caudal thorns and vertebral thin sections from fish of all age classes. The criteria for determining edge type were adapted from Yudin and Cailliet (1990) and Gallagher and Nolan (1999). Edge types were defined as follows: (1) translucent band (or ridge in caudal thorns) forming at the marginal edge; (2) narrow opaque band beginning to form on the marginal edge; and (3) broad opaque band well-formed on the marginal edge. 'Broad' and 'narrow' opaque bands were defined based on the relative width of the previous fully formed opaque band (Cailliet et al. 2006). MIA was performed on vertebral thin sections following the methodology of Conrath et al. (2002). A digital photograph was taken of each specimen and all measurements were made using Image-Pro Plus 5.1 software (Media Cybernetics). Margin width (MW) was divided by penultimate band pair width (PBW) to yield the marginal increment ratio (MIR; Fig. 1). Mean MIR was calculated from at least four 5–8 year-olds during each month of collection. Edge type and mean MIR were both compared against month of collection to determine the seasons of opaque and translucent growth zone deposition and to confirm the annual nature of band pair formation. Differences between months in MIR were assessed using a non-parametric Kruskal–Wallis analysis of variance by ranks test and examined post-hoc with Dunn's test for multiple comparisons (Zar 1999).

Growth estimates

Two different models were fit to observed length-at-age data from males, females, and both sexes combined using the Levenberg–Marquardt routine for nonlinear least squares parameter estimation in the statistical software SPSS (v. 12.0.1 SPSS, Inc). The first model fitted was a three-parameter von Bertalanffy growth function in the form:

$$L_t = L_\infty (1 - e^{-k(t-t_0)})$$

where L_t = total length at age t, L_∞ = theoretical asymptotic length, k = growth rate coefficient, and t_0 = theoretical age at zero length (Quinn and Deriso 1999; Haddon 2001). After fitting the von Bertalanffy model, the theoretical average length of an age zero fish (L_0) was estimated from the model parameters by the equation: $L_0 = L_\infty (1 - e^{kt_0})$, and compared to the independent estimates of length at birth to verify the biological appropriateness of using the von Bertalanffy model (Cailliet and Goldman 2004; Cailliet et al. 2006).

A modified form of the Gompertz growth function was also fit to the data as:

$$L_t = L_0 * e^{G(1 - e^{-gt})}$$

where G is the instantaneous rate of growth at time t, g is the rate of decrease of G, and L_0 is defined as above (Goldman 2004; Bishop et al. 2006). Asymptotic length (L_∞) was estimated by the equation: $L_\infty = L_0 e^G$ using the parameters obtained by fitting the Gompertz model and was then compared to maximum observed length (Mollet et al. 2002).

Model goodness-of-fit was assessed by comparisons of significance level ($P < 0.05$), coefficient of determination (r^2), and residual mean squared error (MSE; Carlson and Baremore 2005; Neer and Thompson 2005), and by the production of reasonable biological estimates (Cailliet et al. 2006). Independent estimates of length at birth and maximum observed length were compared with the values of L_0 and L_∞ predicted by each of the growth functions. Potential differences between males and females were examined for each model through Kimura's (1980) likelihood ratio test as described by Haddon (2001).

Maturity

Size at 50% maturity (L_{50}) was estimated for each sex separately by fitting a logistic model to paired maturity and length data using a weighted non-linear least squares parameter estimation routine in the statistical software R v.2.0.1 (R Dev Core Team 2004). A logistic model of the form:

$$P = \frac{1}{1 + e^{-(a+bx)}}$$

was fit to the data, where P is the proportion of mature individuals in each 1 cm size interval, x is total length (cm), and a and b are parameters. Following parameter estimation, length at 50% maturity was calculated by setting P equal to 0.5 and solving for x. Variance and confidence intervals for length at 50% maturity were calculated after Ashton (1972). Age at 50% maturity (T_{50}) was directly estimated for each sex from paired age and maturity data using the methods above.

Longevity and natural mortality

Longevity (T_{max}) was estimated from the oldest vertebral age estimates observed during this study. Three methods were used to indirectly estimate natural mortality. Hoenig's method (1983) is based on the empirical relationship between mortality and longevity. Hoenig demonstrated that for fishes, the general relationship between the instantaneous rate of mortality (Z) and maximum age is: $\ln(Z) = 1.46 - 1.01*\ln(T_{max})$. Because Hoenig's data came from unexploited or lightly exploited stocks, Z approximates M, the instantaneous rate of natural mortality.

Jensen (1996) demonstrated theoretically that the von Bertalanffy growth coefficient (k) should be a good predictor of M. We used the relationship derived from Pauly's (1980) database ($M = 1.6k$) to obtain a second estimate of M, with approximate confidence intervals for M estimated after Gunderson et al. (2003).

The theoretical relationship between mortality and age at maturity was used to obtain a third estimate of natural mortality with the equation: $M = 1.65/T_{50}$ (Jensen 1996), where T_{50} is age at 50% maturity.

Results

Age estimates and verification

A total of 1356 skates was collected between July 2003 and December 2005. Males ranged in size from 22 to 118 cm TL, and females ranged from 21 cm to 119 cm TL. Vertebral thin sections from 546 specimens were prepared and aged blindly three independent times. Of these, 431 thin sections, or approximately 79%, had low enough APE (< 15%) and adequate quality (>'2') to be considered suitable for use in growth modeling. Acceptable vertebral thin section age estimates came from 200 males, ranging in size from 22 to 114 cm TL, and 231 females ranging in size from 21 to 116 cm TL, during each month of collection (Fig. 3a, b). Age estimates generated from vertebral thin sections and used in growth models ranged from 0 to 15 years for males and 0 to 17 years for females.

Independent estimates of length at birth were generated to verify birthmark increment identification in vertebral thin sections and to assess the adequacy of growth models in describing the early years of growth. The mean size of late-stage embryos, excluding the embryonic tail filament, was approximately 19.8 cm TL (SE = 0.20; n = 39; Hoff[1]) and the smallest observed free-swimming individual in this study was 21 cm TL; therefore the actual length at birth was assumed to fall within 19.8–21 cm TL.

The relationship between TL and VR was nearly linear but was best described by a power function of the form: $TL = 27.166VR^{0.867}$ ($P < 0.0001$; $r^2 = 0.97$; $n = 311$). The relationship between TL and caudal TBL was also best described as a power function, of the

---
[1] G. Hoff. 2006. Personal communication, unpublished data. Alaska Fisheries Science Center, 7600 Sand Point Way NE, Seattle, WA 98115.

Fig. 3 Frequency histograms of vertebral thin section samples used in growth models by (**a**) length and (**b**) month of collection, and (**c**) length frequency histogram of caudal thorn samples

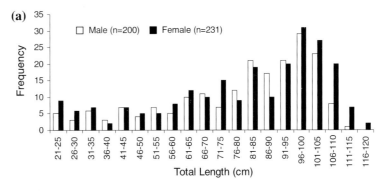

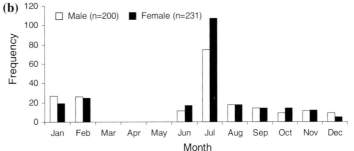

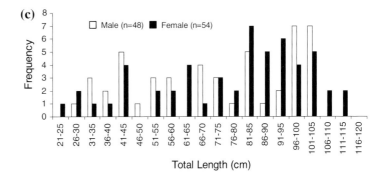

form: TL = 10.342TBL$^{0.9703}$ ($P < 0.0001$; $r^2 = 0.93$; $n = 100$). Through back-calculation, the average TL at vertebral birthmark increment formation (TL$_{BM}$) was estimated as 22.14 cm TL ($n = 310$; 95% CI = 21.87, 22.41), very close to the independent estimates of size at birth (19.8–21 cm TL), thus verifying the correct identification of the birthmark increment.

In total, 102 caudal thorns were aged blindly three independent times and were compared with vertebral thin sections from the same specimens (Fig. 3c). Precision between thorn and vertebral age estimates was generally high (APE = 7.92%, CV = 11.20%), and there was no detectable bias between estimates from the two structures (Fig. 4a). A paired t test demonstrated no significant difference between mean age estimates from caudal thorns and vertebral thin sections after samples with poor quality and APE were excluded ($t = -1.372$, $P > 0.10$, $n = 70$).

Overall precision between the primary reader's three age estimates from vertebral thin sections was generally high (APE = 8.08%, CV = 10.89%, $n = 546$). Precision between the second reader's estimates was slightly lower (APE = 14.49%, CV = 19.19%, $n = 72$), and inter-reader precision was considered acceptable (APE = 9.99%, CV = 14.13%, $n = 72$). Thirty-five percent of the two readers' age estimates were in complete agreement, 76% agreed within 1 year, and 94% of age estimates agreed within 2 years. Furthermore, an age bias plot demonstrated no

Fig. 4 Age-bias graphs showing (**a**) mean caudal thorn age estimates relative to vertebral thin section age estimates, and (**b**) the mean vertebral age estimates of Reader 2 relative to the age estimates of Reader 1. Sample sizes are listed above each age, error bars represent 95% confidence intervals around mean caudal thorn or Reader 2 age estimates, and the 1:1 equivalence line is given

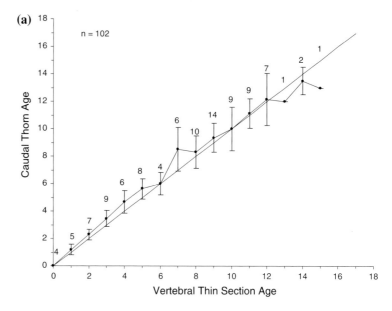

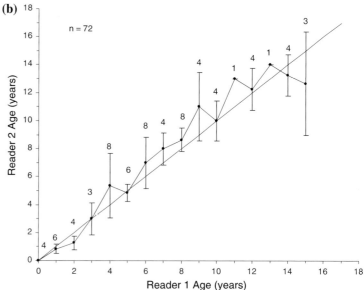

appreciable bias between vertebral age estimates from the two readers (Fig. 4b).

The results of edge analysis and MIA of vertebral thin sections support annual band pair formation (Fig. 5). Translucent bands at the edge (Type 1) were present most frequently during the months of January and February, narrow opaque bands (Type 2) during the months of June through October, and broad opaque bands (Type 3) during November and December, suggesting a seasonal progression in band formation.

A significant difference in MIR between months was detected with a Kruskal–Wallis test (KW $\chi^2 = 43.783$, $P < 0.001$). The lowest mean MIR occurred during the month of February (mean MIR = 0.044, SE = 0.026), after which an increasing trend was apparent, resulting in the highest mean MIR during the month of November (mean MIR = 0.662, SE = 0.071). Significant differences in MIR between the month of February and the months of August ($P < 0.005$), September ($P < 0.005$), November ($P < 0.001$),

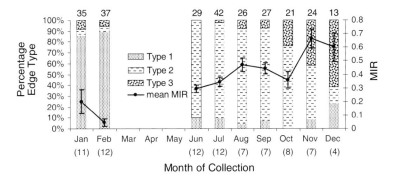

Fig. 5 Percentage edge type (columns) and mean marginal increment ratio (MIR; line) during each month of collection. Edge analysis sample sizes are listed above the columns, and marginal increment analysis sample sizes are listed along the x-axis in parentheses. Error bars represent mean MIR ± 1 SE

and December ($P < 0.005$), and between the months of January and November ($P < 0.01$), were detected with Dunn's post hoc test for multiple comparisons. Thus vertebral translucent band formation appears to occur during the winter months, followed by formation of the opaque band in summer and fall, supporting the assumption that each band pair represents one year of growth.

It was not possible to perform a full marginal increment or edge analysis on caudal thorns, as they were only collected during June and July. However, it should be noted that the majority of these thorns (94%) had a narrow opaque band (Type 2) beginning to form at the edge, suggesting that ridges (translucent bands) are deposited earlier in the year, consistent with our findings for vertebral thin sections.

Growth estimates

Overall, both the von Bertalanffy and Gompertz growth models fit the observed length-at-age data well (Fig. 6; Table 2). Each model provided highly significant ($P < 0.0001$) and reasonable fits with similar MSE and r^2. Independent estimates of length at birth (19.8–21 cm TL) were very close to L_0 (average length at age zero) predicted by the von Bertalanffy model (19–20 cm TL) and the Gompertz model (22–23 cm TL), indicating that both models adequately describe the early years of growth. However, the Gompertz model had slightly lower MSE and higher r^2 than the von Bertalanffy model for both sexes (Table 2). Furthermore, asymptotic length (L_∞) estimated from the Gompertz parameters (111 and 121 cm TL for males and females, respectively) is closest to the observed maximum length of this species (118 and 119 cm TL for males and females respectively.)

Significant differences between the sexes in overall growth were detected with likelihood ratio tests for both the von Bertalanffy model ($\chi^2 = 11.54$, df = 3, $P < 0.01$) and the Gompertz model ($\chi^2 = 16.48$, df = 3, $P < 0.01$). Using likelihood ratio tests on single parameters, significant differences between the sexes were determined for the von Bertalanffy model in terms of the L_∞ parameter ($\chi^2 = 5.78$, df = 1, $P < 0.05$) and k parameter ($\chi^2 = 6.23$, df = 1, $P < 0.05$), and for the Gompertz model in terms of the g parameter ($\chi^2 = 6.55$, df = 1, $P < 0.05$).

Maturity

The smallest observed mature male and female were 85 and 87 cm TL, respectively. The largest immature male and female were 98 and 101 cm TL, respectively. Length at 50% maturity (L_{50}) occurred at 91.75 cm TL in males ($n = 555$; 95% CI = 90.94, 92.57) and at 93.28 cm TL in females ($n = 642$; 95% CI = 92.52, 94.04; Fig. 7a). Females had a slightly later age at maturity (T_{50}) than males (Fig. 7b). Males were estimated to mature at an age of 8.97 years ($n = 200$; 95% CI = 8.60, 9.33) and females matured at 9.71 years ($n = 231$; 95% CI = 9.38, 10.04).

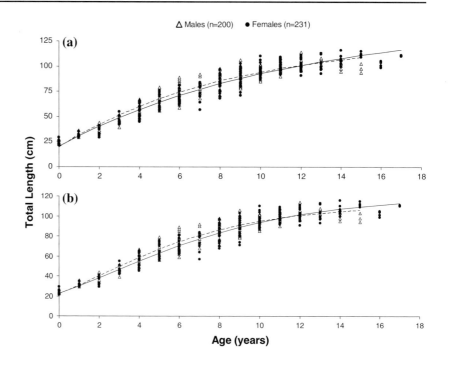

Fig. 6 Three parameter von Bertalanffy (**a**) and Gompertz (**b**) growth functions fit to observed length-at-age data for male (dashed lines) and female (solid lines) *B. parmifera*

Table 2 Parameter estimates and measures of goodness-of-fit from von Bertalanffy and Gompertz growth functions, fit to length-at-age data from males ($n = 200$), females ($n = 231$), and both sexes combined ($n = 431$)

Von Bertalanffy model						
	L_∞ (cm TL)	k (year^{-1})	t_0 (year)	L_0 (cm TL)	r^2	MSE
Males	126.29	0.12	−1.39	19.33	0.916	43.98
	[116.92–135.66]	[0.098–0.14]	[−1.78–(−0.98)]			
Females	144.62	0.087	−1.75	20.48	0.928	46.17
	[131.31–157.93]	[0.070–0.10]	[−2.18–(−1.33)]			
Both sexes	135.39	0.10	−1.60	20.02	0.921	46.05
	[127.44–143.34]	[0.088–0.11]	[−1.90–(1.30)]			
Gompertz model						
	L_0 (cm TL)	g (year^{-1})	G	L_∞ (cm TL)	r^2	MSE
Males	21.90	0.23	1.63	111.26	0.922	40.56
	[19.03–24.76]	[0.20–0.26]	[1.51–1.74]			
Females	22.54	0.19	1.68	120.51	0.935	42.05
	[20.09–24.98]	[0.17–0.21]	[1.59–1.77]			
Both sexes	22.50	0.21	1.64	115.99	0.927	42.67
	[20.62–24.38]	[0.19–0.22]	[1.57–1.71]			

Numbers in brackets are 95% confidence intervals for parameter estimates, and MSE is mean squared error of the residuals for each model fit. Predicted average length at age zero (L_0) and asymptotic length (L_∞) were estimated after fitting the von Bertalanffy and Gompertz functions, respectively

Longevity and natural mortality

Observed estimates of maximum age (T_{max}) were 15 years for males and 17 years for females. Using Hoenig's method (1983), estimates of natural mortality (M) based on T_{max} were 0.25 for females and 0.28 for males. Estimates of M based on the von Bertalanffy growth coefficients (Jensen 1996; Gunderson et al. 2003) were 0.14 for females (approximate 95% CI = 0.11, 0.17) and 0.19 for males (approximate 95% CI = 0.15, 0.23). Estimates of natural mortality using Jensen's method (1996) and

Fig. 7 Maturity ogives for (**a**) length and (**b**) age of male (dashed lines) and female (solid lines) *B. parmifera*

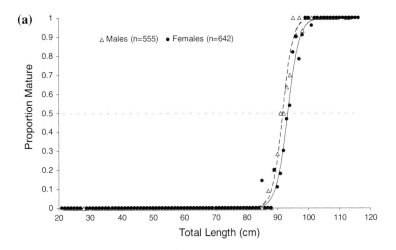

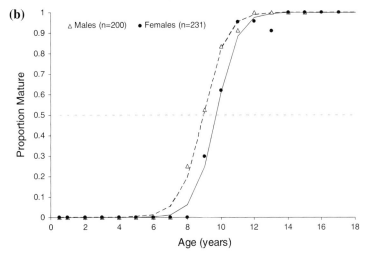

age at 50% maturity were similar: M was 0.17 for females and 0.18 for males.

Discussion

The appropriateness of using both vertebrae and caudal thorns as ageing structures for the Alaska skate is supported by the strong relationships between TL and VR and between TL and TBL. The average TL back-calculated from the birthmark increment in vertebral thin sections was very close to independent estimates of length at birth derived from the smallest free-swimming individuals and from late-stage embryos. Identification of the birthmark increment thus appeared to be accurate and consistent across specimens.

Agreement between age estimates generated from caudal thorns and vertebral thin sections was high, suggesting that thorns may be a suitable alternative to vertebrae as an ageing structure for *B. parmifera*. Thorns are much easier and faster to collect at sea, and require little storage space aboard a vessel and less processing time in the laboratory. Additionally, their removal likely causes minimal mortality, although the potential of thorns as truly non-lethal ageing structures must be studied in more detail. Caudal thorns may have application to other skate age and growth studies, and their usefulness should be tested on a species-by-species basis.

The measures of precision reported here, specifically the values of APE and CV, were consistent with other elasmobranch studies. Campana (2001) reported that CVs from shark age studies rarely fall below 10%; thus the precision of our age estimates was considered acceptable.

Both readers found interpretation of ageing structures to be moderately difficult, and a long training period was required for readers to feel confident in their age estimates. Nevertheless, bias between readers was not evident from visual inspection of the age bias plot.

Edge and marginal increment analyses indirectly validated vertebral age estimates by supporting the assumption of annual band pair deposition in *B. parmifera*. The assumption of predictable and temporally regular band deposition must be tested in elasmobranch ageing studies, as at least one species has been shown to lay down bands inconsistently (Natanson and Cailliet 1990). Edge and marginal increment analyses are among the simplest ways to indirectly validate age estimates, at least for younger fish, since direct methods such as tagging studies and captive rearing often require a great deal of time and funding to obtain results. MIA is generally more successful when applied to specimens that are still in a rapid state of growth (Campana 2001). For this study, MIA was performed on immature specimens from only a few age classes (5–8 years) due to the difficulty associated with quantitatively measuring vertebral growth in mature individuals; however, edge analysis was possible for all age groups. Since edge analysis and MIA yielded similar results, it was assumed that older individuals follow a similar pattern of growth to younger individuals. However, validation of absolute age, using methods such as mark-recapture or captive rearing of fish from hatching, should remain a goal of future research (Campana 2001; Cailliet and Goldman 2004).

Approximately half of the skates included in this analysis were collected through fishery-dependent sources (longliners). Consequently, the sample length distribution is biased, likely due to gear selectivity or to selective retention of larger, more commercially valuable individuals. However, age estimates were generated for a number of very small skates, including young-of-the-year, and both growth models had intercepts that passed very close to the known size at hatching. Thus the skewed length distribution is not believed to present a problem with regard to estimates of growth. The maximum length observed in this study for male and female Alaska skates (118 and 119 cm TL, respectively) is greater than the maximum length previously reported (Ebert 2005). There was little difference between males and females in terms of maximum observed length, although there were generally more large females than males.

The von Bertalanffy and Gompertz models both provided highly significant and reasonable fits to the observed length-at-age data. Many authors have pointed out weaknesses in the von Bertalanffy function and its application to fish growth, yet due to its prevalence in the literature, it still has some utility for comparison with other studies (Haddon 2001). An increasing number of elasmobranch studies are using forms of the Gompertz model to describe growth, especially for batoids (Cailliet and Goldman 2004; Carlson and Baremore 2005; Neer and Thompson 2005). Biologically meaningful parameters can be derived from both the Gompertz function and the von Bertalanffy function, including length at age zero (which approximates length at birth) and asymptotic length. In the case of the Alaska skate, both models appear to describe the early years of growth similarly, passing through the y-axis very close to the independent estimates of length at birth. However, the Gompertz model had the best overall fit to the data, based upon statistical criteria (r^2 and MSE) and reasonable parameter estimates. For both models, the differences in parameter estimates between the sexes appear to be very small based upon visual examination of the curves and are most likely related to the slight difference in maximum observed age for males and females.

Length at 50% maturity (L_{50}) occurred at approximately 78% of the maximum observed TL for both males and females. There was no significant difference between males and females with respect to size at maturity, as evidenced by overlapping confidence intervals for L_{50}. Age at maturity (T_{50}) occurred at approximately 60% of the observed maximum age for both males and females, falling within but near the late end of the continuum for batoids (Zeiner and Wolf 1993; Walmsley-Hart et al. 1999; Francis et al. 2001; Neer and Cailliet 2001; Sulikowski et al. 2003, 2005; Neer and Thompson 2005; McFarlane and King 2006). Maximum observed age also fell within the observed range for skates (Cailliet and Goldman 2004), suggesting that *B. parmifera* has moderate longevity when compared with other species. It is possible that older individuals are present within the population; however, our sample included large individuals, including some larger than the previous maximum reported size. Thus, we feel it is not likely that maximum age would increase substantially with additional sampling.

The methods of Hoenig (1983) and Jensen (1996) produced estimates of natural mortality ranging from 0.18 to 0.28 for males, and from 0.14 to 0.25 for females. These estimates are based on maximum age, von Bertalanffy growth coefficients, and age at maturity observed in this study. In this case, M should be interpreted with caution, as the assumed relationships from which M has been indirectly derived are based primarily on teleost fishes. Future studies should strive to obtain direct estimates of M, possibly through mark-recapture studies or from catch curves.

This study is among the first to describe the biology of *Bathyraja parmifera* in detail. As our knowledge of this species continues to improve, it will be possible to assess demography, stock structure, and population dynamics. Genetic studies have been planned to determine whether multiple populations of *B. parmifera* exist across its range (Michael Canino, Alaska Fisheries Science Center, personal communication) and if this is indeed the case, examination of potential differences in life history parameters among stocks will be necessary.

This species currently experiences relatively low exploitation rates since it is only caught incidentally in the EBS, although it could potentially become the target of a directed fishery in the future, owing to its abundance and large body size. Relative biomass of this species in the EBS has averaged around 400,000 metric tons over the past eight years (Matta et al. 2006). While it is difficult to determine the actual annual catch rate of *B. parmifera* in the EBS, since species composition of the skate catch must be derived from fishery observer estimates, it likely does not exceed 6% of biomass estimates (Matta et al. 2006). Ecosystem model simulations indicate that fishing pressure is a greater source of mortality for *B. parmifera* than predation, as this species has few natural predators in the EBS (Gaichas et al. 2005). Additionally, the mortality of discarded skates is unknown and likely varies with the handling process. Future work may elucidate potential ecosystem effects related to *B. parmifera* removal, sources of natural and human-induced mortality, and levels of fishing mortality required for sustainability.

Charnov (2002) showed that relative age at maturity (T_{50} relative to T_{max}) is one of the three critical life history variables differentiating major vertebrate taxa. Relative age at maturity clearly distinguishes elasmobranchs from teleosts. Stock assessment and management methods that have traditionally been used for teleosts may not be appropriate for elasmobranchs due to the differences in life history patterns. The Alaska skate, like most elasmobranchs, appears to have moderate longevity, late maturity relative to T_{max}, and slow to moderate growth. These life history characteristics, coupled with other aspects of the reproductive biology of the Alaska skate (Matta 2006), suggest that this species could only support a directed fishery with limited catch rates, should one develop (Hoenig and Gruber 1990; Musick et al. 2000; King and McFarlane 2003).

Acknowledgements We wish to thank the Alaska Fisheries Science Center, Seattle, WA, for providing technical support, laboratory space, and equipment. Special thanks go to the NPGOP for making seasonal collection possible. We gratefully acknowledge Christopher Gburski for his expertise in age and growth and Gerald Hoff for generously providing embryo size data. Many thanks to the scientists at the Pacific Shark Research Center, Moss Landing, CA, for providing invaluable age training and advice. We especially appreciate the efforts of David Ebert and James Sulikowski in arranging a special symposium, 'Biology and systematics of skates: what do we really know?' at the 2006 Annual Meeting of the American Elasmobranch Society in New Orleans, LA, and for making these proceedings possible. Lastly, we'd like to thank Dan Kimura, Christopher Gburski, and two anonymous reviewers for their helpful comments regarding earlier versions of this manuscript. This publication is funded by the Joint Institute for the Study of the Atmosphere and Ocean (JISAO) under NOAA Cooperative Agreement No. NA17RJ1232, Contribution #1334, as part of the NOAA Stock Assessment Improvement Plan (SAIP).

References

Adams PB (1980) Life history patterns in marine fishes and their consequences for fisheries management. Fish Bull 78(1):1–10

Ashton WD (1972) The logit transformation, with special reference to its uses in bioassay. Hafner Publishing Company, New York, 87 pp

Beamish RJ, Fournier DA (1981) A method for comparing the precision of a set of age determinations. Can J Fish Aquat Sci 38:982–983

Bishop SDH, Francis MP, Duffy C, Montgomery JC (2006) Age, growth, mortality, longevity and natural mortality of the shortfin mako shark (*Isurus oxyrinchus*) in New Zealand waters. Mar Fresh Res 57:143–154

Brander K (1981) Disappearance of common skate *Raia batis* from Irish Sea. Nature 290:48–49

Cailliet GM, Goldman KJ (2004) Age determination and validation in chondrichthyan fishes. In: Carrier JC, Musick JA, Heithaus MR (eds) Biology of sharks and their relatives. CRC Press LLC, Boca Raton, FL, pp 399–447

Cailliet GM, Smith WD, Mollet HF, Goldman KJ (2006) Age and growth studies of chondrichthyan fishes: the need for consistency in terminology, verification, validation, and growth function fitting. Environ Biol Fish 77(3–4):211–228

Camhi M, Fowler SL, Musick JA, Bräutigam A, Fordham SV (1998) Sharks and their relatives: ecology and conservation. IUCN/SSC Shark Specialist Group, IUCN, Gland, Switzerland and Cambridge, UK, 39 pp

Campana SE (2001) Accuracy, precision, and quality control in age determination, including a review of the use and abuse of age validation methods. J Fish Biol 59:197–242

Carlson JK, Baremore IE (2005) Growth dynamics of the spinner shark (*Carcharhinus brevipinna*) off the United States southeast and Gulf of Mexico coasts: a comparison of methods. Fish Bull 103(2):280–291

Casey JM, Myers RA (1998) Near extinction of a large, widely distributed fish. Science 281:690–692

Charnov EL (2002) Reproductive effort, offspring size and benefit-cost ratios in the classification of life histories. Evol Ecol Res 4:749–758

Committee of Age Reading Experts (2000) Manual on generalized age determination procedures for groundfish. Pacific States Marine Fisheries Commission, 39 pp

Conrath CL, Gelsleichter J, Musick JA (2002) Age and growth of the smooth dogfish (*Mustelus canis*) in the northwest Atlantic Ocean. Fish Bull 100:674–682

Cortés E (2002) Incorporating uncertainty into demographic modeling: application to shark populations and their conservation. Conserv Biol 16(4):1048–1062

Dulvy NK, Reynolds JD (2002) Predicting extinction vulnerability in skates. Conserv Biol 16:440–450

Dulvy NK, Metcalfe JD, Glanville J, Pawson MG, Reynolds JD (2000) Fishery stability, local extinctions, and shifts in community structure in skates. Conserv Biol 14(1):283–293

Ebert DA (2005) Reproductive biology of skates, *Bathyraja* (Ishiyama), along the eastern Bering Sea continental slope. J Fish Biol 66:618–649

Francis RICC (1990) Back-calculation of fish length: a critical review. J Fish Biol 36:883–902

Francis MP, Ó'Maolagáin C, Stevens D (2001) Age, growth, and sexual maturity of two New Zealand endemic skates, *Dipturus nasutus* and *D. innominatus*. N Zealand J Mar Freshw Res 35:831–842

Gaichas S, Matta B, Stevenson D, Hoff J (2005) Bering Sea and Aleutian Islands skates. In: Stock assessment and fishery evaluation report for the groundfish resources of the Bering Sea/Aleutian Islands regions. North Pacific Fishery Management Council, Anchorage, AK, pp 825–857

Gallagher M, Nolan CP (1999) A novel method for the estimation of age and growth in rajids using caudal thorns. Can J Fish Aquat Sci 56:1590–1599

Goldman KJ (2004) Age and growth of elasmobranch fishes. In: Musick JA, Bonfil R (eds) Elasmobranch fisheries management techniques. IUCN Shark Specialist Group/APEC Fisheries Working Group, Singapore, pp 97–132

Gunderson DR, Zimmermann M, Nichol DG, Pearson K (2003) Indirect estimates of natural mortality rate for arrowtooth flounder (*Atheresthes stomias*) and darkblotched rockfish (*Sebastes crameri*). Fish Bull 101:175–182

Haddon M (2001) Modelling and quantitative methods in fisheries. Chapman and Hall/CRC Press, Boca Raton, FL, 406 pp

Hoenig JM (1983) Empirical use of longevity data to estimate mortality rates. Fish Bull 82(1):898–902

Hoenig JM, Gruber SH (1990) Life-history patterns in the elasmobranchs: implications for fisheries management. In: Pratt HL, Gruber SH, Taniuchi T (eds) Elasmobranchs as living resources: advances in the biology, ecology, systematics, and the status of the fisheries. NOAA Technical Report NMFS 90:1–16

Holden MJ (1974) Problems in the rational exploitation of elasmobranch populations and some suggested solutions. In: Harden Jones FR (ed) Sea fisheries research. John Wiley and Sons, New York, pp 117–137

Jensen AL (1996) Beverton and Holt life history invariants result from optimal trade-off of reproduction and survival. Can J Aquat Fish Sci 53:820–822

Kimura DK (1980) Likelihood methods for the von Bertalanffy growth curve. Fish Bull 77(4):765–776

King JR, McFarlane GA (2003) Marine fish life history strategies: applications to fishery management. Fisheries Manag Ecol 10:249–264

Matta, ME (2006) Aspects of the life history of the Alaska skate, *Bathyraja parmifera*, in the eastern Bering Sea. MS Thesis, University of Washington, 92 pp

Matta B, Gaichas S, Lowe S, Stevenson D, Hoff G, Ebert D (2006) Bering Sea and Aleutian Islands skates. In: Stock assessment and fishery evaluation report for the groundfish resources of the Bering Sea/Aleutian Islands regions. North Pacific Fishery Management Council, Anchorage, AK

McFarlane GA, King JR (2006) Age and growth of big skate (*Raja binoculata*) and longnose skate (*Raja rhina*) in British Columbia waters. Fish Res 78:169–178

Mollet HF, Ezcurra JM, O'Sullivan JB (2002) Captive biology of the pelagic stingray, *Dasyatis violacea* (Bonaparte 1832). Mar Freshw Res 53:531–541

Musick JA, Burgess G, Cailliet G, Camhi M, Fordham S (2000) Management of sharks and their relatives (Elasmobranchii). Fisheries 25(3):9–13

Natanson LJ, Cailliet GM (1990) Vertebral growth zone deposition in Pacific angel sharks. Copeia 1990(4):1133–1145

Neer JA, Cailliet GM (2001) Aspects of the life history of the Pacific eagle ray, *Torpedo californica* (Ayres). Copeia 3:842–847

Neer JA, Thompson BA (2005) Life history of the cownose ray, *Rhinoptera bonasus*, in the northern Gulf of Mexico, with comments on geographic variability in life history traits. Environ Biol Fish 73:321–331

Pauly D (1980) On the interrelationships between natural mortality, growth parameters, and mean environmental temperature in 175 fish stocks. ICES J Cons Int Expl Mer 39(2):175–192

Quinn TJ, Deriso RB (1999) Quantitative fish dynamics. Oxford University Press, New York, 542 pp

R Development Core Team (2004) R: A language and environment for statistical computing. R Foundation for Statistical Computing, Vienna, Austria. ISBN 3-900051-07-0, http://www.R-project.org

SPSS for Windows, Rel 12.0.0 (2003) SPSS Inc, Chicago

Stevenson DE (2004) Identification of skates, sculpins, and smelts by observers in north Pacific groundfish fisheries (2002–2003). NOAA Technical Memo NMFS-AFSC-142, 67 pp

Sulikowski JA, Morin MD, Suk SH, Howell WH (2003) Age and growth estimates of the winter skate (*Leucoraja ocellata*) in the western Gulf of Maine. Fish Bull 101:405–413

Sulikowski JA, Tsang PCW, Howell WH (2005) Age and size at maturity for the winter skate, *Leuocoraja ocellata*, in the western Gulf of Maine based on morphological, histological, and steroid hormone analyses. Environ Biol Fish 72:429–441

Walker PA, Heessen HJL (1996) Long-term changes in ray populations in the North Sea. ICES J Mar Sci 53:1085–1093

Walker PA, Hislop JRG (1998) Sensitive skates or resilient rays? Spatial and temporal shifts in ray species composition in the central and north-western North Sea between 1930 and the present day. ICES J Mar Sci 55:392–402

Walmsley-Hart SA, Sauer WHH, Buxton CD (1999) The biology of the skates *Raja wallacei* and *R. pullopunctata* (Batoidea: Rajidae) on the Agulhas Bank, South Africa. South African J Mar Sci 21:165–179

Winemiller KO, Rose KA (1992) Patterns of life-history diversification in North American fishes: implications for population regulation. Can J Fish Aquat Sci 49:2196–2218

Yudin KG, Cailliet GM (1990) Age and growth of the gray smoothhound, *Mustelus californicus*, and the brown smoothhound, *M. henlei*, sharks from central California. Copeia 1990(1):191–204

Zar JH (1999) Biostatistical analysis, 4th edn. Prentice Hall, Upper Saddle River, NJ

Zeiner SJ, Wolf P (1993) Growth characteristics and estimates of age at maturity of two species of skates (*Raja binoculata* and *Raja rhina*) from Monterey Bay, California. In: Branstetter S (ed) Conservation biology of elasmobranchs NOAA Technical Report NMFS 115:87–99

Age and growth of the roughtail skate *Bathyraja trachura* (Gilbert 1892) from the eastern North Pacific

Chanté D. Davis · Gregor M. Cailliet · David A. Ebert

Originally published in the journal Environmental Biology of Fishes, Volume 80, Nos 2–3, 325–336.
DOI 10.1007/s10641-007-9224-7 © Springer Science+Business Media B.V. 2007

Abstract This study provides the first published age estimates for the roughtail skate, *Bathyraja trachura*. Age and growth characteristics of *B. trachura*, a poorly-known deepwater species, were determined from samples collected along the continental slope of the contiguous western United States. A new maximum size was established at 91.0 cm TL. Age was determined using a traditional structure (vertebral thin sections) with widespread application on multiple skate species and a non-lethal structure (caudal thorns) recently used for age analysis on skate species. Caudal thorns were determined not to be a useful ageing structure for this species based on poor precision and significantly lower age estimates when compared to age estimates from vertebral thin sections. The best model for describing growth of *B. trachura* was the two parameter VBGF, assuming annual vertebral band deposition and using length-at-age data. Although females grew slower and reached a larger maximum size than males, their growth was not statistically different (ARSS; $P = 0.90$); therefore, data were pooled ($L_\infty = 99.38$, $k = 0.09$). Annual band deposition was found to be a reasonable assumption for this species, but has yet to be validated. The maximum age estimated for *B. trachura* was 20 years for males and 17 years for females using vertebral thin sections.

Keywords Vertebral thin sections · Caudal thorns · Age estimation · Elasmobranch · Skate

Introduction

In Californian waters, skates are directly targeted and taken incidentally as bycatch in commercial fisheries (Martin and Zorzi 1993). Landing records often report skate species as "unidentified skate," thus preventing an assessment of the potential effects of fishing on population dynamics (Zorzi et al. 2001). Determining the impact of fishing mortality on exploited skate populations is further hindered by a lack of species-specific life history information (Holden 1972; Martin and Zorzi 1993; Dulvy and Reynolds 2002). This is of concern because some skates have life history characteristics (e.g. slow growth, late age at maturity, low fecundity, and moderate longevity) that make them susceptible to overfishing (Holden 1972; Stevens et al. 2000). If current trends continue, fishing effort on skates is likely to increase, subjecting these vulnerable fishes to possible overexploitation.

C. D. Davis (✉) · G. M. Cailliet · D. A. Ebert
Moss Landing Marine Laboratories, Pacific Shark Research Center, 8272 Moss Landing Rd., Moss Landing, CA 95039, USA
e-mail: cdavis@mlml.calstate.edu

Fishing mortality has been associated with the declines of many skate populations worldwide. Walker and Hislop (1998) reported that relative abundance of skate species in north-east Atlantic waters was once dominated by the late-maturing blue skate, *Dipturus batis*. However, because of fishery exploitation, the skate assemblage is now dominated by the early-maturing thorny skate, *Amblyraja radiata*. Fishing mortality has also contributed to population declines in the northwest Atlantic of the winter skate, *Leucoraja ocellata*, barndoor skate, *D. laevis*, blue skate, *D. batis* (Johnson 1979; Brander 1981; Casey and Myers 1998). Knowledge of life history parameters and how they influence population dynamics of individual species is needed to better understand the impact of fishing mortality and to better develop appropriate management strategies.

Fishery biologists incorporate age and growth information into models to analyze the population dynamics of individual species (Campana 2001). Early chondrichthyan age and growth research focused on skates because of their abundance, hardiness and ability to survive in captivity, but was limited to nearshore species (Ishiyama 1951; Holden 1972). Since these early studies fishing pressure in some areas, especially deeper waters, has increased motivating research into the life history of these skates (Casey and Myers 1998; Cailliet and Goldman 2004).

Sharks, skates and rays lack otoliths, therefore other calcified structures such as vertebral centra, neural arches, spines and caudal thorns are used to estimate age (Holden and Meadows 1962; Cailliet et al. 1983; Cailliet and Tanaka 1990; Gallagher and Nolan 1999; Cailliet and Goldman 2004). These structures exhibit a banding pattern of opaque and translucent bands that have been found in many species to form during summer and winter months, respectively. However, this assumption must be validated for each to ensure accurate estimates of age (Campana 2001; Cailliet and Goldman 2004).

Bathyraja trachura is a poorly-known eastern North Pacific (ENP) skate with a range of 213–2550 m (most common below 600 m) and may be vulnerable as bycatch in trawl fisheries (Ishihara and Ishiyama 1985; Ebert 2003). Information about the genus *Bathyraja* (Rajiformes: Arhynchobatidae) in the ENP region is limited to taxonomic guides and a few descriptions of egg cases and advanced embryos (Cox 1963; Ishihara and Ishiyama 1985). *Bathyraja trachura* occurs from the Bering Sea to northern Baja California (Ebert 2003; Love et al. 2005). It is a medium-sized species (89.0 cm maximum total length (TL_{max})) with a dorsal surface that is plum brown or slate gray in color.

The objectives of this study were to provide estimates of age and describe growth characteristics for *B. trachura* in ENP waters. Specifically, we estimated size-at-age for *B. trachura* using age estimates from vertebral thin sections and caudal thorns. We attempted to validate age estimates using centrum edge and marginal increment ratio (MIR). Growth models were generated using length- and weight-at-age data for each sex. Results of this project provide previously unknown but critical life history information for the formulation of an effective management plan for *B. trachura* in the ENP.

Materials and methods

Specimen collection

Skates were obtained from along the Pacific coast of the contiguous United States between 48.6° and 33.35° North latitude (Fig. 1). Samples were collected in summer and fall 2002–2003 during the Northwest Fisheries Science Center annual groundfish surveys. Additional samples were collected during winter and spring 2004 from commercial fishery landings via the Pacific States Marine Fisheries Commission – West Coast Groundfish Observer Program.

Sex was determined for each specimen, vertebrae and caudal thorns were removed, and biological information was recorded. Total length (TL), disc length (DL), and disc width (DW), were measured to 0.1 cm and total weight (kg) was recorded following Hubbs and Ishiyama (1968). The first eight vertebrae and first six caudal thorns were removed from each specimen and frozen prior to analysis.

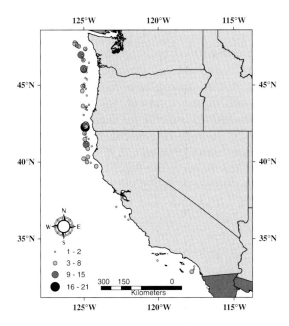

Fig. 1 Map of study area indicating distribution of trawl stations from commercial vessels via the NWFSC and West Coast Groundfish Observers ($n = 231$). Size of points represents the number of specimen obtained per haul

Preparation and evaluation of ageing structures

Vertebral columns were cleaned of extraneous tissue with a scalpel, neural, and haemal arches were removed and individual centra were separated. To be a useful aging structure, vertebral centra must grow predictably with TL. Mean vertebral centrum diameter was calculated across two perpendicular axis to 0.1 mm. A linear regression was used to determine the relationship between vertebral growth and somatic growth.

Vertebrae were embedded in a polyester casting resin. A 0.3 mm thin section containing the nucleus was removed using a Buehler Isomet low speed saw with paired 10 cm Norton diamond-edged blades. Thin sections were embedded on slides with Cytoseal 60, polished using 1200 grit wet sandpaper, and viewed under a dissecting scope with transmitted light. A pilot study comparing the application of band enhancement techniques following Gruber and Stout (1983) and Gallagher and Nolan (1999) proved that unstained thin sections provided the best band clarity.

All trunk centra from 11 specimens were removed and used to verify consistency of age estimates throughout the vertebral column. Age estimates of the first five vertebrae from the anterior part of the column were compared to age estimates of vertebrae from the last five vertebrae from the posterior part of the column. A paired t-test was used to determine if mean age estimates differed significantly (Zar 1999).

Caudal thorns were manually trimmed of extraneous tissue (Goldman 2004). Whole thorns were submerged in 3% trypsin for 48–72 h to remove remaining tissue (Gallagher and Nolan 1999). A subsample of caudal thorns were partially embedded in a fiberglass resin and sectioned laterally. A banding pattern was not apparent on sectioned thorns therefore, age was determined from whole caudal thorns for remaining samples. Whole thorns were placed on slides at 30°–50° angle and viewed under a dissecting scope with transmitted light.

To determine if caudal thorns would be a useful ageing structure, structural growth, consistency of age estimates, and the potential for thorn replacement were evaluated. Caudal thorn base diameter, measured along anterior-posterior thorn axis, and height was measured from proto-thorn tip to caudal thorn base using dial calipers to 0.1 mm. Measurements were plotted against TL to determine the relationship between caudal thorn size and body size. A pilot study comparing the application of band enhancement techniques following the methods of Gruber and Stout (1983) and Gallagher and Nolan (1999) concluded that unstained thin sections provided the best band clarity. Age estimates determined during the pilot study were not included in final analysis.

All caudal thorns were removed from the tail of 11 specimens and used to evaluate consistency of age estimates along the tail. Age estimates from the first five anterior most caudal thorns were compared to age estimates from the last five posterior most caudal thorns. A paired t-test was used to determine if mean age estimates differed (Zar 1999). To determine the potential for thorn replacement, caudal thorns were counted on each specimen and plotted against TL. A linear regression was used to determine the relationship between thorn count and somatic growth.

Age determination and validation

The birthmark in each vertebral centrum was identified as the change in angle of the corpus calcareum and was located on thin sections through all size classes (Fig. 2). Banding patterns were not enhanced using these staining methods, therefore thin sections were processed unstained.

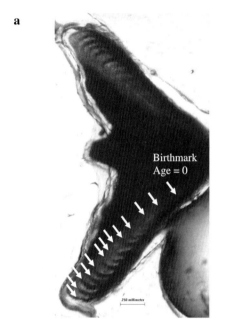

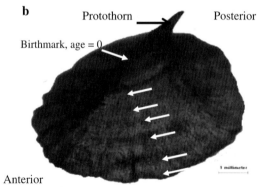

Fig. 2 Description of banding pattern for vertebral thin sections from a 59.8 cm TL female (**a**) and caudal thorns from a 46.5 cm TL male (**b**). On vertebral thin section the change in angle of the corpus calcerum signifies the birthmark (age = 0). On caudal thorns, the birthmark is a prominent ridge which encompasses the thorn below the protothorn. Subsequent white arrows identify opaque bands. (vertebral thin section age = 12, magnification 3.2×; caudal thorn age = 6, magnification = 1.25×)

Each opaque and translucent band pair was considered to represent one year of growth.

A sample of 100 caudal thorns was selected for analysis using stratified random sampling representing all size classes. Four samples were damaged, leaving a total of 96 caudal thorns (51 females, 45 males) for age analysis. The birthmark, identified on caudal thorns through all size classes, was identified as a distinct ridge at the base of the protothorn (Fig. 2). The protothorn forms the tip of the caudal thorn and lacks growth bands. Each opaque and translucent band pair was considered to represent one year of growth

Age estimates were determined using three rounds of independent age estimates by one reader without advanced knowledge of length, season of capture, or sex of the sample. If agreement was not achieved among these three estimates, a fourth read was completed. If agreement was not achieved by the fourth read, samples were removed from analysis. A clarity grade was assigned to each sample based on criteria adapted from Officer et al. (1996). Samples receiving a poor clarity grade were removed from analysis.

Age estimates were evaluated for reader precision and structural bias. Precision among age estimates was assessed using average percent error (APE) (Beamish and Fournier 1981), coefficient of variation (CV) and the index of precision (D) (Chang 1982). Percent agreement (PA) was calculated to determine precision of age estimated between rounds and was evaluated for exact agreement, agreement within one year, and within two years of age (Cailliet and Goldman 2004). Age-bias plots were used to determine the potential bias of age estimates within and between each independent read and between structures (Campana et al. 1995).

Final age estimates were compared between vertebrae and caudal thorns from the same animal. Precision and bias were calculated as previously described to evaluate variation of age estimates between structures. A paired t-test was used to determine if there was a significant difference between mean age estimates (Zar 1999).

Validation of vertebral band deposition was attempted using edge analysis and marginal increment ratio (MIR) (Hyndes et al. 1992; Campana 2001). Edge analysis is an optical classification of a thin sections' outermost band as opaque or translucent. Alternately MIR is a measured ratio representing the relative completion of the newest deposited band compared to the previously completed band pair and was calculated as: MIR = MW/PBW, where MW is margin width and PBW is previous band width (Conrath et al. 2002). The resulting ratios were plotted against month of collection to determine the periodicity of band deposition (Cailliet and Goldman 2004). Equality of variances was determined using Cochran's test, following which an ANOVA was used to test for seasonality of band deposition (Zar 1999).

Growth

The von Bertalanffy (VBGF) and Gompertz growth functions were fitted to length- and weight-at-age data for both sexes (Cailliet et al. 2006). The parameter estimates for each function were estimated using SigmaPlot version 8.0 (SPSS Inc., 2002). To determine if growth differed between sexes, analysis of the residual sums of squares (ARSS) was calculated (Haddon 2001).

The VBGF was fitted to length-at-age data and was calculated as: $l_t = L_\infty(1 - e^{-k(t-t_o)})$ where l_t is the predicted length at age t, L_∞ is the maximum length predicted by the equation, k is the growth coefficient, t_o is the theoretical age at which length is zero (Beverton and Holt 1957). Fabens (1965) incorporated known size at birth to better reflect biological reality. A two parameter VBGF with a fixed length at birth (L_o) was calculated as: $l_t = L_\infty - (L_\infty - L_o) e^{(-kt)}$ where l_t is the predicted length at age t, L_∞ is the maximum length predicted by the equation, k is the growth coefficient and L_o is set to the known length at birth (19.0 cm). A von Bertalanffy growth function fitted to weight-at-age data following Fabens (1965) and Ricker (1979) was also applied: $W_t = W_\infty(1 - e^{-k(t-t_o)})^3$ where W_∞ is the theoretical asymptotic weight, and values for k, t, t_0 are the same as previously described.

The Gompertz function (Ricker 1979) was fitted to weight and length-at-age data: $W_t = W_\infty e^{(-ke^{-gt})}$ where W_t is age, t is estimated age, g is the instantaneous growth coefficient, k is a dimensionless parameter and the other parameters are as previously described. To evaluate goodness-of-fit, the standard estimate of the error (SEE) and plots of standardized residuals were evaluated (Cailliet et al. 1992). Model selection was based on statistical fit (r^2), convenience, and biological relevance.

Results

Sample collection

Vertebral centra of 231 specimens (102 females, 129 males) and caudal thorns from 100 specimens (54 females, 46 males) were used for age estimation. The smallest female was 14.5 cm TL and male was 16.0 cm TL. The largest female was 86.5 cm TL and male was 91.0 cm TL (Fig. 3).

Preparation and evaluation of ageing structures

Vertebrae were determined to be a useful ageing structure based on the positive linear relationship between their size and TL (Fig. 4). There was no significant difference in this relationship between sexes ($F_{0.05,1,206}$ = 1.665, P = 0.198); therefore, values were pooled for analysis. A positive linear relationship was identified between TL and centrum diameter (n = 231, y = 0.071x–0.2402, r^2 = 0.92, P < 0.001).

Whole vertebral columns were removed from 11 specimens to determine if age was consistent throughout the structure. However, clear age estimates were only available in anterior and posterior vertebrae from nine specimens. There was no significant difference between age estimates from anterior and posterior vertebral thin sections (n = 9, $t_{(0.05(2),8)}$ = 0.748, P = 0.47). All age estimates were based on anterior vertebral centra.

The utility of caudal thorns as ageing structures was not demonstrated (Fig. 5). Caudal thorn size increased with body size, there was no evidence of thorn replacement, and ages were consistent along the tail. A logarithmic curve represented the best fit between thorn height and total length

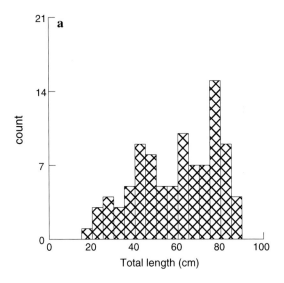

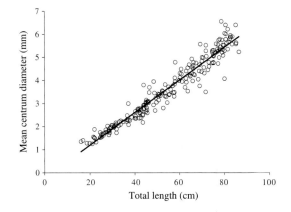

Fig. 4 Relationship between mean vertebral centrum diameter and total length for sexes combined ($n = 231$, $y = 0.071x - 0.2402$, $r^2 = 0.92$)

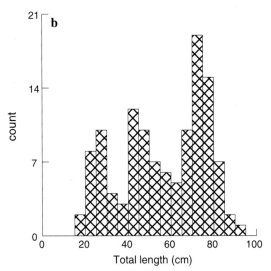

Fig. 3 Size frequency histogram of specimens used in age analysis; males (**a**;$n = 110$, a) and females (**b**;$n = 86$, b)

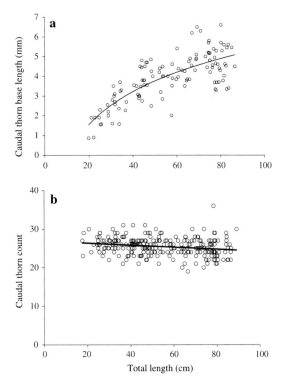

Fig. 5 Best fit relationships between caudal thorn base length and body size (**a**; $y = 2.4073\text{Ln}(x) - 5.652$, $r^2 = 0.60$, $n = 100$) and number of caudal thorns and body size (**b**; $y = -0.0266x + 26.91$, $r^2 = 0.04$, $n = 248$) are depicted for *Bathyraja trachura*

($n = 100$, $y = 0.6737\text{Ln}(x) + 0.3329$, $r^2 = 0.24$) and between thorn base length and total length ($n = 100$, $y = (2.4073 \times \text{Ln}(x))-5.652$, $r^2 = 0.60$) (Fig. 5a). Meristic counts of caudal thorns (from 248 additional specimens not aged in this study) indicated that caudal thorn count and TL were not significantly related ($y = -0.0266x + 26.91$, $r^2 = 0.04$, $P < 0.001$) (Fig. 5b). The presence of healed scars suggests that thorn replacement does not occur in this species. Additionally, there was no significant difference between age estimates from anterior and posterior caudal thorns ($t_{(0.05(2),7)} = 1.62$, $P = 0.149$); age was estimated from anterior caudal thorns.

Fig. 6 Age bias plots of age estimates between independent rounds of (**a**) vertebral age estimates (*n* = 197) and (**b**) caudal thorn (*n* = 96) band counts. The 45° line represents 1:1 agreement of age estimates between rounds

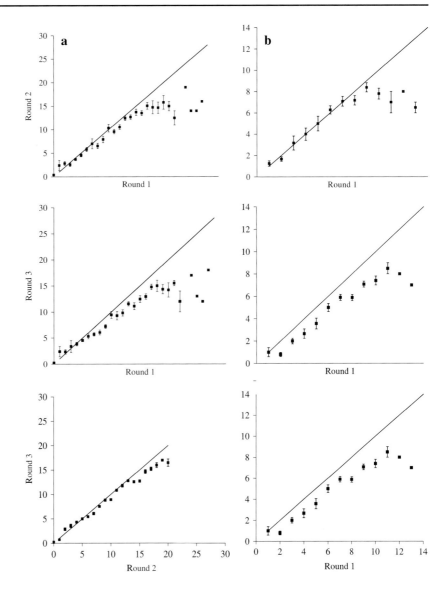

Age determination and validation

Final age estimates were determined from 197 vertebral samples which provided the best band clarity. Analysis of vertebral band clarity grades indicated that visibility of bands was variable and inconsistent between samples. Approximately 9.6% were determined to be of poor clarity (grade 5) and were not included in analysis.

Calculations of bias and precision indicated the first round of vertebral band counts was biased toward older age estimates. Average percent error and coefficient of variation were acceptable between the three independent rounds of band counts (APE: 13.6%, CV: 18.6%, D: 9.3). Percent agreement within ±2 years was 76.1% and 67.5% respectively, for first vs. second and first vs. third rounds of band counts. Greater agreement was found between the second and third round of band counts (86.8%). Age-bias plots indicated that the first round of band counts were biased toward older age estimates (Fig. 6a).

Caudal thorn samples that received poor clarity grades (2.0%) were removed from analysis and final age was determined for 96 samples. Analysis of clarity grades indicated that visibility of bands were variable and inconsistent between samples. Approximately 2.2% were determined to be of poor clarity (grade 5) and were not included in analysis.

Precision and bias estimates indicated that the first round of band counts had the greatest variation. Average percent error and coefficient of variation values for caudal thorns were highest when all rounds of band counts were combined (APE: 17.7%, CV: 24.6%, D: 12.3). Percent agreement calculations also indicated that the second and third round of band counts were most similar, 82.2% within two band counts. Age bias plots indicated that the first round of band counts produced greater bias when compared to later rounds, indicating that reader ability improved over time (Fig. 6b).

Vertebral thin sections were determined to be a better structure than caudal thorns for age estimation of *B. trachura*. Age estimates between the structures were consistent until age 7 and then became quite variable (Fig. 7). Age estimates compared between structures produced unacceptable precision values and bias (APE: 29.9%, CV: 40.4%, D: 28.5). There was a significant difference between ages estimated from vertebral thin sections and those estimated from caudal thorns ($t_{0.05(2)17}$ = 3.003, P = 0.007). Age estimates of vertebral thin sections were more precise, and were used for validation and growth analyses. The maximum ages determined by estimates from vertebral thin sections were 17 years (females) and 20 years (males).

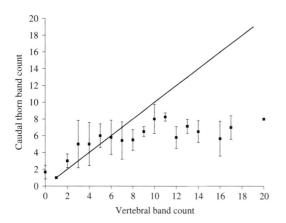

Fig. 7 Age bias plot of caudal thorn and vertebral age estimates (n = 74). The 45° line represents 1:1 agreement of age estimates between structures

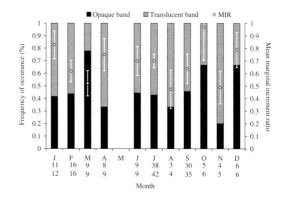

Fig. 8 The frequency of occurrence (FO; n = 153) of opaque and translucent bands and mean Marginal Increment Ratio (MIR; n = 139) are plotted by month. Black bars indicate frequency of opaque bands, grey bars depict frequency of translucent bands, and white diamonds represent mean MIR values. Numerical values reported below month indicate sample size of MIR (top) and FO (bottom)

Band deposition could not be validated using the methods applied in this study. About 35% of thin sections were removed from edge analysis because of poor clarity of the thin section. Edge analysis did not show a significant difference between the occurrence of opaque and translucent bands among months ($\chi^2_{0.05,3}$ = 1.06, P < 0.75; Fig. 8). Likewise, seasonality of band formation could not be validated using MIR. About 65% of thin sections were of optimal clarity and used for MIR analysis. There was no significant difference between MIR values among months ($F_{0.05,10,128}$ = 0.75, P = 0.67; Fig. 8). Although validation was not achieved using edge and MIR analysis, growth models were derived assuming annual deposition of one opaque and translucent band.

Growth

The model providing the most biologically reasonable fit for the data was the two parameter VBGF (Fig. 9). Females reached a larger maximum size (L_∞ = 101.53) with a slightly lower growth coefficient (k = 0.08) than males (L_∞ = 100.17, k = 0.09). However, there was no significant difference between female and male growth ($F_{0.05,1,307}$ = 0.196, P = 0.90). Therefore, data were combined and the resulting growth parameters for

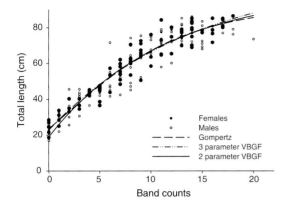

Fig. 9 Growth functions fitted to length-at-age data for combined sexes. Note VBGF = von Bertalanffy growth function

the pooled two parameter VBGF were $L_\infty = 101.25$ cm and $k = 0.09$.

Discussion

Our study expanded the known size range of *B. trachura* to include a new TL_{max}, one adult male was measured at 91.0 cm TL. Prior to this study the maximum size was 89.0 cm TL (Ishihara and Ishiyama 1985; Craig 1993). Samples were obtained from commercial fisheries and fishery-independent surveys, but neither source used trawling gear that could sample the entire depth range of this species; it is possible that larger specimen exist.

Caudal thorns were not determined to be a valid structure for age estimation of *B. trachura*. Caudal thorns of *B. trachura* did not grow linearly with body size, suggesting that their growth slowed with increased size and age estimation of larger individuals would be difficult (Francis and Maolagáin 2005). Caudal thorn height was difficult to relate to TL because of protothorn erosion in larger specimens. Because thorns originate from dermal dentical scales and, unlike vertebrae, do not support mass; thorn growth may not be inherently linked to somatic growth (Gallagher et al. 2005). Caudal thorns were originally considered as a possible ageing structure for this study because of the consistency of age estimates throughout the structure. Subsequent evaluation of precision proved that age estimates from thorns were not precise. Maximum ages determined by caudal thorns were less than half the maximum age determined by vertebral thin sections.

Gallagher and Nolan (1999) first used caudal thorns as an ageing structure for four species of *Bathyraja* from the Falkland Islands: *B. brachyurops*, *B. griseocauda*, *B. scaphiops*, and *B. albomaculata*. Caudal thorns were found to be suitable ageing structures that provided precise age estimates and ease of band interpretation. Since their publication caudal thorns often have proven to be a difficult structure to age, producing poor precision (Francis and Maolagáin 2005) and may be complicated by the slowing of thorn growth with increased somatic growth (Perez 2005). Perez (2005) found evidence of thorn replacement in *B. kincaidii*, indicating that not every thorn depicted actual age for this species.

Seasonality of band pair deposition in vertebral thin sections could not be validated using MIR and edge analysis. Diminished clarity of the banding pattern was caused by poor calcification of the vertebral edge, distortion of edge quality caused by overpolishing or inability to acquire a clear photographic image of the region. This lack of clarity limited the number of samples included in MIR analysis, potentially restricting detection of band periodicity.

Although vertebral band deposition was not validated, standard age determination methods were applied. Vertebral thin sections have been used for ageing more than 67 elasmobranch species (Cailliet and Goldman 2004; Perez 2005; Licandeo et al. 2006; McFarlane and King 2006). Among the studies, annual band deposition was validated for 54 species including seven skate species (Cailliet and Goldman 2004). These findings support the assumptions used to determine age for *B. trachura*. However, caution is necessary when using unvalidated age estimates for management purposes because of the inherent possibility for over or underestimation of age (Campana 2001; Cailliet and Goldman 2004).

In this study, multiple growth functions were used to model growth and the criteria used for growth model selection were statistical fit, convenience, and biological relevance (Cailliet et al 2006, Table 1). Growth functions fitted to weight-at-age data provided lowest variance about the mean, however W_∞ was underestimated by both

Table 1 Growth functions and estimated parameters for *Bathyraja trachura*. Length and weight-at-age data are represented for sexes combined. Note: SEE = standard error of the estimate, VBGF = von Bertalanffy growth function, t_o = size at birth, k and g = growth coefficients, L_o (W_o) = size(weight) at birth, L_∞ (W_∞) = maximum size(weight)

Parameter	Length-at-age			Weight-at-age	
	3 Parameter VBGF	2 Parameter VBGF	Gompertz	VBGF	Gompertz
L_∞ (W_∞)	112.11	101.25	92.62	(3.23)	(3.22)
$L_o(W_o)$	22.85	19.0	23.59	–	(0.02)
k	0.06	0.09	1.37	0.19	5.07
t_o	–3.45	–	2.20	0.70	7.70
g	–	–	–	0.19	0.21
SEE	5.36	5.55	5.26	0.49	0.54
r^2	0.93	0.92	0.92	0.79	0.74

functions (W_{max} = 5.0 kg). The two-parameter VBGF, fitted to length-at-age data, estimated a reasonable L_∞ and was the only length model that did not overestimate L_o; making it the most biologically reasonable choice. Since the three-parameter VBGF is the most commonly applied model for elasmobranch species (Cailliet and Goldman 2004), these values were used for comparison with the published literature.

The predicted growth parameters for *B. trachura* are not consistent with the assumption that larger batoids live longer and grow slower than smaller batoids. *Bathyraja trachura* is a medium sized species (91 cm) with a growth coefficient (k = 0.06) that is smaller than some smaller relatives such as *R. erinacea* (k = 0.35; Waring 1984) but not some others like *Raja clavata* (k = 0.05; Brander and Palmer 1985). Maximum age of *B. trachura* was determined to be 20 years, which is older than smaller species but not younger than all larger species such as *Amblyraja radiata* (16 years; Sulikowski et al. 2005).

Acknowledgements We thank Guy Fleisher, Keith Bosley, Erica Fruh, Aimee Keller, Victor Simon, and Dan Kamikawa NMFS/Northwest Fisheries Science Center for collecting samples during their yearly groundfish cruises in the eastern North Pacific. We also thank John Cusick and observers at the West coast Groundfish Observer Program for collecting samples throughout the year. We are grateful for the students and staff at Pacific Shark Research Center (PSRC) at Moss Landing Marine Laboratories who helped process and prepare samples for analysis, especially Joe Bizzarro, Colleena Perez, and Wade Smith. Funding for this research was provided by NOAA/NMFS to the National Shark Research Consortium and Pacific Shark Research Center, the National Sea Grant College Program of the U.S. Department of Commerce's National Oceanic and Atmospheric Administration under NOAA Grant no. NA04OAR4170038, project number R/F–199, through the California Sea Grant College Program and in part by the California State Resources Agency. Samples were collected under IAUCUC protocol #801.

References

Beamish RJ, Fournier DA (1981) A method for comparing the precision of a set of age determinations. CA J Fish Soc 112(6):735–743

Beverton RJH, Holt SJ (1957) On the Dynamics of exploited fish populations. Chapman and Hall, New York

Brander K (1981) Disappearance of common skate *Raja batis* from Irish Sea. Nature (5801) 290:48–49

Brander K, Palmer D (1985) Growth rate of *Raia clavata* in the northeast Irish Sea. J Cons Int Explor Mer 42:125–128

Cailliet GM, Tanaka S (1990) Recommendations for research needed to better understand the age and growth of elasmobranchs. In: HL Pratt, SH Gruber, T Taniuchi (eds) Elasmobranchs as living resources: advances in the biology, ecology, systematics, and the status of fisheries, vol 90. NOAA Tech. Rep., pp 505–507

Cailliet GM, Goldman KJ (2004) Age determination and validation in chondrichthyan fishes. In: JC Carrier, JA Musick, MR Heithaus (eds) Biology of sharks and their relatives, Chap. 14. CRC Press LLC, Boca Raton, Florida, pp 399–447

Cailliet GM, Smith WD, Mollet HF, Goldman KJ (2006) Age and growth studies of chondrichthyan fishes: the need for consistency in terminology, verification, validation, and growth function fitting. Enviro Bio Fish 77:211–228

Cailliet GM, Martin LK, Kusher D, Wolf P, Welden BA (1983) Techniques for enhancing vertebral bands in age estimation of California elasmobranchs. In: Prince E Pulos L (eds) Proc. Int. Tunas, Billfishes, Sharks. Spec. Sci. Rep./Fish. NMFS 8:179–188, pp 157–165,

Cailliet GM, Mollet HF, Pittenger GG, Bedford D, Natanson LJ (1992) Growth and demography of the Pacific angel shark (*Squatina californica*), based upon tag returns off California. Aust J Mar Freshwater Res 43:1313–1330

Campana SE (2001) Accuracy, precision and quality control in age determination, including a review of the use and abuse of age validation methods. J Fish Biol 59:197–242

Campana SE, Annand MC, McMillan JI (1995) Graphical and statistical methods for determining the consistency of age determinations. Trans Amer Fish Soc 124:131–138

Casey JM, Myers RA (1998) Near extinction of a large, widely distributed fish. Science 281:690–691

Chang WYB (1982) A statistical method for evaluating the reproducibility of age determination. Can J Fish Aquat Sci 39:1208–1210

Conrath CL, Gelsleichter J, Musick JA (2002) Age and growth of the smooth dogfish (*Mustelus canis*) in the northwest Atlantic Ocean. Fish Bull 100:674–682

Cox KW (1963) Egg-cases of some elasmobranchs and a cyclostome from Californian waters. Calif Fish and Game 64:271–289

Craig JC (1993) The systematics of the *Bathyraja* and *Rhinoraja* (Rajidae) species of the Bering Sea and adjacent areas in the North Pacific ocean. Masters Thesis, Texas A and M University, Texas, p 180

Dulvy NK, Reynolds JD (2002) Predicting extinction vulnerability in skates. Cons Bio 16(2):440–450

Ebert DA (2003) Sharks, rays, and chimaeras of California. University of California Press. Bentley, California, p 284

Fabens AJ (1965) Properties and fitting of the von Bertalanffy growth curve. Growth 29:265–289

Francis MP, Maolagáin CÓ (2005) Age and growth of the Antarctic skate, *Amblyraja georgiana*, in the Ross Sea. CCAMLR Science 12:183–194

Gallagher M, Nolan CP (1999) A novel method for the estimation of age and growth in rajids using caudal thorns. Can J Fish Aquat Sci 56:1590–1599

Gallagher M, Nolan CP, Jeal F (2005) The structure and growth processes of caudal thorns. J Northwest Atlantic Fish Sci 35(31):125–129

Goldman KG (2004) Age and growth of elasmobranch fishes In: Musick JA, Bonfil R (eds) Elasmobranch fisheries management techniques. APEC Fisheries Working Group, Singapore, pp 97–132

Gruber SH, Stout RG (1983) Biological materials for the study of age and growth in a tropical marine elasmobranch, the lemon shark, *Negaprion brevirostris* (Poey). US Dep Commer, NOAA Tech Rep NMFS 8:193–205

Haddon M (2001) Modeling and quantitative methods in fisheries. CRC Press LLC, Boca Raton, Florida, p 404

Holden MJ (1972) The growth of *Raja brachyura*, *R. clavata*, and *R montagui* as determined from tagging data. J Cons Int Explor Mar 34(2):161–168

Holden MJ, Meadows PS (1962) The structure of the spine of the spur dogfish (*Squalus acanthias*) and its use for age determination. J Mar Biol Ass UK 42:179–197

Hyndes GA, Loneragan NR, Potter IC (1992) Influence of sectioning otoliths on marginal increment trends and age and growth estimates for the flathead *Platycephalus speculator*. Fish Bull 90:276–284

Hubs CL, Ishiyama R (1968) Methods for the taxonomic study and description of skates (Rajidae). Copeia 3:48–3491

Ishihara H, Ishiyama R (1985) Two new North Pacific skates (Rajidae) and a revised key to *Bathyraja* in the area. Jpn J Ichthyol 32(2):143–179

Ishiyama R (1951) Age determination of *Raja hollandi* Jordan & Richardson inhabiting the waters of the East China Sea. Bull Jap Soc Sci Fish 16(12):119–124

Johnson AG (1979) A simple method for staining the centra of teleost vertebrae. Northeast Gulf Sci 3:113–115

Licandeo RR, Lamilla JG, Rubilar PG, Vega RM (2006) Age, growth, and sexual maturity of the yellownose skate *Dipturus chilensis* in the south-eastern Pacific. J Fish Biol 68:488–506

Love MS, Mecklenburg CW, Mecklenburg TA, Thorstein LK (2005) Resource inventory of marine and estuarine fishes of the West coast and Alaska: a checklist of North Pacific and Arctic ocean species from Baja California to the Alaska-Yukon border. U.S Department of the Interior, U.S. Geological survey Biological Resources Division, Seattle, Washington, p 276

Martin L, Zorzi GD (1993) Status and review of the California skate fishery. In: Branstetter S (ed) Conservation Biology of Elasmobranchs. NOAA Tech Rep NMFS 115:39–52

McFarlane GA, King JR (2006) Age and growth of big skate (*Raja binoculata*) and longnose skate (*Raja Rhina*) in British Columbia waters. Fish Res 78:169–178

Officer RA, Gason AS, Walker TI, Clement JG (1996) Sources of variation in counts of growth increments in vertebrae from gummy, *Mustelus antarcticus*, and school shark, *Galeorhinus galeus*, implications for age determination. Can J Fish Aquat Sci 53:1765–1777

Perez CR (2005) Age, growth, and reproduction of the sandpaper skate, *Bathyraja kincaidii* (Garman, 1908) in the eastern North Pacific. Masters Thesis, California State University and Moss Landing Marine Laboratory, California, p 100

Ricker WE (1979) Growth rates and models. In: Fish Physiology vol. 3. Academic Press Inc., New York, pp 677–742

Stevens JD, Bonfil R, Dulvy NK, Walker PA (2000) The effects of fishing on sharks, rays, and chimaeras (chondrichthyans), and the implications for marine ecosystems. ICES J Mar Sci 57:476–494

Sulikowski JA, Jurek J, Danley PD, Howell WH, Tsang PCW (2005) Age and Growth estimates of the thorny skate (*Amblyraja radiata*) in the western Gulf of Maine. Fish Bull 103:161–168

Walker PA, Hislop JRG (1998) Sensitive skates or resilient rays? Spatial and temporal shifts in ray species composition in the central and north-western North Sea between and the present day. ICES J Mar Sci 55:392–402

Waring GT (1984) Age, growth and mortality of the little skate off the northeast coast of the United States. Trans Amer Fish Soc 113:314–321

Zar JH (1999) Biostatistical analysis, 4th edn. Prentice Hall, New Jersey, p 663

Zorzi GD, Martin LK, Ugoretz J (2001) Skates and Rays. In: Leet WS, Dewees CM, Klingbeil R, Larson EJ (eds) California's living marine resources: a status report. California Department of Fish and Game, California, pp 257–261

Age and growth of big skate (*Raja binoculata*) and longnose skate (*R. rhina*) in the Gulf of Alaska

Christopher M. Gburski · Sarah K. Gaichas · Daniel K. Kimura

Originally published in the journal Environmental Biology of Fishes, Volume 80, Nos 2–3, 337–349.
DOI 10.1007/s10641-007-9231-8 © Springer Science+Business Media B.V. 2007

Abstract In 2003, big skates, *Raja binoculata*, and longnose skates, *Raja rhina*, were the target of a commercial fishery around Kodiak Island in the Gulf of Alaska (GOA) for the first time. The sudden development of a fishery for these species prompted the need for improved life history information to better inform fishery managers. Due to the selective nature of the skate fishery, mostly larger individuals were captured. Back-calculation from skate vertebral measurements was used to estimate size-at-age for younger skates. Because back-calculated age-length data within individuals were highly correlated, bootstrap resampling methods were used to test for differences between male and female growth curves. Results from bootstrapping indicated that differences between male and female growth were statistically significant for both species. This investigation indicates that growth of big skates in the GOA (max size 178 cm total length, max age 15 years) is similar to that in California, but different from that in British Columbia. For longnose skates, our GOA results agree with those reported in British Columbia, but were considerably older (max size 130 cm, max age 25 years) than those reported in California, which may not be surprising because longnose skates in the present study were generally larger. This life history information suggests that both big and longnose skates are at risk of unsustainable exploitation by targeted fisheries.

Keywords Alaska skate fisheries · Age determination · Back-calculation · Rajidae

C. M. Gburski (✉) · S. K. Gaichas · D. K. Kimura
National Marine Fisheries Service, Alaska Fisheries Science Center, 7600 Sand Point Way NE, Seattle, WA 98115, USA
e-mail: christopher.gburski@noaa.gov

Introduction

In the Gulf of Alaska (GOA) the most common skate species are the big skate, *Raja binoculata*, longnose skate, *Raja rhina*, Aleutian skate, *Bathyraja aleutica*, Bering skate, *Bathyraja interrupta*, and the Alaska skate, *Bathyraja parmifera* (Gaichas et al. 2003, 2005). The range of the big skate extends from the Bering Sea to southern Baja California in depths ranging from 2 to 800 m (Love et al. 2005). The longnose skate has a similar biogeographical range, from the southeastern Bering Sea to Baja California from 9 to 1,069 m depths (Love et al. 2005). While these two species have wide depth ranges, they are generally found in continental shelf waters (0–200 m) in the GOA. The greatest biomass of skates is found in depths less than 100 m and is dominated by the big skate (Fig. 1). Longnose

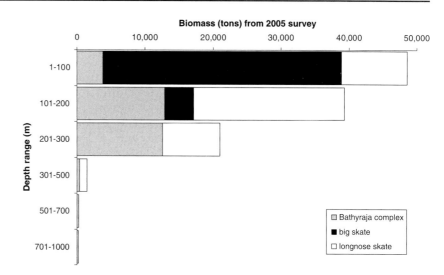

Fig. 1 Biomass at depth for major GOA skate species (big, longnose, and *Bathyraja* spp) (Gaichas et al. 2003, 2005)

skate is the dominant skate species at depths of 100–200 m, and *Bathyraja* spp are dominant in deeper waters (200 to >1,000 m) (Fig. 1). In the eastern Bering Sea, *Bathyraja* spp dominate the skate biomass in continental slope waters (Ebert 2005; Gaichas et al. 2003, 2005).

In 2002, markets for skates in the GOA developed (Bang and Bolton, Alaska Fisheries Inc, personal communication), and the resource became economically valuable in 2003 when the ex-vessel price became equivalent to that of Pacific cod, *Gadus macrocephalus*. In 2003, vessels began retaining and delivering skates as target species in federal waters partly because the market for skates had improved and partly because Pacific cod could be retained as bycatch in a skate 'Other species' target fishery, even though directed fishing for cod was seasonally closed (Gaichas et al. 2003, 2005). The result was a dramatic increase in skate landings around Kodiak Island in the GOA (Fig. 2; Gaichas et al. 2003, 2005). Moreover, in recent years (2003–2005) over half the annual catch was composed of *R. binoculata* and *R. rhina* (Gaichas et al. 2003, 2005).

In response to skate fisheries in other regions, age, and growth studies have been applied by Sulikowski et al. (2003, 2005a) for more informed management of the winter skate, *Leucoraja ocellata*, and the thorny skate, *Amblyraja radiata*, in the western Gulf of Maine (Sulikowski et al. 2005b). Two previous studies, Zeiner and Wolf (1993) and McFarlane and King (2006) studied age and growth of big and longnose skates in California waters and off the west coast of British Columbia, Canada, respectively, and a third age and growth study of the longnose skate was completed by Thompson (2006) on the US west coast. The purpose of the current study is to provide biological information on the age and growth of big and longnose skates in the GOA, in response to the development of the directed skate fishery.

Materials and methods

Specimen collection and preparation

Big and longnose skates were obtained from the directed fishery by longline in the surrounding waters of Kodiak Island, Alaska (57°45′N and 153°29′W), from February 2003 to June 2003 and in February 2004 to May 2004 (Fig. 2). Six thoracic vertebrae were excised from each skate by the Alaska Department of Fish and Game (ADFG) port sampling program in Kodiak, stored frozen and shipped to the Alaska Fisheries Science Center (AFSC) for later examination. All skates were identified, sexed, and measured for either total or pectoral length (±1 cm) at the port.

In this study, a total of 100 big skates (43 males and 57 females) and 103 longnose skates (48 males and 55 females) were examined. At the

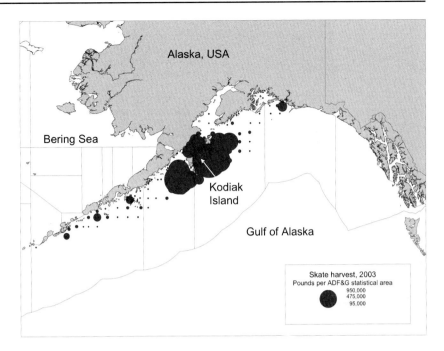

Fig. 2 Skate catch from Alaska Department of Fish and Game (*ADFG*) fish ticket database in 2003, data courtesy of Michael Ruccio, ADFG. Vertebrae used for our study were collected from skates in the waters surrounding Kodiak Island

AFSC, excess vertebral tissues, along with the neural and haemal arches, were removed from each vertebra with a scalpel. Individual vertebrae were stored in vials containing 70% ethanol. After initial processing, vertebrae were embedded in polyester resin (Artificial Water^{TM}) and mounted in a slotted block attachment on an Isomet^{TM} low speed saw. Each vertebra was cut along the sagittal (i.e., longitudinal) axis at a speed of ~200 rpm using two Norton^{TM} 4-in. diamond blades. Sections were mounted on a microscope slide with UV-light cured Loctite^{TM} 349 adhesive. After curing, each mounted thin section (≈0.30 mm thick) was sanded and polished using an Ecomet^{TM} table sander and 800 to 1,200 grit sandpaper. Mounted thin sections were viewed under a dissecting microscope to determine the optimal thickness for readability by examining the degree of translucence through the thin section. Section thickness was adjusted with further polishing.

Skate ageing methodology

Growth bands were examined under a Leica^{TM} MZ 95 dissecting microscope with reflected light from a fiber optic light source (Fig. 3). Mineral oil was applied to the thin sections to enhance bands. Opaque and translucent bands were followed from one corpus calcareum 'arm' across the intermedialia to the corpus calcareum 'arm' on the opposite side (Cailliet and Goldman 2004). Growth bands, which could be followed in this manner were interpreted as annual marks. The angle of each growth band changes where the intermedialia intersects the corpus calcareum and was the region which was used for counting (Fig. 3a). An hourglass-shaped constriction at age zero was assumed to correspond to the event of hatching from the egg case (Fig. 3a). The next translucent band was considered the first year (Cailliet and Goldman 2004). Thin sections were assigned a readability code according to the following scale: 1-clear thin section; 2-a single age can be generated with variable level of confidence; or 3-very difficult, where the reader can only assign a minimum age or age range.

Precision and bias

Band counts were made independently by two readers for each specimen without prior knowledge of total length or sex. Inter-reader precision was calculated by using the coefficient of variation (CV) (Chang 1982) and percent agreement between readers. Discrepancies between the two readers were resolved using a teaching (double)

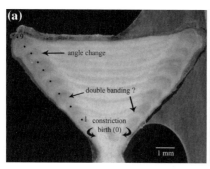
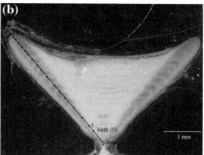

Fig. 3 (a) Big skate vertebral thin section with an estimated age of 9 years, showing angle change, potential double banding (*checks*), first year and constriction used to locate the birth mark (age 0). (b) Longnose skate vertebral thin section showing *annular rings* with an estimated age of 17 years. Back-calculation measurements are indicated along the *line* from focus to the outer-most edge

microscope. Age bias plots were constructed for each species to evaluate any potential bias between reader 1 age estimates and the resolved ages. A matched pairs *t*-test was performed between reader 1 and reader 2 ages and reader 1 and resolved ages.

Back-calculation, growth parameter estimation, and sample bootstrapping

Back-calculation was used to estimate the size-at-age for all age classes, since small skates were not targeted by the commercial fishery. Vertebral measurements (V_i) were measured from the focus of the V-shaped vertebral section, along the corpus calcareum, to the distal side of each presumed annual mark (Fig. 3b). The simple Dahl-Lea method (Francis 1990) was used, which assumes an isometric relationship between linear measurements in the skate vertebra and skate total length:

$$L_i = (V_i/V_A)L_A,$$

where L refers to skate total length, V refers to the vertebral measurement, i is the presumed ith annual ring, and A is the age at capture at the time of formation of the Ath presumed annual ring.

Although back-calculation allowed us to easily generate large amounts of age-length data, it must be recognized that these data were highly correlated within each of the individual specimens. To test differences between male and female growth curves, bootstrap resampling methods were used (Efron 1982). Length-at-age data were fit with a von Bertalanffy growth function:

$$L_t = L_\infty(1 - \exp(-K(t - t_0))),$$

where L_t is the total length at time t (age in years), L_∞ the theoretical asymptotic length, K the Brody growth coefficient, and t_0 is the theoretical age at zero length. The usual *F*-statistic to test between the male and female von Bertalanffy growth curves (Kimura 1980, 1990; Quinn and Deriso 1999) uses probabilities, which are usually drawn from the *F*-distribution. To bootstrap the null distribution for this statistic, the male and female data were combined into one sample. The same sample sizes that were present in the original male and female samples were randomly sampled, with replacement. When a specimen was selected, all back-calculated age-length data from that specimen were also selected. Note that each of the bootstrapped samples actually contained a mixture of males and females (e.g., if $n = 43$ specimens are selected, this would include both males and females). The *F*-statistic was calculated in the same way as the original male and female samples, and was bootstrapped 10,000 times to approximate the sampling distribution of the *F*-statistic under the null hypothesis that there is no difference between male and female growth curves. The proportion of these bootstrapped *F*-statistics greater than or equal to the original *F*-statistic calculated for testing between the male and female growth curves was determined. This proportion equals the estimated probability of the *F*-statistic for testing between male and female growth curves, and is more realistic than the probability calculated from the *F*-distribution.

Estimation of the natural mortality rate (M)

Hoenig's (1983) equation is commonly used in fishery stock assessment work to estimate the natural mortality rate from maximum age.

Natural mortality was estimated using the equation:

$$M = 4.306/t^{1.01},$$

where *M* is the natural mortality rate on an annual basis and *t* is the maximum age measured in years.

Results

Sample analysis

The size range of collected specimens varied from 84 to 141 cm TL for male big skates ($n = 43$) and 67–178 cm TL for female big skates ($n = 57$), and 100–129 cm TL for male longnose skates ($n = 48$) and 98–140 cm TL for female longnose skates ($n = 55$) (Fig. 4).

A similar isometric relationship was observed between vertebral radius and total length for both the big skate ($\hat{\beta} = 22.1$) and for the longnose skate ($\hat{\beta} = 21.5$) (Fig. 5). Isometry was also supported by statistical tests, which indicated the hypothesis of a slope intercept of zero could not be rejected for either species ($P = 0.66$ for big skate; $P = 0.15$ for longnose skate) (Fig. 5).

For big skates, a comparison of age estimates between the two readers showed a 38% total agreement (±0) with a slight overall bias where the second reader tended to overestimate the primary reader's ages. The inter-reader CV was 9.35% ($n = 50$). For longnose skates, a comparison of age estimates between the two readers showed a 14.3% total agreement (±0) with a much greater bias where the second reader tended to underestimate the primary reader's ages. The inter-reader CV was 11.85% ($n = 49$). A matched pairs *t*-test (Zar 1999) showed the overall bias between reader 1 and reader 2 was not significant for big skates ($t = 1.07$, $df = 46$, and $P = 0.29$) or longnose skates ($t = -1.93$, $df = 46$, and $P = 0.06$).

Discrepancies between reader 1 and reader 2 were investigated and potential ageing criteria differences were resolved. An age bias plot between reader 1 and resolved ages showed that differences in ageing criteria between readers was successfully resolved (Fig. 6). Because there was no significant age bias between readers, the primary reader's ages were used as final ages in this study. Verification was not performed since vertebrae were only sampled from the directed skate fishery over a 4–5-month period of time. Based on results of other skate age and growth studies, we assumed that each opaque and translucent band pair for big and longnose skates represented 1 year's growth.

Observed ages ranged from 4 to 15 years for big skates and 13–25 years for longnose skates, indicating that the maximum age of longnose skates may be 10 years greater than the maximum age of big skate (Table 1, Fig. 4). Due to the lack of small skates, back-calculation was used to estimate the size and age of skates that were missing from our samples. The inclusion of this technique greatly increased the sample sizes available for fitting the von Bertalanffy growth curves (Table 2). Average back-calculated length-at-age data are plotted along with the von Bertalanffy growth curves estimated from the individual back-calculated data (Fig. 7). The von Bertalanffy growth parameters for big skates were $L_\infty = 153.3$ cm TL, $K = 0.1524$ year^{-1} and $t_0 = -0.632$ year ($R^2 = 0.88$, males) and $L_\infty = 247.5$ cm TL, $K = 0.0796$ year^{-1} and $t_0 = -1.075$ year ($R^2 = 0.82$, females) (Table 2). For longnose skates, the von Bertalanffy growth parameters were $L_\infty = 168.8$ cm TL, $K = 0.0561$ year^{-1} and $t_0 = -1.671$ years ($R^2 = 0.92$, males) and $L_\infty = 234.1$ cm TL, $K = 0.0368$ year^{-1} and $t_0 = -1.993$ years ($R^2 = 0.94$, females) (Table 2). Plots of average length-at-age (Fig. 7) suggest that female skates of both species continue to grow after the length at maturity of 110 cm for big skate and 60 cm for

Fig. 4 Frequency histograms of total length and age for sampled big and longnose skates, sexes combined

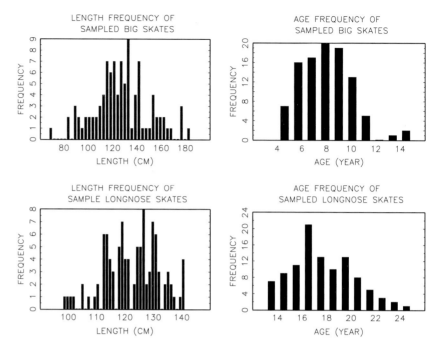

longnose skate estimated by Zeiner and Wolf (1993) and Martin and Zorzi (1993). Bootstrapping the F-statistic indicates that the difference between male and female growth curves is highly significant for big skate ($F = 28.96$, $P < 0.0004$) and longnose skate ($F = 45.93$, $P < 0.0001$) (Table 2).

Based on maximum age (Hoenig 1983) the instantaneous natural mortality rate for big skates was estimated to be 0.28, and the natural mortality rate for longnose skates was 0.17.

In order to compare ageing results from specific regional studies, we plotted and compared the size-at-age from growth curves estimated by Zeiner and Wolf (1993) for California, by Thompson (2006) for the US west coast longnose skates, by McFarlane and King (2006) for British Columbia and in our study for the GOA (Table 3, Fig. 8). Since Zeiner and Wolf (1993) used the logistic growth curve when fitting data from big skates, these data were plotted in Fig. 8 but were omitted from Table 3 to avoid confusion. Thompson (2006) forced the von Bertalanffy curve through an estimated length at birth. It is clear from all four studies that female big and longnose skates grow larger than males (Table 3). This is also reflected in estimated L_∞ (Table 3, Fig. 8).

Discussion

The relationship between TL and vertebral radius demonstrated that vertebrae grow isometrically with respect to the rest of the body in the big skate ($n = 100$) and the longnose skate ($n = 103$). However, sampling restrictions for both species limited the body size range (67–178 cm TL for the big skate and 98–140 cm TL for the longnose skate) from which vertebral age estimates could be obtained. The isometric relationship between total length and vertebral radius was useful for back-calculation and consequently expanding the size range and number of samples. Back-calculation is a common practice used to supplement the lack of small individuals from a data set (Cailliet and Goldman 2004). The accuracy of ages produced using this technique was illustrated by the back-calculated size at hatching for both big and longnose skates, which compared well with available size at birth data from specimens previously collected in the GOA (Gerald Hoff, Alaska Fisheries Science Center, personal communication 2006), California[1] and data reported by Zeiner and Wolf (1993).

---

[1] K. Cox and C. Hitz, unpublished data.

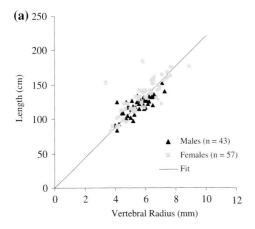

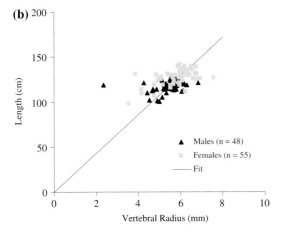

Fig. 5 The isometric relationship between vertebral radius and skate total length for big skate (a) and longnose skate (b), sexes combined. For the big skate, the slope and SE were $\hat{\beta} = 22.1, \mathrm{SE}(\hat{\beta}) = 0.25$; for longnose skate the slope and SE were $\hat{\beta} = 21.5, \mathrm{SE}(\hat{\beta}) = 0.25$. All regressions were through the origin

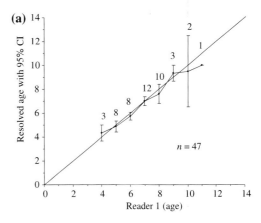

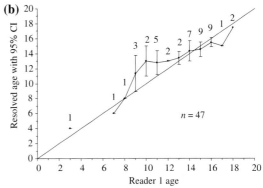

Fig. 6 Age bias graphs for (a) big skate ($n = 47$) and (b) longnose skate ($n = 47$) vertebral band counts by reader 1 and corresponding mean resolved ages. Each *error bar* represents the 95% confidence interval for the mean resolved age to all fish assigned a given age by reader 1. The *diagonal line* represents the one-to-one equivalence line. Sample sizes are given above each corresponding age

The primary criterion for determining that a growth ring is an annual mark was the consistency of growth zones across the thin section (Cailliet and Goldman 2004). Technically, some species of skates can be difficult to age due to faint annuli or inconsistent growth patterns. This is suggested in recent age and growth studies for the longnose skate (Thompson 2006) and the Alaska skate, *B. parmifera* (Matta 2006). Thompson (2006) indicated a CV of 13.75% for the longnose skate, while Matta (2006) described a CV of 10.89% for the Alaska skate. Variation in growth patterns have been found within the *Raja* spp by McFarlane and King (2006), who observed that thin sections of big skate centra produced a clearer pattern of dark bands than longnose skate centra. Precision from inter-reader age comparisons for the big skate ($n = 50$) resulted in a 38% (±0) agreement and a CV of 9.35% versus the longnose skate ($n = 49$) with a 14.3% (±0) agreement and a CV of 11.85%. The CV's from vertebral studies of sharks have frequently exceeded 10% (Campana 2001), and thus we considered our level of precision acceptable. In our study, the level of precision and the presence of bias are likely due to differences in growth band interpretation between readers. The higher agreement and lower CV for the big skate may be attributed to readability differences between species. Growth bands became constricted approaching the vertebral thin section edge for the longnose skate, whereas big skate thin

Table 1 Maximum and mean total lengths and ages observed for GOA big and longnose skates

Species	Sex	Sample size	Max length (cm)	Max age (year)	Mean length (cm)	Mean age (year)
R. binoculata	Male	43	141	15	119.6	8.02
R. binoculata	Female	57	178	14	132.4	8.09
R. rhina	Male	48	129	25	116.6	17.50
R. rhina	Female	55	140	24	127.7	18.6

Table 2 Back-calculated sample sizes and estimated von Bertalanffy parameters for big and longnose skates in the GOA

Species	Sex	Back-calculated sample size	L_∞ (cm TL)	K	t_0 (year)	F	P
R. binoculata	Male	388	153.3	0.1524	−0.632	28.96	0.0004
R. binoculata	Female	518	247.5	0.0796	−1.075		
R. binoculata	Combined	906	189.6	0.1145	−0.835		
R. rhina	Male	888	168.8	0.0561	−1.671	45.93	0.0001
R. rhina	Female	1,083	234.1	0.0368	−1.993		
R. rhina	Combined	1,971	203.8	0.0437	−1.868		

F-statistics refer to the standard F-statistic for testing between male and female growth curves, and P is the bootstrapped probability of the F-statistics

sections remained more consistent in growth band spacing. Thompson (2006) indicated this band constriction for the longnose skate led to

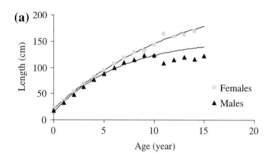

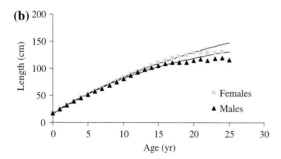

Fig. 7 Average total length-at-age for male and female big skate (a) and longnose skate (b) with von Bertalanffy growth curves. Big skate: males (L_∞ = 153.3 cm TL, K = 0.1524 year^{-1}, t_0 = −0.632 year, R^2 = 0.88), females (L_∞ = 247.5 cm TL, K = 0.0796 year^{-1}, t_0 = −1.075 year, R^2 = 0.82). Longnose skate: males (L_∞ = 168.8.3 cm TL, K = 0.0561 year^{-1}, t_0 = −1.671 year, R^2 = 0.92), females (L_∞ = 234.1 cm TL, K = 0.0368 year^{-1}, t_0 = −1.993 years, R^2 = 0.94)

difficulties in age estimation for this species. Similarly, Cailliet and Goldman (2004) suggest that ageing vertebrae from older animals can be problematic due to bands being tightly grouped. Overall, the big skate age estimates had greater inter-reader precision, with a higher agreement, lower CV and less bias compared to the longnose skate. The matched pairs t-test showed bias between reader 1 and resolved ages not to be significant (P > 0.05), and thus reader 1 ages were used as the final ages to estimate growth parameters

Brander and Palmer (1985) suggest that seasonal changes including environmental conditions, migrations, and spawning can affect growth. For the bull shark, *Carcharhinus leucas*, it was implied that ring deposition may be attributed to the spring pupping season (Neer et al. 2005). Checks, sometimes associated with presumed annual marks, may be due to these seasonal changes or may be randomly occurring. Opaque growth bands are wide and are typically laid down during the summer or productive growth periods (Cailliet and Goldman 2004). Translucent growth bands are narrower, signifying less growth during the winter months or less productive growth periods. Longnose skates appear to have more constricted growth patterns compared to big skates, which may be the result of potential differences between the two species in depth preference or reproductive events. Generally,

Biology of Skates

Table 3 A comparison of three age and growth studies of big skates and four age and growth studies of longnose skates, which took place in four geographic regions

Species	Region/source	Sex	Length range TL (cm)	Max age (year)	L_∞ (cm)	K	t_0 (year)
R. binoculata	California	Male	17.5–132.1	11	N/A	N/A	N/A
R. binoculata	California	Female	22.7–160.7	12	N/A	N/A	N/A
R. binoculata	British Columbia	Male	16.6–183.6	25	233.0	0.05	−2.10
R. binoculata	British Columbia	Female	16.9–203.9	26	293.5	0.04	−1.60
R. binoculata	GOA	Male	84–141	15	153.3	0.152	−0.632
R. binoculata	GOA	Female	67–178	14	247.5	0.080	−1.075
R. rhina	California	Male	35.9–132.2	13	96.7	0.25	0.73
R. rhina	California	Female	30.3–106.8	12	106.9	0.16	−0.30
R. rhina	West coast	Male	19–129	20	207.2	0.042	0
R. rhina	West coast	Female	18–142	22	180.9	0.051	0
R. rhina	British Columbia	Male	18.6–122.0	23	131.5	0.07	−2.17
R. rhina	British Columbia	Female	18.4–124.6	26	137.2	0.06	−1.80
R. rhina	GOA	Male	100–129	25	168.8	0.056	−1.67
R. rhina	GOA	Female	98–140	24	234.1	0.037	−1.99

growth zones, which did not follow the horizontal pattern across the thin section, or were faint, were dismissed as non-annual marks in our study. The attributes contributing to the growth pattern of each species could be the focus of a future study.

The lack of age validation presents a limitation to this study. However, based on previous studies of other elasmobranchs, including the longnose skate, we believe that one opaque and translucent band pair is deposited annually. Thompson (2006) performed MIA on the longnose skate and found annual periodicity of band pair formation and evidence that opaque bands typically form in the winter and spring. Other skate studies support annual periodicity in band pair formation including Sulikowski et al. (2003, 2005a) for the winter skate, Sulikowski et al. (2005b) for the thorny skate and Matta (2006) for the Alaska skate. Conrath et al. (2002) performed edge analysis on the smooth dogfish, *Mustelus canis*, and found

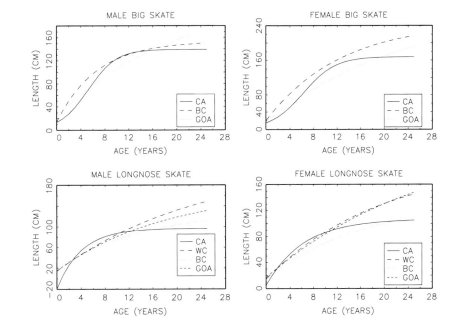

Fig. 8 Age-total length growth curves from four studies of male and female big and longnose skates [*GOA* Gulf of Alaska, *BC* British Columbia, *CA* California, *WC* US west coast (Cape Flattery, WA to Cape Mendocino, CA, USA)]. Growth curves are all von Bertalanffy except for the CA big skate data, which was fit with a logistic curve

MIA conclusive for juvenile-size animals. Annual periodicity was also indicated by the marginal increment analysis for the salmon shark, *Lamna ditopis* (Goldman and Musick 2006).

Big skate growth rates were different between the sexes in our study. Longnose skate growth rates were more similar between the sexes. Females of both species had slower growth rates than males. McFarlane and King (2006) also reported a slightly slower growth rate for females when compared to males for big and longnose skates. In other elasmobranchs, like the salmon shark, females tend to exhibit larger maximum sizes and slower growth rates than males (Goldman and Musick 2006). Female big and longnose skates may be at a higher risk to over-fishing than males because they grow more slowly to larger sizes. The larger body size of big and longnose skates compared to other skate species may make them more vulnerable to over-fishing (Dulvy and Reynolds 2002).

When examining the L_∞ parameter estimates, it should be understood that this is an *average* theoretical maximum size. Some individuals may never attain the age required to reach this maximum size, and large individuals may exceed L_∞ because it is an average. It should be noted that for our study, the von Bertalanffy parameters were estimated from the individual data, while the mean length at each back-calculated age appears on the plots. The result is that the von Bertalanffy curve does not follow the means at older ages because these are based on fewer data points. This was evident in the age and growth of the dusky shark, *Carcharhinus obscurus*, where the lack of larger specimens prevented an accurate estimation of L_∞ (Simpfendorfer et al. 2002).

The natural mortality rate for longnose skates ($M = 0.17$) from our study was somewhat lower than Thompson's (2006) natural mortality rate for longnose skates ($M = 0.26$). The big skate natural mortality rate ($M = 0.28$) derived from the maximum age was somewhat higher than that of the longnose skate. Since the big skate and the longnose skate have not been previously aged in Alaskan waters, this study presents the first estimates of natural mortality based on maximum age from that region.

Comparing skate growth data from California, the US West Coast, British Columbia, and GOA (present study)

The four age and growth studies of big and longnose skates have been largely regional in nature (Zeiner and Wolf 1993; McFarlane and King 2006; Thompson 2006; our study). Each has had their set of limitations. For instance, the study by Zeiner and Wolf (1993) in waters off the coast of California (Santa Cruz to Monterey) lacked the larger body sizes, especially for longnose skates. In our study, the GOA commercial fishery selected for larger specimens and inadequately sampled smaller sizes, resulting in fewer young specimens. The maximum age of longnose skates off California (Zeiner and Wolf 1993) was much younger compared to GOA longnose skates, most likely because more large-bodied longnose skates were collected in the GOA. The maximum ages of longnose skates from British Columbia (Dixon entrance to the northwest coast of Vancouver Island) by McFarlane and King (2006) and from the GOA were very similar and longnose skates from the US west coast (Coos Bay, Oregon to the US/Mexico Border) by Thompson (2006) were only slightly younger.

Comparisons between the four studies suggest differences in life history parameters. These differences could be truly geographical in nature, or they could be the result of differences in sampling or age estimation criteria between the studies. For example, an 11 years difference in maximum ages existed between British Columbia (26 years, ≈190 cm TL, female) and GOA (15 years, 141 cm TL, female) big skates. Since the GOA sampling for big skate included large individuals, it was expected that GOA maximum ages would be similar to McFarlane and King's (2006) age results. Limitations in our study's sample size ($n = 100$) when compared to McFarlane and King (2006; $n = 242$) may have contributed to the discrepancy in maximum ages. Additionally, a 60 cm difference in maximum length between our study and McFarlane and King (2006) may have still resulted in age differences. Differences between British Columbia and GOA maximum ages may potentially be related to the different ageing methodologies between the studies. McFarlane and King (2006) estimated the position of the first 3 years by using vertebral measurements from younger specimens and applying them to older specimens even when presumed annual marks were not clearly visible. The GOA big skate ages resembled those of Zeiner and Wolf (1993), who employed similar ageing criteria where bands were assumed to be annual only when visible on the thin section. Their maximum age of 12 years (161 cm TL) compared well with our maximum age of 15 years (141 cm TL). In our study, big skate centra produced clearer growth zones compared to longnose skates, though a majority of thin sections from both species were assigned a readability code of 2 by the primary age reader. Clarity differences also existed between specimens within species.

Maximum ages of longnose skate collected from the US west coast, British Columbia, and the GOA appear reasonably consistent. The maximum age of longnose skate collected off the US west coast (22 years) by Thompson (2006) was similar to our results from the GOA (25 years). Ageing methodologies of Thompson (2006) and our study were similar due to technical assistance provided to both studies by Moss Landing Marine Laboratories. Our maximum ages for longnose skates caught in the GOA were much older compared to those from California (13 years) (Zeiner and Wolf 1993). Longnose skates caught in the GOA were generally larger compared with those caught off California (Table 3). Studies of other skate species also suggests the possibility that maximum ages for longnose skates can be older than previously reported (Frisk et al. 2001, 2004).

Age and growth studies have been found useful in the management of skate fisheries in the northeast Atlantic Ocean. The thorny skate, *A. radiata*, in the western Gulf of Maine has decreased in biomass, requiring biological data to assist managers in rebuilding these stocks (Sulikowski et al. 2005b). Sulikowski et al. (2003, 2005a) suggests the susceptibility of the winter skate, *L. ocellata*, to over-fishing, particularly when large individuals are selectively removed from the population. Body size can be a predictor of vulnerability since most skate fisheries select for larger individuals, and many skate species mature at a large body size (Dulvy and Reynolds 2002). Comparing age and growth of the same species on a regional level may help fisheries managers monitor populations and identify potential detrimental effects of fishing pressure on different stocks.

It is important to better understand the life history traits of big and longnose skates through age and growth studies because these species are being targeted by commercial fisheries in the GOA (Gaichas et al. 2003, 2005). Future studies and long-term monitoring by the AFSC will help ensure that these skate populations can be managed properly; in particular, age validation will be performed to verify annual vertebral band pair periodicity for big and longnose skates in Alaska waters. Skates in other parts of the world have been over-fished (Brander 1981; Dulvy et al. 2000; Sulikowski et al. 2003, 2005a). This study is a direct response to the initiation of a directed skate fishery to avoid a similar situation in the GOA. The growth characteristics of these species suggest that they could not sustain high levels of catch (Hoenig and Gruber 1990; Musick et al. 2000; King and McFarlane 2003), and this study will help fisheries managers determine appropriate strategies of management and conservation for big and longnose skates.

Acknowledgments We would like to thank Michael Ruccio, Ken Goldman, Bill Bechtol, Kally Spalinger and Willy Dunne at the ADFG for providing skate vertebrae, and Rob Swanson of the NMFS Kodiak Lab for facilitating collections at the processing plants and for initial data review. Wade Smith and many others at Moss Landing Marine Laboratories provided much assistance in vertebral ageing and processing techniques as well as a background in elasmobranch life history. We also thank Beth Matta, a graduate student at the University of Washington, who worked concurrently on her *B. parmifera* thesis and collaborated with us on all phases of this project. We thank Josie Thompson at Oregon State University for allowing us to use data from her thesis. Finally, we thank Betty Goetz, Jerry Hoff, Beth Matta, James Sulikowski, Dave Ebert and Greg Cailliet for reviewing the manuscript.

References

Brander K (1981) Disappearance of common skate *Raia batis* from Irish Sea. Nature 290:48–49

Brander K, Palmer D (1985) Growth rate of *Raja clavata* in the Northeast Irish Sea. ICES J Cons 42(2):125–128

Cailliet GM, Goldman KJ (2004) Age determination and validation in chondrichthyan fishes. In: Carrier J, Musick JA, Heithaus MR (eds) Biology of sharks and their relatives. CRC Press LLC, Boca Raton, FL, pp 399–447

Campana SE (2001) Accuracy, precision and quality control in age determination, including a review of the use and abuse of age validation methods. J Fish Biol 59:197–242

Chang WYB (1982) A statistical method for evaluating the reproductivity of age determination. Can J Fish Aquat Sci 89:1208–1210

Conrath CL, Gelsleichter J, Musick JA (2002) Age and growth of the smooth dogfish (*Mustelus canis*) in the northwest Atlantic Ocean. Fish Bull US 100:674–682

Dulvy NK, Metcalfe JD, Glanville J, Pawson MG, Reynolds JD (2000) Fishery stability, local extinctions, and shifts in community structure in skates. Conserv Biol 14(1):283–293

Dulvy NK, Reynolds JD (2002) Predicting extinction vulnerability in skates. Conserv Biol 16:440–450

Ebert DA (2005) Reproductive biology of skates, *Bathyraja* (Ishiyama), along the eastern Bering Sea continental slope. J Fish Biol 66:618–649

Efron B (1982) The jackknife, the bootstrap, and other resampling plans. Society of industrial and applied mathematics. CBMS NSF Monogr 38:92

Francis RICC (1990) Back-calculation of fish length: a critical review. J Fish Biol 36:883–902

Frisk MG, Miller TJ, Fogarty MJ (2001) Estimation and analysis of biological parameters in elasmobranch fishes: a comparative life history study. Can J Fish Aquat Sci 58:969–981

Frisk MG, Miller TJ, Dulvy NK (2004) Life histories and vulnerability to exploitation of elasmobranchs: inferences from elasticity, perturbation, and phylogenetic analysis. J Northwest Atl Fish Sci 34:1–19

Gaichas S, Ruccio M, Stevenson D, Swanson R (2003) Stock assessment and fishery evaluation of skate species (*Rajidae*) in the Gulf of Alaska. North Pacific Fishery Management Council 605 W. 4th Ave., Anchorage, AK 99501

Gaichas S, Sagalkin N, Gburski C, Stevenson D, Swanson R (2005) Gulf of Alaska skates. In Stock assessment and fishery evaluation report for the groundfish resources of the Gulf of Alaska, North Pacific Fishery Management Council 605 W. 4th Ave., Anchorage, AK 99501

Goldman KJ, Musick JA (2006) Growth and maturity of salmon sharks (*Lamna ditropis*) in the eastern and western North Pacific, and comments on back-calculation methods. Fish Bull US 104:278–292

Hoenig SJ (1983) Empirical use of longevity data to estimate mortality rates. Fish Bull US 82:898–902

Hoenig JM, Gruber SH (1990) Life-history patterns in the elasmobranchs: implications for fisheries management. In: Pratt HL, Gruber SH, Taniuchi T (eds) Elasmobranchs as living resources: advances in the biology, ecology, systematics, and the status of the fisheries. NOAA Technical Report NMFS 90, pp 1–16

Kimura DK (1980) Likelihood methods for the von Bertalanffy growth curve. Fish Bull US 77:765–776

Kimura DK (1990) Testing nonlinear regression parameters assuming heteroscedastic, normally distributed errors. Biometrics 46:697–708

King JR, McFarlane GA (2003) Marine fish life history strategies: applications to fishery management. Fish Manage Ecol 10:249–264

Love MS, Mecklenberg CW, Mecklenberg TA, Thorsteinson LK (2005) Resource inventory of marine and estuarine fishes of the West Coast and Alaska: a checklist of North Pacific and Arctic Ocean species from Baja California to the Alaska-Yukon Border. US Department of the Interior, US Geological Survey, Biological Resources Division, Seattle, Washington, 98104, OCS Study MMS 2005-030 and USGS/NBII 2005-001

Martin L, Zorzi GD (1993) Status and review of the California skate fishery. US Dept Commer NOAA Tech Rep 115:39–52

Matta ME (2006) Aspects of the life history of the Alaska skate (*Bathyraja parmifera*) in the eastern Bering Sea. MS Thesis, University of Washington, 92 pp

McFarlane GA, King JR (2006) Age and growth of big skate (*Raja binoculata*) and longnose skate (*Raja rhina*) in British Columbia waters. Fish Res 78:169–178

Musick JA, Burgess G, Cailliet G, Camhi M, Fordham S (2000) Management of sharks and their relatives (Elasmobranchii). Fisheries 25(3):9–13

Neer JA, Thompson BA, Carlson JK (2005) Age and growth of (*Carcharhinus leucas*) in the northern Gulf of Mexico: incorporating variability in size at birth. J Fish Biol 67:370–383

Quinn TJ, Deriso RB (1999) Quantitative fish dynamics. Oxford University Press, Oxford, 542 pp.

Simpfendorfer CA, McAuley RB, Chidlow J, Unsworth P (2002) Validated age and growth of the dusky shark (Carcharhinus obscurus) from western Australian waters. Mar Freshw Res 53:567–573

Sulikowski JA, Dorin MD, Suk SH, Howell WH (2003) Age and growth estimates of the winter skate (*Leucoraja ocellata*) in the western Gulf of Maine. Fish Bull US 101:405–413

Sulikowski JA, Tsang PCW, Howell WH (2005a) Age and size at sexual maturity for the winter skate (*Leucoraja ocellata*), in the western Gulf of Maine based on morphological, histological and steroid hormone analysis. Environ Biol Fish 72:429–441

Sulikowski JA, Kneebone J, Elzey S, Jurek J, Danley DD, Howell WH, Tsang PCW (2005b) Age and growth estimates of the thorny skate (*Amblyraja radiata*) in the western Gulf of Maine. Fish Bull US 103:161–168

Thompson JE (2006) Age, growth and maturity of the longnose skate (*Raja rhina*) on the US West coast and sensitivity to fishing impacts. Unpublished MS Thesis, Oregon State University, Corvallis, OR, 147 pp

Zar JH (1999) Biostatistical analysis. Prentice Hall, Upper Saddle River, NJ, pp 161–164

Zeiner SJ, Wolf P (1993) Growth characteristics and estimates of age at maturity of two species of skates (*Raja binoculata* and *Raja rhina*) from Monterey Bay, California. US Dept Commer NOAA Tech Rep 115:87–99